MICRO TOTAL ANALYSIS SYSTEMS

MESA Monographs

MICRO TOTAL ANALYSIS SYSTEMS

Proceedings of the μTAS '94 Workshop,
held at MESA Research Institute,
University of Twente, The Netherlands,
21–22 November 1994

Edited by

A. VAN DEN BERG
and
P. BERGVELD

MESA Research Institute,
University of Twente,
The Netherlands

SPRINGER-SCIENCE+BUSINESS MEDIA, B.V.

A C.I.P. Catalogue record for this book is available from the Library of Congress

ISBN 978-94-010-4072-3   ISBN 978-94-011-0161-5 (eBook)
DOI 10.1007/978-94-011-0161-5

Printed on acid-free paper

Preface

The MESA Research Institute of the University of Twente was created in 1990 through the joining of the research unit Sensors and Actuators with the department of Microelectronics. The multidisciplinary institute, with participation from the faculties of Electrical Engineering, Applied Physics and Chemical Technology, was recently recognized as a Centre of Excellence by the Dutch Science Foundation. It is fully equipped with modern Clean Room facilities (1000 m^2) and a number of research laboratories. The objective of MESA is to perform research and development of systems in modern information technology, and on the units on which they are based: the microstructures that process and transduce signals. The institute gradually expanded during the past few years till some 125 persons in 1994. Given the wide variety of research subjects within MESA, it has been decided to start a **MESA Monographs** series, appearing on a more or less regular, yearly basis. In this way, after some time a good overview of research topics under investigation at MESA will be obtained.

The first volume of this series coincides with the Proceedings of *μTAS '94*, the first Workshop on Micro Total Analysis Systems, held on November 21-22 at the University of Twente in Enschede, The Netherlands. μTAS has recently been defined as the first strategic research orientation of MESA, aiming at synergetic collaboration between the different disciplines present in MESA. The use of silicon-based microsystems for chemical analysis is one of the most promising concepts in the recent developments in Micro System Technology (MST). It is expected that chemical sensors will be increasingly integrated in μTAS, systems in which all steps in a chemical determination from sampling to detection and data treatment are integrated in one miniature instrument. μTAS offer a variety of advantages over conventional analysis systems such as improved analytical performance, reduced reagent and power consumption, small size, possibility of new and more complicated functions, higher reliability and lower fabrication costs. Application of μTAS may be found in fields like process industry, environmental monitoring, medical diagnostics, aeronautics, automotive industry, etc.

The aim of *μTAS '94* is to present the state of the art of research and development in the field of miniature instruments for (bio)chemical analysis by leading research institutes and industries. The academic and industrial worlds are brought together to stimulate R&D co-operation and discuss present and future opportunities for μTAS. In this Proceedings, apart from short papers of all posters presented at the Workshop, contributions are given of all invited speakers, covering the following subjects:

- Basic concepts of μTAS
- Technologies and components
- Industrial applications
- Gas and liquid analysis systems

The organisers wish to thank the Dutch Science Foundation (NWO) and the University of Twente for financial support of the Workshop and hope you will enjoy the first **MESA Monograph: μTAS**.

Enschede, August 1994.

Jan Fluitman
Director of MESA

Albert van den Berg and Piet Bergveld
Organisers of *μTAS '94*

Contents

THE CHALLENGE OF DEVELOPING µTAS

Piet Bergveld
MESA Research Institute, University of Twente
P.O. Box 217, 7500 AE Enschede, The Netherlands

0. Abstract.

By comparing the developments taken place in the field of consumer electronics, where large scale equipment has been miniaturized due to the integration of electronic functions in one and the same chip, it is made clear how in the same way the present large scale equipment for chemical analysis might be miniaturized to portable µTAS systems. The progress in silicon sensor technology as well as in the micro systems technology for liquid handling systems makes this miniaturisation possible, but for the practical realisation of µTAS systems a standardisation for the interconnection of the system components has to be developed.

1. Introduction.

After the first world war, more and more citizens came in the possession of a wireless radio set, which consisted in those days of a separate power supply, a so-called detector and a separate loudspeaker. It took at least one decade before the manufacturers integrated the different parts into one radio as a nice piece of furniture, primarily available for the upper class. These radio's were relatively large and heavy, due to the application of large vacuum tubes, transformers and detection coils. Only after world war two, the vacuum tubes became smaller and the radio's needed less energy, resulting in the development of portable radio's, of which initially the batteries formed the heaviest part. After the invention and application of the transistor, the energy need decreased tremendously and many types of transistorised radio's became available. The first types still used relatively large batteries as in use for electric torches, but as soon as integrated circuits became common, the transistor radio's came down in size and weight and soon became pocket radio's. Nowadays the whole radio consists of only one integrated circuit, to be supplied by one miniature battery, mounted in a matchbox-sized housing, to which an ear-phone can be connected.

The same story as given above can be told about to the development of many other electronic products, such as recorders, TV sets, computers etc. Beside the miniaturization of the equipment, the number of built-in functions greatly increased. These developments have been accepted world-wide as a logical progress in technological possibilities.

1

A. van den Berg and P. Bergveld (eds.), Micro Total Analysis Systems, 1–4.
© 1995 *Kluwer Academic Publishers.*

2. From single analyser to auto-analysers.

Already in the twenties many instruments existed or were under development in the field of analytical and clinical chemistry for the measurement of one specific parameter of a sample solution, mostly a concentration Various calorimeters, photometers and a pH measuring apparatus are well known examples The more parameters that could be analysed, the more equipment came into existence In order not to divide samples in ever more portions to be analysed separately by the available equipment with their specific method of measurement, in the fifties the tendency arose to combine the different methods in one and the same apparatus In this way, again as a logical consequence of the possibilities, multi analysers were built, completely automated with respect to sample manipulation, measurement, registration, data processing etc The first auto-analysers were pieces of room filling equipment, but due to the introduction of, among other things, integrated circuit technology and the interrelated data processing and mechatronic control actions, many parts of the system could be miniaturized as in the case of the radio's mentioned above At present the equipment contains much more functions, is highly automated and is still relatively small compared to the original designs Nevertheless most of the equipment is too large to be applied outside a central laboratory and moreover needs skilled personnel for a proper operation The question may arise whether, as in the case of electronic equipment as mentioned in the introduction, the development of auto-analysers will also show progress in further miniaturization This might ultimately result in pocket size equipment, for application in the field of environmental or bedside monitoring, operated by relatively untrained personnel

3. The development of chemical sensors.

Since about 1970, the construction of miniature sensors has been a promising field of research The reason is that at this time the possibilities of applying silicon technologies for the construction of sensors became available It was expected, as in the case of transistors, that many types of sensors could be produced in this way in large quantities and for a very low price Especially university research groups took this challenge and indeed showed that the application of silicon and related technologies is very useful for the manufacturing of various types of physical and chemical sensors The industry discovered however that it is not the chip price which determines the costs of the sensor, but the encapsulation which can hardly be standardised due to the wide range of specific applications Therefore it took at least 20 years of development before the first solid state sensors became commercially available, especially the chemical- and biosensors At present different types of electrochemical and optical sensors are available, but the application is rather traditional The construction of the sensors is mainly in the form of dipstick and flow-through systems as miniature equivalents of conventional systems using the larger types of sensors The main advantage of the smaller sensors is that they can handle smaller samples in shorter times due to their miniature design and faster response With only some exceptions the sensors have, like their conventional counterparts, the same needs for regular calibration to compensate for parameter drift This limits the application to laboratory type measurements whereas for instance long-term in vivo application is still not possible

In·the same time that the modern sensor technology was developed, the micromachining technology, especially focusing on silicon as the basic material, came of age Micro valves, -pumps, -mixers, -motors, etc were developed in silicon, using a large variety of operational principles Since these parts are, beside the sensors, the essential parts of the large-scale auto-analysers, the question arises whether the combination of all the silicon parts can or can not lead to the development of pocket-size auto-analysers At present the technology for making miniature sensors and liquid-handling systems in silicon is about at the same level of possibilities, which is in fact a unique coincidence Furthermore the necessary measuring and control electronics to operate the different components can, where necessary, be integrated in the same piece of silicon, whereas in addition the application of neural networks and fuzzy electronics seems very useful Due to the fact that all necessary elements of an analyser can be made in the same technology on one and the same substrate, these systems are called micro Total Analysis Systems

4. Reality and application of μTAS.

In the previous section the possibility of making μTAS was given on a technology-driven base in analogy with the development of the ever smaller consumer electronics mentioned in the introduction The question can be put forward whether or not this is a realistic view Most probably the first systems will consist of a hybrid construction, where only parts of the system are really integrated, but connected with each other by intermediate conventional tubes For a complete integration the tubing should be made in a solid, planar substrate, such as a silicon wafer, by means of etched channels and cavities This type of "micro-plumbing" still has to be developed, which not only needs a specific technology, but above all a standardisation Only in that case the different parts like sensors, pumps, mixers etc can be manufactured as standard modules, to be mounted on top of the pretubed substrate in a connection pattern typical for a certain application in about the same way as electronic components are mounted on a printed circuit board This standardisation should be an international activity An artist impression of such a modular system is given in figure 1

4

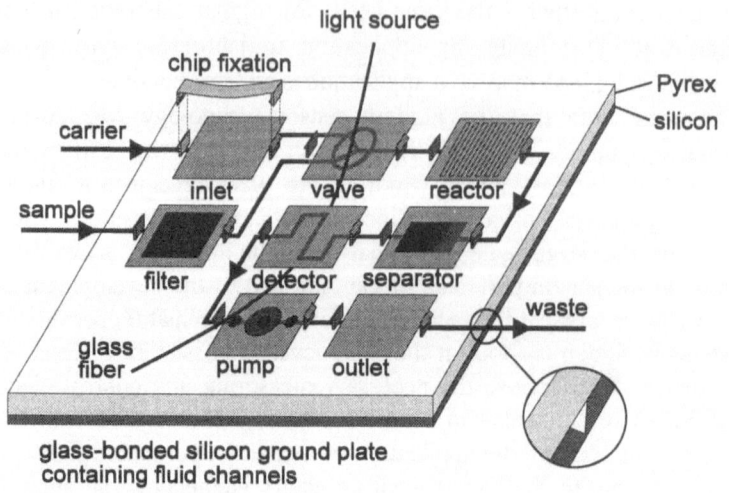

Fig. 1. Artist's impression of modular μTAS with standard "floorplan"

With respect to the application of μTAS it should be noted that it can not be expected that the new miniaturised systems will replace the standard auto-analysers in the central laboratory, although this might have been suggested in the previous section The μTAS is aimed to be applied outside the central laboratory for bedside use, environmental monitoring, etc Even in those applications it should not be expected that the systems replace the sampling techniques and laboratory tests μTAS should be seen as screening test equipment for a quick first analysis of the most important parameters of a sample The advantage of the system should be that it is portable and has a fast response, combined with such a high degree of automation and intelligence that it can be operated by rather unskilled personnel To come back to the history of the development of consumer electronics as described in the introduction, the existence of μTAS beside the large auto-analysers might be compared with the existence of transistor radio's, camcorders etc , beside respectively HiFi tuners and professional video cameras

5. Conclusion.

Both, the present availability of technologies for producing miniature sensors and liquid-handling systems and the needs for portable stand-alone analysers, make it possible to develop a new class of auto-analysers, μTAS, to be applied outside a central laboratory It will be necessary to standardise the interconnection schemes for the different parts, in order to develop μTAS as modular systems, to be composed on specific application specifications

µ-TAS: MINIATURIZED TOTAL CHEMICAL ANALYSIS SYSTEMS

Andreas Manz, Elisabeth Verpoorte, Daniel E. Raymond, Carlo S. Effenhauser, Norbert Burggraf, H. Michael Widmer
Corporate Analytical Research, Ciba-Geigy Ltd ,
CH-4002 Basel, Switzerland

Abstract

The miniaturized total chemical analysis system is a concept for on-line monitoring combining classical analytical techniques and photolithographically defined micro structures. Examples of silicon and glass micro structures for flow-injection analysis, capillary liquid chromatography and capillary electrophoresis are given. The results obtained indicate faster separations, dramatically reduced reagent consumption, and access to novel types of analysis techniques.

1. Introduction

The continuous monitoring of a chemical parameter, usually the concentration of a chemical species, is of increasing importance in chemical production, environmental analysis, medical diagnostics, and the auto industry. In almost all cases, the chemical compound of interest is to be found in sample matrices containing a host of other compounds, including species which could potentially interfere in an analysis. Several strategies have been developed to specifically distinguish an analyte against chemically complex backgrounds. Those based on the use of chemical sensors strive to maximize the chemical selectivity of a determination. In contrast, approaches incorporating separation processes focus on separation efficiency, to lessen the selectivity requirements placed on the detector. Common to all analytical methodologies is the goal of sufficient sensitivity to measure over the entire range of relevant concentrations.

A van den Berg and P Bergveld (eds), Micro Total Analysis Systems, 5–27
© 1995 *Kluwer Academic Publishers*

Selective Chemical Sensors. The classical example of an almost ideal chemical sensor is the glass pH electrode. The difference in physical properties and chemical behavior of the proton (or H_3O^+) relative to other chemical species is responsible for the tremendous selectivity. Even a poorly designed sensor based on this sensing principle can often be successfully implemented! The preceding statements also apply to certain humidity and gas sensors. Generally speaking, however, the application of such sensors under industrial conditions might involve a variety of problems [1].

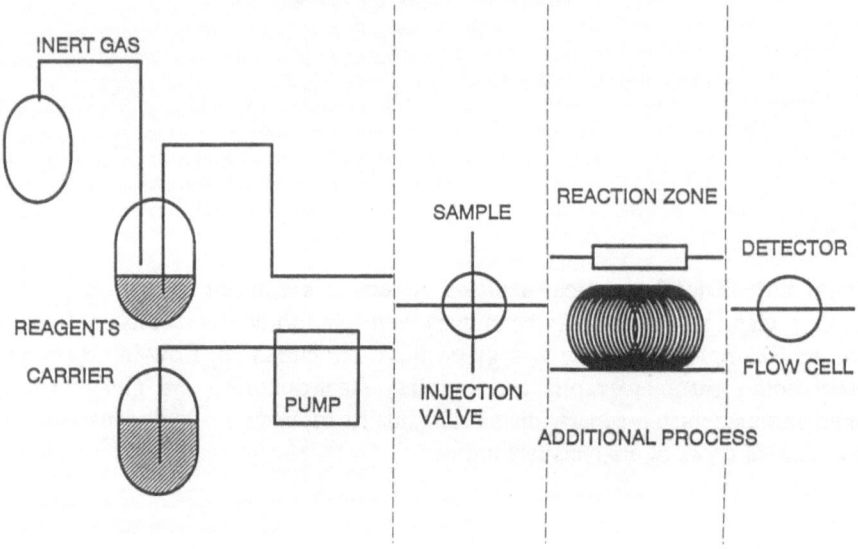

Fig. 1. General arrangement of a typical flow-injection analysis (FIA) system.

An ideal sensor is specific, that is, it transduces the concentration of the chemical compound of interest, to the exclusion of all others, into an electrical or optical signal. In addition, the sensor can be immersed directly into a sample solution or stream (see Figure 5b). These are the goals of many sensor technologists, but so far results regarding the selectivity and lifetime of most sensors have been unconvincing. There are a variety of ionic and biologically active species measurable under special matrix conditions. However, the development of a chemical sensor for a particular type of analyte usually necessitates a significant effort in terms of time, with months to years being the norm. With the exception of pH electrodes, sensors are not employed for monitoring within chemical production. The development of novel transducers, such as the ion sensitive field effect transistor (ISFET) or integrated optics, have to date not improved the status of sensors in this regard. At present, most chemical sensors are implemented using flow systems or robotics for sample handling and sample pretreatment. Future developments may change the present situation, but applications in industry require a more versatile and rugged technology.

Flow Injection Analysis. A possibility to circumvent lifetime and stability problems with sensors is provided by 'flow injection analysis' (FIA). The samples only periodically contact the sensor or detector, perhaps for a few seconds, according to a desired measuring interval typically 20 sec to 1 min in duration. In contrast to automated robot sampling systems, the samples here are transported by flowing carrier liquids. Figure 1 shows a general arrangement of an FIA system with the essential components: propelling unit (pump or gas-pressure unit), injection system, tube zone (linear and/or coiled, with reactor/mixing chamber), detection system (flow-cell) and recorder. FIA also lends itself very well to the adaptation to flow analysis of the majority of traditional wet chemical analysis methods reported over the past decades.

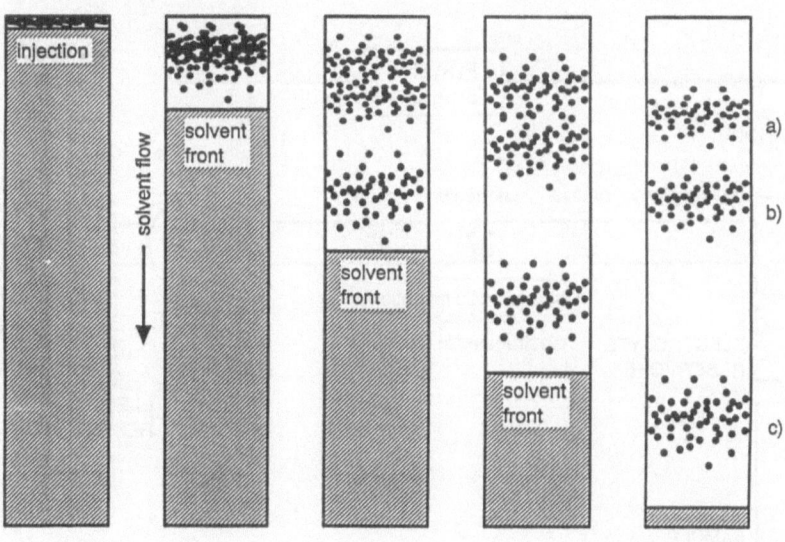

Fig. 2. Principle of a chromatographic separation, starting with the injection and depicted at various stages of a typical separation.

8

For detailed information about FIA consult the literature in [2,3].

Separation Techniques. Another extremely powerful and generally applicable set of chemical analytical techniques includes chromatography and electrophoresis. Both techniques exploit the fact that physical properties of similar chemical compounds are similar but not identical. To adapt to a novel analyte requires only a few days to weeks, in contrast to chemical sensors.

Figure 2 shows the principle underlying chromatography. Viewing the figure from left to right, a solvent, the so-called mobile phase, is forced through a thin film, a capillary, or a packed bed of particles by means of gravity, capillary forces, high pressure pumps or electric fields. The sample is injected into this stream as a small volume rectangular plug at one end of the system. During elution, the individual components of the sample interact with the so-called stationary phase, which covers surfaces in the system. The components exhibit different overall elution rates, depending on the interaction equilibria established by the individual components during the run. The detector can be positioned at one end of the system, measuring continuously (signal = f(time)), or perform a scan of the system at a designated time (signal = f(distance)). The many

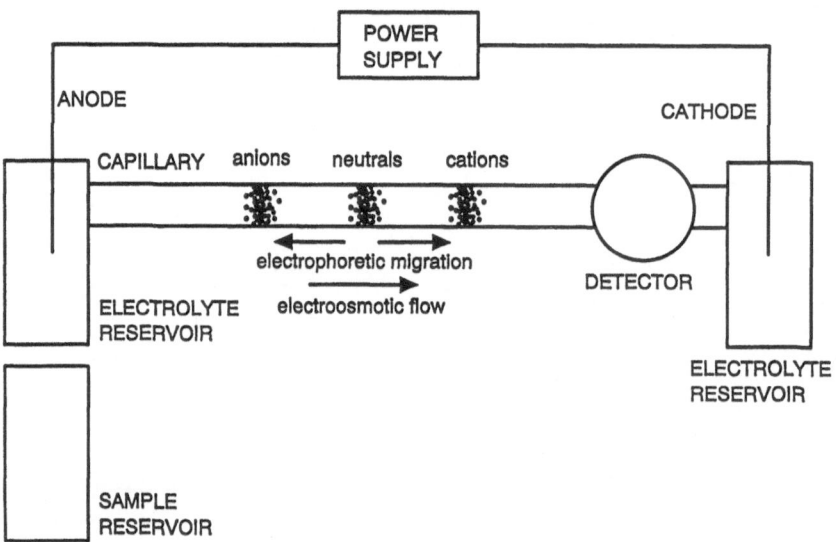

Fig. 3. Arrangement of a typical capillary electrophoresis (CE) experiment. The capillary and the anode are moved to the sample reservoir for injection.

9

different types of chromatography may be classified according to: 1) setup (column or thin layer chromatography), 2) the mobile phase (gas, supercritical fluid or liquid chromatography), 3) separation mechanism (polar or non-polar gas chromatography, adsorption, distribution, reverse phase, ion-exchange, affinity or gel permeation liquid chromatography) and 4) volume (analytical, preparative, capillary, packed column chromatography). Further details can be found in references [4, 5].

Figure 3 shows the general setup for capillary electrophoresis (CE). The driving force is the electrical field generated by the power supply. Typical field strengths are around 200 to 500 V/cm. The separation mechanism is the difference in the electrophoretic mobilities of the sample components. Even though anions move toward the anode, the net displacement is toward the cathode, due to the electrokinetic phenomenon known as electroosmosis. Electroosmosis causes bulk electrolyte flow as a result of the layer of counterions at the charged capillary surface being brought into motion toward one electrode upon application of an electric field. As in chromatography, there are different modes of operation in CE, including capillary, bed, free flow, or gel electrophoresis, and isotachophoresis or isoelectric focusing. For detailed information consult the literature cited [6, 7].

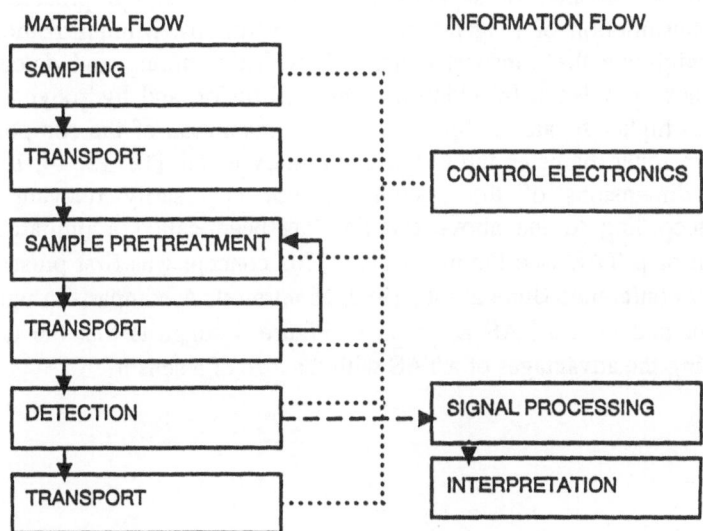

Fig. 4. General flow chart of a quantitative chemical analysis of a single component.

2. Concept

Total Chemical Analysis System. The two groups of chemical analytical techniques presented above can be used as 'stand alone techniques', that is, an apparatus, standing in a lab, operated by a qualified technician. Most of these methods are precise and reproducible, but also time consuming. For monitoring purposes, the instrument needs to be fully automated, from sampling through information evaluation. The state-of-the-art strategy is the so called 'Total Chemical Analysis System' (TAS), which periodically transforms chemical information into electronic information. Sampling, sample transport, any necessary chemical reactions, chromatographic separations, and detection are automatically performed in a flowing stream (see Figure 4) [8]. Because the sample pretreatment serves to eliminate most of the interfering chemical compounds, the detector or sensor in a TAS need not be highly selective. Furthermore, calibration can be incorporated into the system. This approach presents a possibility for the 'real-time' monitoring of rapidly changing industrial sample compositions (e.g., river water, chemical reaction mixtures, cultivation media in bioprocesses). Some examples of TAS are given in [9,10,11].

Miniaturized Total Chemical Analysis Systems. Main disadvantages of TAS, especially for liquid phase samples, include slow transport of the sample in the analysis system, poor separation speed in the case of liquid chromatography or electrophoresis, a lack of selectivity (sensors) or separation efficiency (separation processes), and a considerable consumption of reagents and carrier solutions. Miniaturization of flow manifolds to minimize the transport distance between sampling and detection point addresses all these problems. In addition, simple diffusion and hydrodynamic theory clearly points to higher separation speed and better resolution of the components with smaller capillary inner diameter and a shorter capillary length [12, 13, 14, 15, 16]. The overall outer dimensions of the system are not necessarily relevant. A TAS, miniaturized according to the above criteria, has been called a miniaturized total analysis system or μ-TAS (see Figure 5) [17]. This concept was first presented at the Transducers '89 conference (June 25-30, 1989, Montreux). A comparison with an ideal chemical sensor and with a TAS as given in Figure 5 suggests that the μ-TAS is a hybrid combining the advantages of a TAS with the size of a sensor.

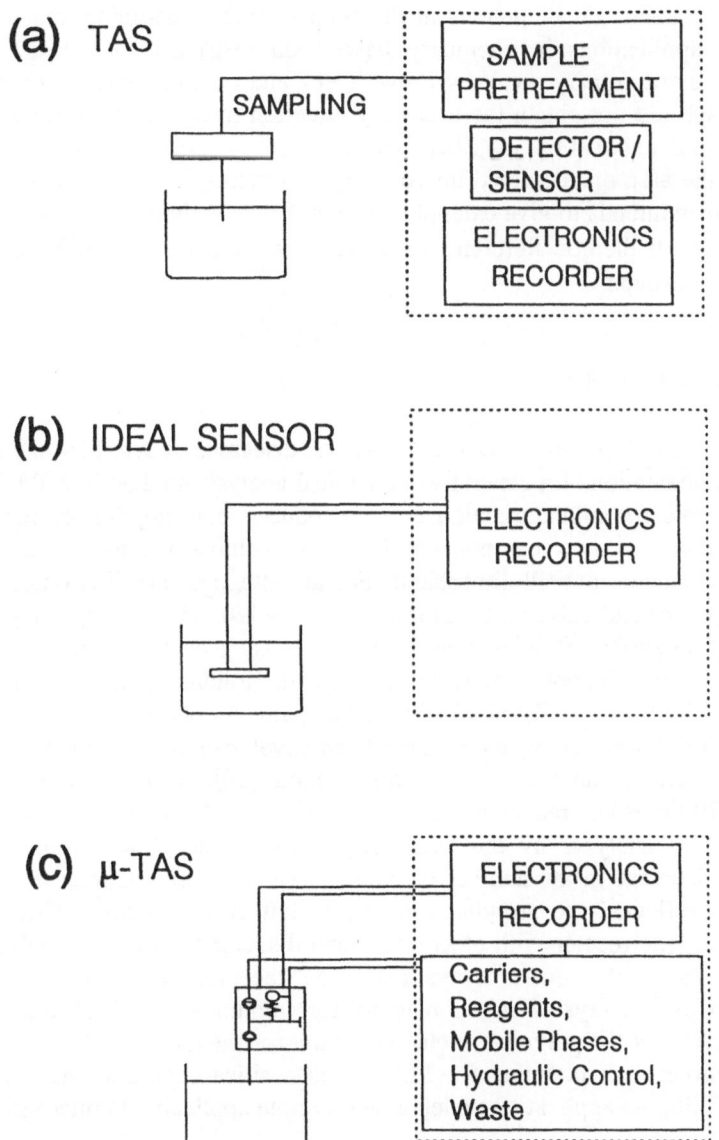

Fig. 5. Schematic diagram of (a) a total chemical analysis system (TAS) (b) an ideal chemical sensor and (c) a miniaturized total chemical analysis system (μ -TAS).

Despite the dramatic developments in the field of micromachining and micro system technology, applications in chemistry have been restricted to sensors, valves and pumps. The extraordinary work by Steve Terry and co-workers on an integrated gas chromatograph did not obtain the necessary attention from the scientific community at the time of its publication in 1979 [18, 19]. It has, however, stimulated renewed research in the area of integrated micro analysis systems over the last few years. This paper primarily intends to give examples from our own activities using micro structures in μ-TAS or parts thereof. References should include most of the published work from other research groups.

3. Experimental Results

Example I: Flow-Injection Analysis (FIA). As discussed above, FIA is a good choice for the implementation of a general wet chemical analysis method in a TAS [2, 3]. The main arguments for miniaturization are the reduced consumption of reagents, more efficient mixing of two components in small volume reactors, shorter transport distances and smaller overall dimensions. For the integration of FIA onto planar micro structures, pumps and valves play an important role [20, 21]. Among early activities in the realm of integrated fluidics combined with sensing elements are a micro titration unit by B. van der Schoot (University of Twente, Holland) [22, 23] and a blood gas analyzer [24] by S. Shoji (Tohoku University, Japan). A concept for using stacked chips for a three-dimensional flow manifold has been developed in our lab [25, 26, 27], and is discussed in detail in another paper of this volume [28]. A phosphate analyzer of this type uses 270 times less reagents than a conventional FIA. A similar system have been used as an ion analyzer by the University of Neuchâtel (Switzerland) [29]. More recently, an increase in the number of publications describing silicon micro structures for FIA has reflected the growing research activities in this area. These include an integrated enzyme reactor with electrochemical detector for glucose analysis [30, 31], detection cells for biosensor arrays based on receptors and potentiometric signal transduction [32], arrays of micro reactors for immunoassays [33] and miniaturized reaction vessels for the polymerase chain reaction for the rapid DNA analysis [34]. The use of electroosmotic pumping for FIA in fused silica capillaries has been reported recently [35, 36], an application which is in principle applicable in micro structures [37, 38].

Example II: Continuous Sample Preparation. For the primary sample preparation steps in a TAS, continuous techniques are clearly advantageous, e.g., filtration, dialysis, SPLITT [39] or free-flow electrophoresis (FFE) [40]. A special version of continuous zone electrophoresis in quartz channel structures has been presented by Mesaros et al.

recently [41]. The general principle of FFE is given in Figure 6. A pumped flow (carrier) in a first dimension and an electric field in a second dimension allows for the continuous separation of sample components being continuously introduced at the top of the flat separation bed. Conventional FFE is a preparative method used for the isolation of various biological components [40]. Typical separation bed dimensions are 10 cm wide, 50 cm long and 0.4 - 1 mm deep. Ion exchange membranes are used to isolate the electrodes from the separation bed. FFE has been miniaturized and integrated onto a silicon micro structure by reducing all dimensions by a factor of 10 (Figure 7a). The ion exchange membranes have been replaced by arrays of narrow 'V'-groove channels each 10 μm deep and 12 μm wide at the base (Figure 7b). It is possible to achieve a continuous separation of small ions according to their charge. A voltage drop of 50 V gave a reasonable resolution of ions differing by one charge unit with a response time of 2 to 5 minutes (see Figure 8) [42]. The fraction of interest can be isolated and transferred to a next pretreatment step or a detection unit. Furthermore, it is possible to separate small ions from biological cells. Increased voltages should allow for higher separation efficiency and a response time of a few seconds.

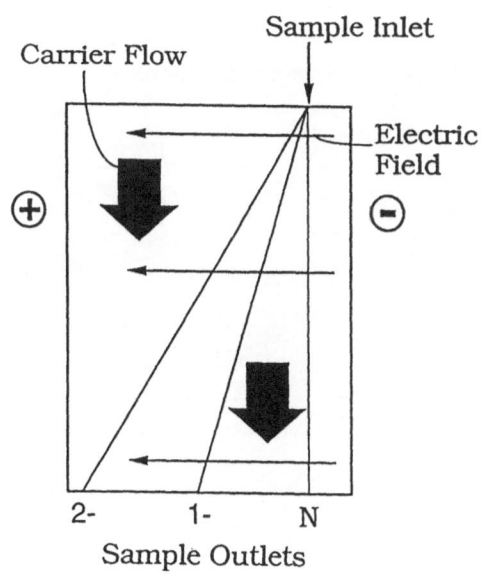

Sample Outlets

Fig. 6. Schematic illustrating the concept of free-flow electrophoresis (FFE). '2-', '1-' and 'N' denote the positions at which di-anion, mono-anion and neutral components will exit the system.

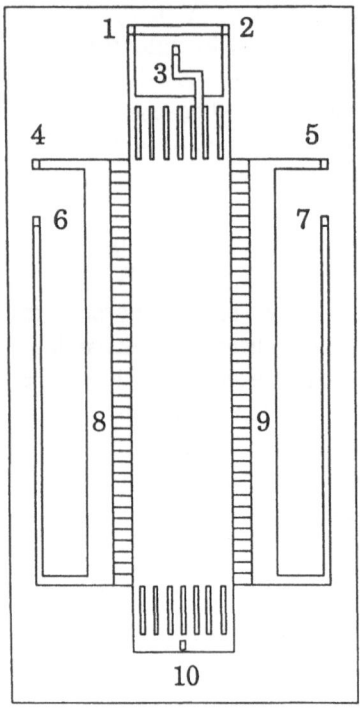

Fig. 7. (a) Silicon FFE device layout. 1 and 2: carrier buffer inlets, 3: sample inlet, 4 and 5: side bed inlets, 6 and 7: side bed outlets, 8 and 9: side beds containing platinum electrodes, 10: outlet. (b) Electron micrograph of a detail of the silicon FFE structure. Carrier and buffer inlet channels and side channels can be seen on the left and right, respectively [42].

a)

b)

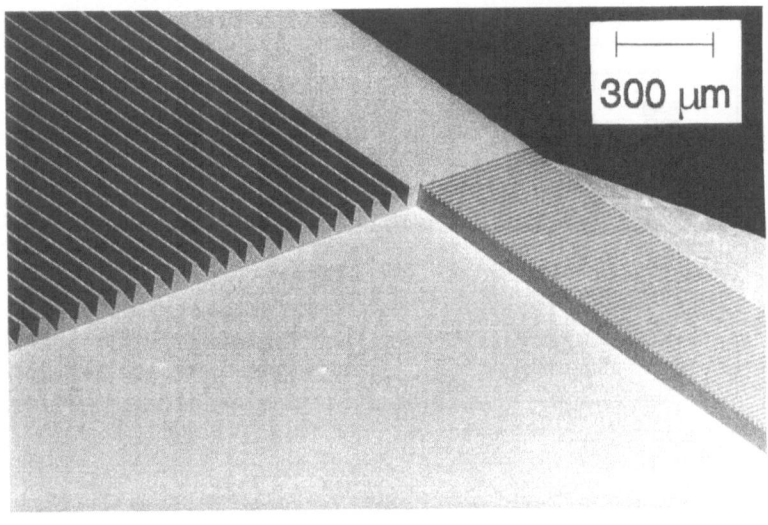

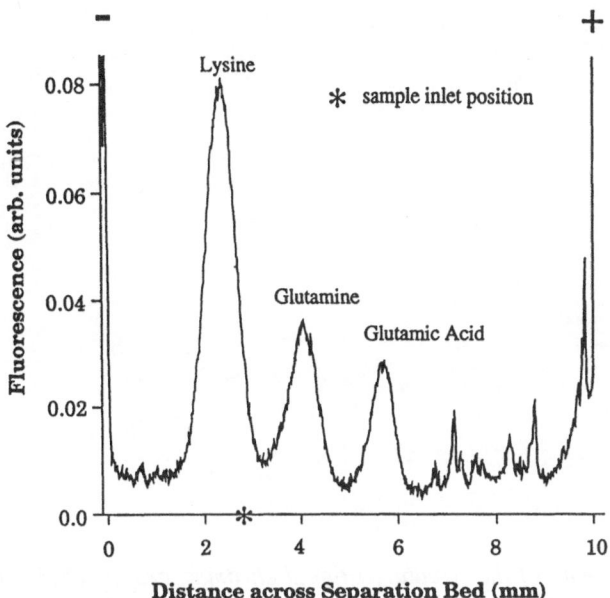

Fig. 8. Separation profile of rhodamine-B isothiocyanate labeled lysine, glutamine, and glutamic acid, obtained 24 mm along the separation bed using the FFE chip shown in Fig. 7 with 50 V applied [42]. Sample: 0.05 M labeled amino acids at 200 nL/min, carrier: pH 7 phosphate buffer (9.5 x 10 $^{-4}$ Ω $^{-1}cm^{-1}$) at 5 μ L/min, side bed electrolyte: pH 7 phosphate buffer and 0.25 M sodium sulfate (35.5 x 10 $^{-3}$ Ω $^{-1}cm^{-1}$) at 15 μ L/min. Residence time to the point of detection was 140 s.

Example III: Liquid Chromatography. According to theory, liquid chromatography in open capillaries should allow for shorter analysis times with increased separation efficiency as compared to standard packed-column high-pressure liquid chromatography (HPLC). Optimum capillary dimensions include inner diameters of less than 5 μm and capillary lengths of a few cm. The key problems of injection and detection volumes of less than a few pL has been addressed using split or stopped flow injection and small volume detector cells [15, 43, 44]. A more elegant way of defining the pL detector cells has been presented by Hitachi Ltd. [45]. In this case, a capillary of 1.5 nL volume integrated onto a silicon chip meets an electrochemical detector giving 1.2 pL of detection volume. In the case of absorption measurements the optical path length has to be as long as possible for a satisfactory sensitivity. A silicon made 'U'-shaped detector cell showed the expected sensitivity of 1 mm optical path length at volumes of 4-15 nL [46].

16

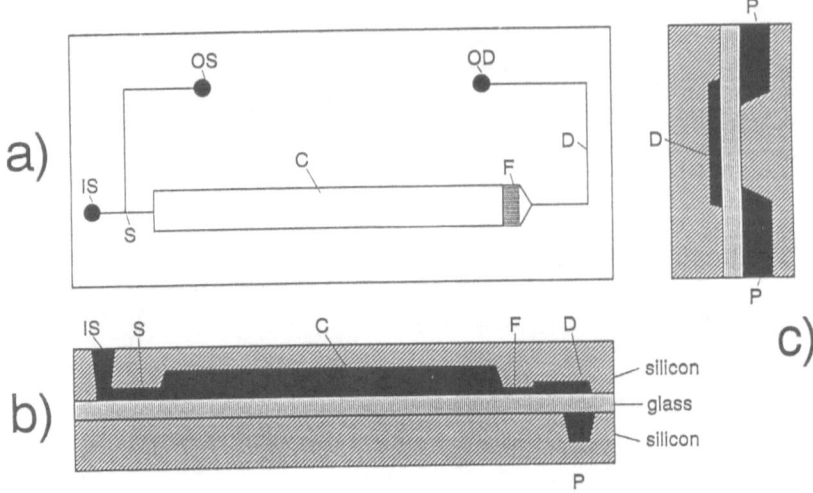

Fig. 9. Schematic layout of the structured liquid chromatography chip (LC). Shown are (a) the layout of the channel system in silicon, (b) a cross-section along the separation channel axis, and (c) a cross-section along the detector cell axis. IS: inlet for sample peak and mobile phase, S: split injector, OS: outlet for the rejected portion of the sample peak and mobile phase, C: separation channel (column), F: frit, D: optical detector cell, OD: outlet, P: positioning grooves for optical fiber.

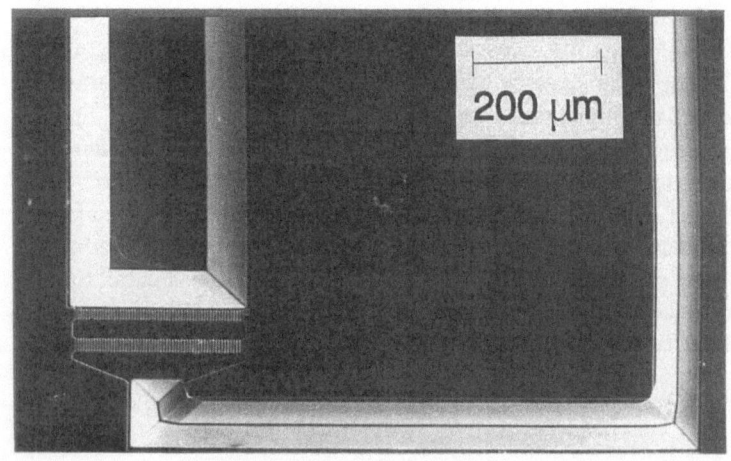

Fig. 10. Electron micrograph of the silicon LC device showing part of the separation channel, frit and optical detector cell [47].

To avoid extra-column band broadening, a split injector, a packed column, a frit and an optical detector cell have been integrated onto a silicon chip (see Figure 9). Internal volumes are 0.5 μL for the empty column, and 3.5 nL for the total of the dead volumes on the chip. A single capillary of the frit had a defined volume of only 59 fL (see Figure 10). Preliminary results indicated that these HPLC chips could be operated at pressures up to 120 atm, with the limiting factor not the chip itself, but the interface with the peripheral system. It was possible to pack the capillary column on the assembled chip, and a separation of two components could be obtained [47].

Recently, using glass micro structures for electroosmotically driven liquid chromatography was reported [48], as well as using multichannel array thin-layer chromatography with diffuse reflectance infrared spectroscopy detection [49]. Open-tubular liquid chromatography integrated onto silicon has been successful [50].

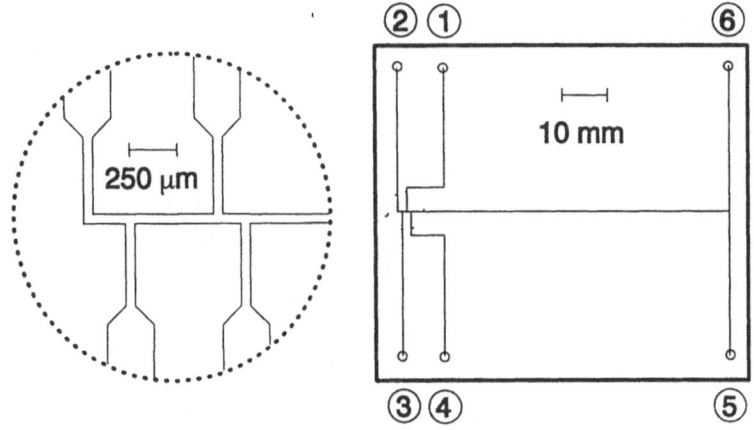

Fig. 11. Layout of the capillary electrophoresis (CE) glass micro structure. The horizontal channel was used for the separation. 1: sample inlet, 2: carrier electrolyte inlet, 3: modifier electrolyte inlet, 4: sample outlet, 5 and 6: outlets.

Example IV: Capillary Electrophoresis. Theoretical and experimental work indicated a major breakthrough in miniaturized capillary electrophoresis (CE). Short analysis times are easily obtained and high voltages applied to the capillaries allow for a good separation efficiency [16]. The first results obtained using glass structures were presented in June 1991 at the HPLC '91 conference (Basel) [51] and at the Transducers '91 (San Francisco) [52, 53]. CE is well suited to the concept of μ-TAS [54]. For a perfect injection and detection, the separation efficiency (number of theoretical plates)

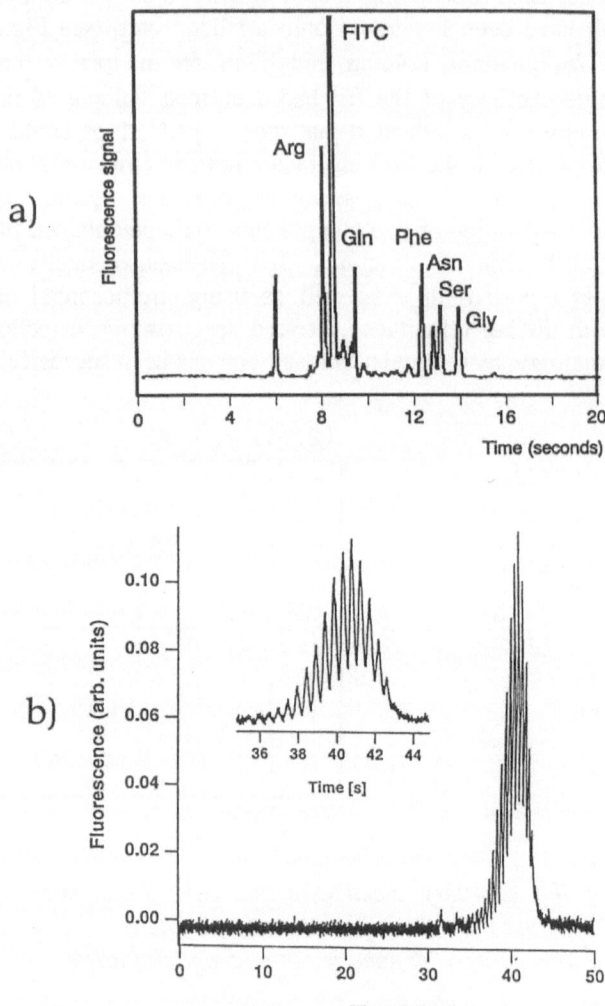

Fig. 12. Electropherograms obtained using the CE structure shown in Fig. 11. (a) Separation of FITC-labeled amino acids in zone CE [56]. Sample: 10 μ M each amino acid, carrier electrolyte: 20 mM boric acid and 100 mM Tris buffer pH 9.0. Separation length 24 mm using 1060 V/cm. (b) Separation of FITC-labeled phosphorothioate oligonucleotide mixture PS pd(T)$_{10-25}$ [60]. Gel: 10% T non-crosslinked polyacrylamide matrix, buffer: 100 mM Tris, 100 mM boric acid, 2 mM EDTA, 7 M urea at pH 8.5. Separation length 38 mm using 2000 V/cm.

is proportional to the applied voltage, regardless of the length of the capillary. The analysis time is dramatically reduced for short capillaries [55, 56] and capillaries can be operated in parallel [57]. The capillary cross-section has to be adapted to the known, but not desired, heating power generated by the electrical current. At the same time, the requirements for the injection and detection volumes are less strict (50-200 pL) in comparison to capillary HPLC (1-5 pL) [17, 58, 59]. Figure 11 shows a layout for a micro structure in glass. Optimized experiments yielded separation efficiencies of 150,000 and 200,000 in free electrolyte and in non-crosslinked polyacrylamide gels, respectively (see Figure 12) [56, 60]. That means, the height equivalent to a theoretical plate is only 200-300 nm. Separations usually take less than a minute. The injection, electrokinetically performed in a double 'T' arrangement, did not discriminate according to mobilities [61] and allowed the definition of 100 pL injection volumes of truly representative sample. The carrier electrolyte consumption was 600,000 times less than compared to standard HPLC, and the analysis time was shorter by a factor of 50 to 100.

It has been shown that silicon micro structures are not acceptable for CE in channels of micrometer dimensions, because the insulation allows for voltages of only several hundred V, which is far too low for a reasonable separation efficiency [62]. Polymer micro structures have been used by B. Ekström, Pharmacia AB (Sweden) [63], more complex manifolds in glass have been tested by D.J. Harrison at the University of Alberta [64] and millisecond to minute separations although with poor separation efficiencies have been published by M. Ramsey, Oak Ridge National Lab [65, 66].

Example V: Synchronized Cyclic Capillary Electrophoresis (SCCE). A novel concept for capillary electrophoresis column switching has been possible using the small-volume connection capabilities of a planar glass micro structure. Similar to an electromotor, the applied voltages, and therefore the ionic current, are rotated by 90 degrees relative to a layout of 4-fold axial symmetry. A detailed procedure is given in Figure 13. It allows for elimination of too rapid and too slow components, and at the same time, an increasing separation resolution of the remaining components [67, 68]. During one cycle, the ions migrate in an electric field corresponding to twice the voltage applied and therefore show a higher values for plate numbers and resolution. Experimental results using a glass device with 80 mm separation capillary circumference clearly demonstrate the principle of SCCE (see Figure 14). After 10 cycles, 1,300,000 theoretical plates were obtained using only 2 kV applied voltage [69]. The main disadvantage at the moment is the decrease in peak area caused by a loss of material at the intersections and by photo bleaching of fluorescently labeled sample.

20

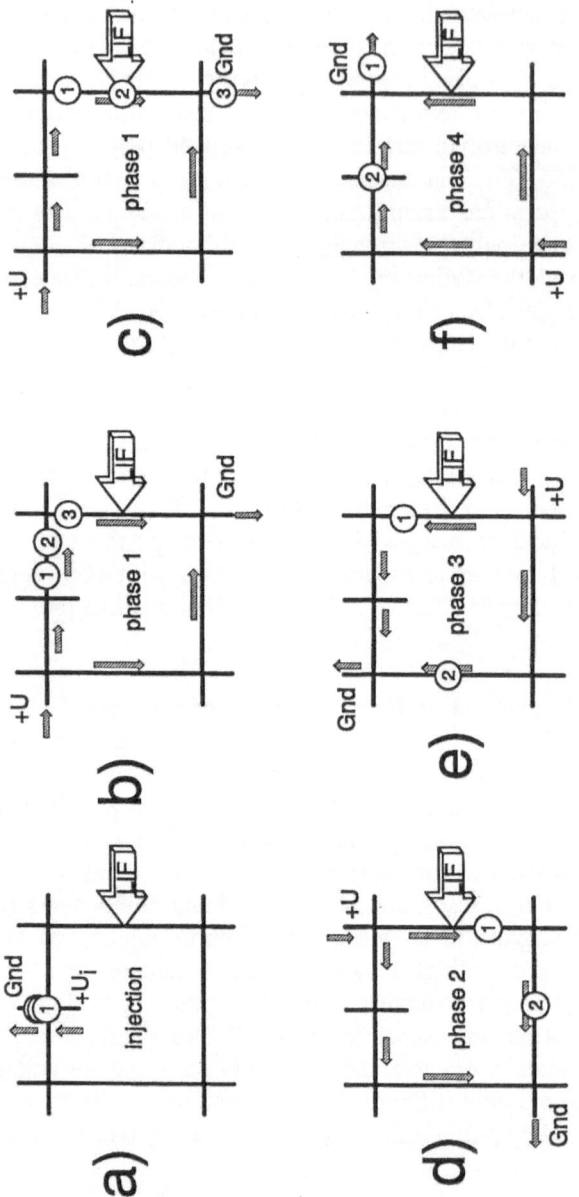

Fig. 13. Principle of synchronized cyclic capillary electrophoresis (SCCE). Three sample components are symbolized by the circled numbers 1, 2 and 3. The voltage switching process is synchronized to component 2. (a) injection phase, (b) during phase 1, (c) at the end of phase 1, (d) phase 2, (e) phase 3, and (f) end of the cycle.

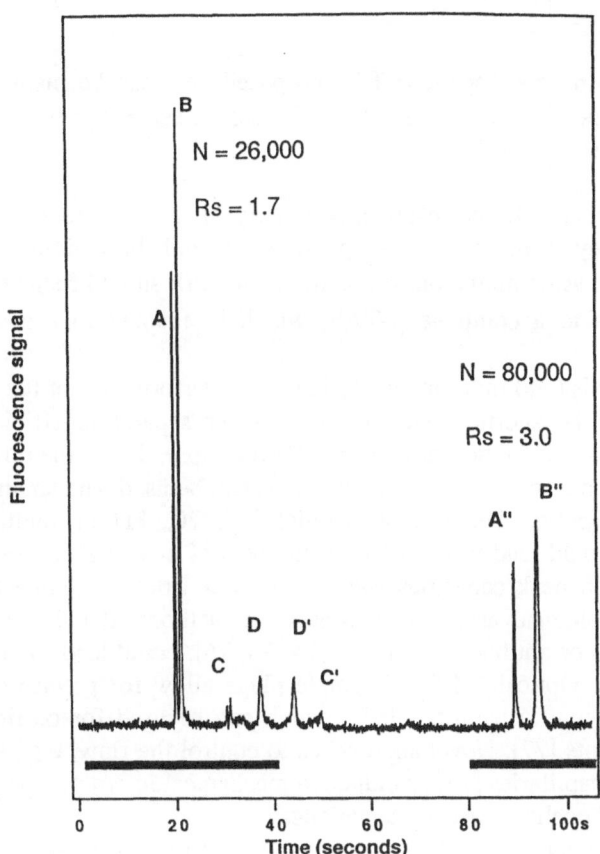

Fig. 14. Electropherogram obtained using the SCCE structure given in Fig. 13 [69].
Separation of a fluorescein-isothiocyanate sample and its degradation products.
Carrier electrolyte: 20 mM boric acid, 100 mM Tris buffer pH 9.0. Length of a full
cycle 80 mm, voltage applied 2 kV across the diagonal. A, B, C and D: components
detected during phase 1 and 2 (see Fig. 13) at a separation length of 20 mm, D' and
C': slow components in backward motion detected during phases 3 and 4, A" and B":
components detected during the second cycle at a separation length of 100 mm. The
separation efficiency is given in N: number of theoretical plates and Rs: resolution.

4. Outlook

The micro structures needed for the μ-TAS proposed here can be manufactured with ease in silicon or glass. The fabrication of small research series can be carried out in 1 to 3 months for a reasonable price. Mass production would clearly lead to even cheaper devices.

The micro structures and the corresponding functional models are not yet industry-proof. However, they demonstrate the possibilities and limitations of individual elements of a μ-TAS. A continuation of this research work should focus on combining the various elements to a complete μ-TAS, which is adapted to a specific on-line analysis problem.

The publications in this field indicate clearly improved performance of the physical side of the analysis, that is, shorter analysis times, better separation efficiencies and a dramatically reduced consumption of reagents. Furthermore, due to the minute volumes for internal connections, new types of combinations can be used, and small samples can be analyzed with success. The use of parallel [57, 70, 71] or multi-dimensional arrangements [72] would lead to even larger numbers of analyses per unit time, or to dramatically increased peak capacities (separation of > 1,000 components [73]). The trend to combine biological assays with separation methods, that is, protein protein interactions, enzymes or antibodies with CE [74, 75, 76], could lead to novel concepts for chemical sensing. Optically defined sample plugs allow for precise small volume injections for millisecond separations [55], which can be used for on-line or in-vivo monitoring experiments [77]. Novel approaches to control the flow, e.g., radial control of electroosmosis in capillaries [78] or inductive mechanical micro pumps [79, 80], will allow access to novel cyclic separation techniques.

Finally, this development will lead to a complete integration of an analysis lab onto a single chip. This is possible because chemical information needs almost no space and very little time. An increasing number of publications deal with detection of individual molecules [81, 82, 83].

The number of interested and contributing research labs is steadily increasing. This conference 'μ-TAS' held in Enschede documents the status of activities in many of the leading labs worldwide. The public interest is growing, and a broad evaluation of the concept will take place in the years to come. Though this technology is a few years away from a commercial product, the benefits clearly indicate that it will play an important role in the analytical chemistry of the 21st century.

Acknowledgment

We would like to thank our colleagues contributing to the fruitful background of this research in Ciba-Geigy Ltd., Basel, Switzerland (Aran Paulus, James C. Fettinger, Stefan Haemmerli, Gregor Ocvirk, Wietske K. Duursma, Roman Bolinger, Franz von Heeren, Eithne Dempsey, Michael Busch), at the University of Alberta, Edmonton, Canada (D. Jed Harrison, Zhonghui Fan, Kurt Seiler, Karl Fluri, Paul Glavina), at the Université de Neuchâtel, Switzerland (Bart H. van der Schoot, Sylvain Jeanneret, Volker Gass) and in the Hitachi Central Research Laboratory, Tokyo, Japan (Yuji Miyahara and K. Sato). Hans Lüdi, Niklaus Graber, Morten B. Garn, Pierre Bataillard, Sietse Wouters, Nils Blom, Alfredo Bruno (all Ciba-Geigy), Nico F. de Rooij (Université de Neuchâtel), Hiroyuki Miyagi, J. Miura, Y. Watanabe (all Hitachi), Wilhelm Simon, Werner E. Morf (both ETH Zürich), Steven Terry, Marc Boillat, Vladimir Vaganov (all ICSensors Inc., Milpitas, California) and Rudolf Wildi (Baumer IMT, Greifensee, Switzerland) contributed by interesting discussions on various aspects, technical, conceptual and financial support. This project is partly supported by the European project BCR (Measurement and Testing, EU, Brussels), by the Swiss national projects KWF ('Kommission zur Förderung der wissenschaftlichen Forschung', Bern) and SNF ('Schweizerischer Nationalfonds', Bern).

References

[1] N.Graber, H.Lüdi, H.M.Widmer, Sensors & Actuators **B1** (1990) 239.

[2] M.Valcarel, M.D.Luque de Castro, Flow injection analysis, principles and applications, Ellis Horwood Ltd., Chichester (England) 1987.

[3] J.Ruzicka, E.H.Hansen, Flow injection analysis, Wiley, New York 1988.

[4] L.S.Ettre, W.H.McFadden, Ancillary techniques of gas chromatography, Wiley, New York 1969.

[5] L.R.Snyder, J.J.Kirkland, Introduction to modern liquid chromatography, Wiley, New York 1979.

[6] S.F.Y.Li, Capillary electrophoresis, Principles, practice and applications, Elsevier, Amsterdam 1992.

[7] P.Jandik, G.Bonn, Capillary electrophoresis of small molecules and ions, VCH Publishers, New York 1993.

[8] H.M.Widmer, Trends Anal. Chem. **2** (1983) 8.

[9]   H.M.Widmer, J.F.Erard, G.Grass, Int. J. Environ. Anal. Chem. **18** (1984) 1.

[10] M.Garn, P.Cevey, M.Gisin, C.Thommen, Biotechnol. Bioeng. **34** (1989) 423.

[11] A.Giorgetti, N.Periclés, H.M.Widmer, K.Anton, P.Dätwyler, J. Chromatogr. Sci. **27** (1989) 318.

[17] A.Manz, N.Graber, H.M.Widmer, Sensors and Actuators **B1** (1990) 244-248.

[12] A.Manz, D.J.Harrison, E. Verpoorte, H. M. Widmer, Advances in Chromatography **33** (1993) 1-66.

[13] M.J.E.Golay in *Gas Chromatography* (V.J.Coates et al. eds.), Academic Press, New York, 1958.

[14] J.H.Knox, M.T.Gilbert, J. Chromatogr. **186** (1979) 405.

[15] S.Müller, D.Scheidegger, C.Haber, W.Simon, J. High Resolution Chromatogr. **14** (1991) 174.

[16] C.A.Monnig, J.W.Jorgenson, Anal. Chem. **63** (1991) 802.

[18] S.C.Terry, 'A gas chromatography system fabricated on a silicon wafer using integrated circuit technology', Ph.D. dissertation, Stanford University, 1975.

[19] S.C.Terry, J.H.Jerman, J.B.Angell, IEEE Trans. Electron. Devices **ED-26** (1979) 1880.

[20] J.Ruzicka, E.H.Hansen, Anal. Chim. Acta **161** (1984) 1.

[21] P.Gravesen, J.Branebjerg, O.S.Jensen, MME '93, Workshop Digest, Swiss Foundation for Research in Microtechnology, Neuchâtel 1993, pp. 143-164.

[22] B.van der Schoot, P.Bergveld, Sensors & Actuators **8** (1985) 11.

[23] W.Olthuis, B.H.van der Schoot, F.Chavez, P.Bergveld, Sensors & Actuators **17** (1989) 279.

[24] S.Shoji, M.Esashi, T.Masuo, Sensors & Actuators **14** (1988) 101.

[25] A.Manz, J.C.Fettinger, E.Verpoorte, S.Haemmerli, H.M.Widmer, Micro System Technologies 91, R.Krahn and H.Reichl, eds., VDE-Verlag Berlin and Offenbach (1991) 49-54.

[26] J.C.Fettinger, A.Manz, H.Lüdi, H.M.Widmer, Sensors and Actuators **17B** (1993) 19-25.

[27] E.M.J.Verpoorte, B.H.van der Schoot, S.Jeanneret, A.Manz, N.F.de Rooij, American Chemical Society, Symposium Series (1994), in press.

[28] B.H.van der Schoot, E.M.J.Verpoorte, S.Jeanneret, A.Manz, N.F.de Rooij, in this volume (1994).

[29] B.H.van der Schoot, S.Jeanneret, A.van den Berg, N.F.de Rooij, Anal. Methods Instrum. 1 (1993) 38-42.

[30] l.Karube, K.Yokoyama, Sensors & Actuators, B13-14 (1993) 12-15.

[31] Y.Murakami, T.Takeuchi, K.Yokoyama, E.Tamiya, I.Karube, M.Suda, Anal. Chem. 65 (1993) 2731-2735.

[32] L.Bousse, R.J.McReynolds, G.Kirk, T.Dawes, P.Lam, W.R.Bemiss, J.W.Parce, Transducers '93 Digest of Technical Papers, ed. Institute of Electrical Engineers of Japan, Tokyo ISBN 4-9900247-2-9 (1993) 924-926.

[33] M.I.Song, K.Iwata, M.Yamada, K.Yokoyama, T.Takeuchi, E.Tamiya, I.Karube, Anal. Chem. 66 (1994) 778-781.

[34] M.A.Northrup, M.T.Ching, R.M.White, R.T.Watson, Transducers '93 Digest of Technical Papers, ed. Institute of Electrical Engineers of Japan, Tokyo ISBN 4-9900247-2-9 (1993) 399-402.

[35] S.Liu, P.K.Dasgupta, Anal. Chim. Acta 268 (1992) 1-6.

[36] P.K.Dasgupta, S.Liu, Anal. Chem. 66 (1994) 1792-1798.

[37] K.Seiler, Z.Fan, K.Fluri, D.J.Harrison, Anal. Chem. (1994), in press.

[38] W.Kaplan, H.Elderstig, C.Vieider, Proceedings of the MEMS '94 conference, Jan. 25-28, Oiso, Japan (1994) 63-68.

[39] S.Levin, G.Tawil, Anal. Chem. 65 (1993) 2254-2261.

[40] K.Hannig, H.G. Heidrich, *Free-Flow Electrophoresis*, GIT Verlag, Darmstadt (Germany) 1990, 115 pp.

[41] J.M.Mesaros, G.Luo, J.Roeraade, A.G.Ewing, Anal. Chem. 65 (1993) 3313-3319.

[42] D.Raymond, A.Manz, H.M.Widmer, Anal. Chem. 66 (1994), in press.

[43] A.Manz, W. Simon, J. Chromatogr. Sci. 21 (1983) 326-330.

[44] A.Manz, W. Simon, Anal. Chem. 59 (1987) 74-79.

[45] A.Manz, Y.Miyahara, J.Miura, Y.Watanabe, H.Miyagi, K.Sato, Sensors and Actuators B 1 (1990) 249-255.

[46] E.Verpoorte, A.Manz, H.Lüdi, A.E.Bruno, F.Maystre, B.Krattiger, H.M.Widmer, B.H.van der Schoot, N.F.de Rooij, Sensors and Actuators B 6 (1992) 66-70.

[47] G.Ocvirk, E.Verpoorte, A.Manz, M.Grasserbauer, H.M.Widmer, Anal. Methods
      Instrum. (1994), in press.

[48] S.C.Jacobson, R.Hergenröder, L.B.Koutny, J.M.Ramsey, Anal. Chem. 66 (1994)
      2369-2373.

[49] S.P.Bouffard, J.E.Katon, A.J.Sommer, N.D.Danielson, Anal. Chem. 66 (1994) 1937-
      1940.

[50] S.Cowen, D.Craston, E.Wasson, An on-chip miniature liquid chromatography system
      with on-line electrochemical detection, oral presentation at the Deauville Conference,
      Montreux (Switzerland), May 16-20, 1994.

[51] A.Manz, D.J.Harrison, E.M.J.Verpoorte, J.C.Fettinger, H.Lüdi, H.M.Widmer, J.
      Chromatogr. 593 (1992) 253-258.

[52] A.Manz, D.J.Harrison, J.C.Fettinger, E.Verpoorte, H.Lüdi, H.M.Widmer,
      Transducers '91, Digest of Technical Papers, IEEE 91CH2817-5, New York ISBN
      0-87942-586-5 (1991) 939-941.

[53] D.J.Harrison, A.Manz, P.G.Glavina, Transducers '91, Digest of Technical Papers,
      IEEE 91CH2817-5, New York ISBN 0-87942-586-5 (1991) 792-795.

[54] D.J.Harrison, K.Flury, K.Seiler, Z.Fan, C.S.Effenhauser, A.Manz, Science 261
      (1993) 895-897.

[55] A.W.Moore, J.W.Jorgenson, Anal. Chem. 65 (1993) 3550-3560.

[56] C.S.Effenhauser, A.Manz, H.M.Widmer, Anal. Chem. 65 (1993) 2637-2642.

[57] X.C.Huang, M.A.Quesada, R.A.Mathies, Anal. Chem. 64 (1992) 967-972.

[58] A.Nann, I.Silvestri, W.Simon, Anal. Chem. 65 (1993) 1662-1667.

[59] S.Sloss, A.G.Ewing, Anal. Chem. 65 (1993) 577-581.

[60] C.S.Effenhauser, A.Paulus, A.Manz, H.M.Widmer, Anal. Chem. 66 (1994), in press.

[61] C.S.Effenhauser, Anal. Methods Instrum. 1 (1994) 172-176.

[62] D.J.Harrison, P.G.Glavina, A.Manz, Sensors & Actuators B10 (1993) 107-116.

[63] B.Ekström, G.Jacobson, O.Öhman, H.Sjödin, Patent: WO 91/16966 (1990).

[64] Z.Fan, D.J.Harrison, Anal. Chem. 66 (1994) 177-184.

[65] S.C.Jacobson, R.Hergenröder, L.B.Koutny, R.J.Warmack, J.M.Ramsey, Anal. Chem.
      66 (1994) 1107-1113.

[66] S.C.Jacobson, R.Hergenröder, L.B.Koutny, J.M.Ramsey, Anal. Chem. 66 (1994) 1114-1118.

[67] N. Burggraf, A.Manz, N. F. de Rooij, H. M. Widmer, Anal. Methods Instrum. 1 (1993) 55-59.

[68] N.Burggraf, A.Manz, C.S.Effenhauser, E. Verpoorte, N.F. de Rooij, H.M.Widmer, J. High Resolut. Chromatogr. 16 (1993) 594-596.

[69] N.Burggraf, A.Manz, E.Verpoorte, C.S.Effenhauser, H.M.Widmer, N.F.de Rooij, Sensors & Actuators (1994), in press.

[70] S.Takahashi, K.Murakami, T.Anazawa, H.Kambara, Anal. Chem. 66 (1994) 1021-1026.

[71] K.Ueno, E.S.Yeung, Anal. Chem. 66 (1994) 1424-1431.

[72] A.V.Lemmo, J.W.Jorgenson, Anal. Chem. 65 (1993) 1576-1581.

[73] A.T.Andrews, Electrophoresis, Theory, techniques and biochemical and clinical applications, Clarendon Press, Oxford 1986 (2nd ed.).

[74] D.Wu, F.E.Regnier, Anal. Chem. 65 (1993) 2029-2035.

[75] N.M.Schultz, R.T.Kennedy, Anal. Chem. 65 (1993) 3161-3165.

[76] K.Shimura, B.L.Karger, Anal. Chem. 66 (1994) 9-15.

[77] B.L.Hogan, S.M.Lunte, J.F.Stobaugh, C.E.Lunte, Anal. Chem. 66 (1994) 596-602.

[78] M.A.Hayes, I.Kheterpal, A.G.Ewing, Anal. Chem. 65 (1993) 2010-2013.

[79] G.Fuhr, B.Wagner, Transducers '93 Digest of Technical Papers, ed. Institute of Electrical Engineers of Japan, Tokyo ISBN 4-9900247-2-9 (1993) 88-92.

[80] T.Schnelle, R.Hagedorn, G.Fuhr, S.Fiedler, T.Müller, Biochim. Biophys. Acta 1157 (1993) 127-140.

[81] A.Castro, F.R.Fairfield, E.B.Shera, Anal. Chem. 65 (1993) 849-852.

[82] T.T.Perkins, D.E.Smith, S.Chu, Science 264 (1994) 819-822; and: T.T.Perkins, S.R.Quake, D.E.Smith, S.Chu, Science 264 (1994) 822-826

[83] F.Flam, report on a presentation by Eigen and Rigler, Science 265 (1994) 32.

CHANCES OF µTAS IN ANALYTICAL CHEMISTRY

W.E. van der Linden
MESA Research Institute, University of Twente,
P.O.Box 217, 7500 AE Enschede, The Netherlands

0. Abstract

A brief overview is given of various aspects that may be of importance in the assessment of the potentials of micro total analysis systems (µTAS) from the point of view of the user. It is concluded that some intrinsic features of µTAS such as low reagent consumption, portability, reliability and robustness may make them attractive for the employment in environmental, clinical and process monitoring. If the systems can be produced at relatively low costs it can be expected that there are large scale applications in these domains.

1. Introduction

The use of the word 'chances' in the title presupposes a future. The question whether there is a future for micro total analysis systems (µTAS) in chemical analysis can be answered only when the question is seen in the broader context of the developments of chemical analysis in general. These developments should be discussed in terms of demand and supply where the demand is determined by the prevailing problem domains and the supply by the technological and scientific achievements:

problem domains
> = environmental monitoring
> = clinical monitoring
> = quality control in production processes

technological/scientific achievements
> = possibilities of miniaturization
> = data evaluation
> = new sensing principles
> = new chemical compounds selectively complexing analytes

Both lists can be easily extended and made more specific. Especially with regard to the problem domains several studies have been made. In the following sections these points will be discussed further. Subsequently the basic question about the chances of µTAS in chemical analysis will be reconsidered.

A. van den Berg and P. Bergveld (eds.), Micro Total Analysis Systems, 29–35.

2. Problem domains

For all three fields mentioned the "continuous" measuring function, indicated as monitoring, is essential. Because many analyzing systems cannot operate in a real continuous way, it is important to assess the frequency at which samples should be analyzed. This will depend on the role the analytical results have to play. When, for instance, the monitoring of the river Rhine is concerned, one may be interested in the variation of the content of sodium- and potassium chloride, ammonia, phosphate, nitrate, etc. Normally these compounds show a gradual variation so that the analysis of two or three times test samples a day is sufficient. However, when the analysis system is set up as a warning system aiming at the timely signaling of the appearance of a toxic pollutant, availability of results on the minutes scale is often a prerequisite because immediate action may be required. In general, it is important to assess the correlation between measurement result and the actual state of a process. The various aspects of interest are nicely combined in the concept of *measurability* as introduced by Van der Grinten [1]:

$$
m = \exp\left[-\frac{\left(T_d + \frac{T_a}{2} + \frac{T_g}{3}\right)}{T_x} \right]\left[1 - \frac{\sigma_a \sqrt{T_e}}{\sigma_x \sqrt{T_x}} \right]
$$

where m = measurability; T_d = delay time (time lag between sampling and analytical result); T_a = time between successive samples; T_g = grab time (time during which the sampling is performed); T_x = time constant of the process; T_e = time constant of measuring device; s_a = standard deviation of measuring method and s_x = standard deviation of process value to be measured.
As can be directly seen, the measurability varies between 0 and 1, where the value 0 means that the analytical results don't have any relation to the actual status of the process and 1 is the maximum value attainable.

To get a high measurability T_d, T_a and T_g should be small compared to T_x, and s_a small with respect to s_x. Usually the effect of T_g can be neglected, whereas the value of T_a, which can be simply selected by the operator, is taken equal to T_d which is determined by the total analytical system. This means that the next test sample is offered to the analytical system as soon as the analytical result of the previous test sample is available. As can be seen T_x is the key factor with regard to time. It is defined as the time span (DT) over which a reasonable correlation exist between two successive measurements in a 'time-series'(fig.1A). T_x can be evaluated from the *auto-covariance function* (G(DT) of this time-series Fig.1B)[2,3].

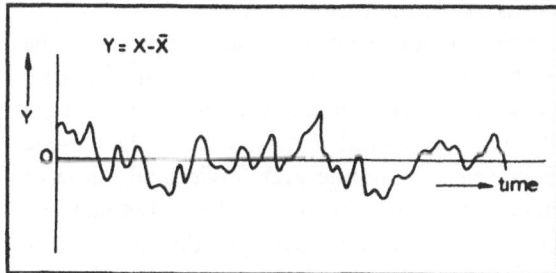 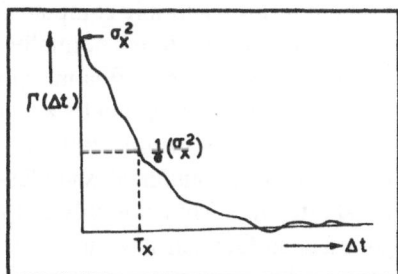

Figure 1 A) Time series is deviation from mean   *B) Autocovariance function of time-series*

One of the possible advantages of μTAS may be the reduction in the delay time, T_a, and the better reproducibility (smaller s_a) normally observed in flow systems where all the steps in the analysis are reproducibly timed as long as the flow rate is constant.

Another important aspect that is common to the three problem domains, is the difficulty of calibration in combination with continuous monitoring systems. For real continuously operating monitors like, *e g.* ion-selective electrodes directly inserted in the medium to be analyzed, calibration is virtually impossible without taking the sensor out of the measuring device. This may lead to irreversible alterations of the condition of the sensor or the sensor surface. This is another good reason for developing μTAS because they probably will operate batch-wise so that it is relatively easy to insert standards in between each series of samples to be analyzed.

It is good to realize that the problem domains as defined above are generally characterized by a great complexity with regard to the composition of the samples. Moreover, very often the samples will contain not only dissolved material but also solid particles may be present with inherent risks of clogging of the analyzing equipment, etc. Dissolved gas may also cause problems, especially when the sample is heated and gas-bubble formation may occur. These problems may require special precautions to be taken in combination with μTAS such as filtration and degassing of the solutions. However, there are many more aspects to be considered which can be better discussed for the various domains separately.

Environmental monitoring

In the environment one may be interested in some single components like NO_3^-, Cl^-, NH_4^+, Cu^{2+}, etc. but nowadays there is an increasing need to know the form in which elements like copper are present, i.e. not only free metal ion concentrations but also the concentrations of the complexes and other compounds in which the metal is present and the inertness or the lability of these compounds. The reason for this need for more detailed information is the difference in toxicity of the various forms in which copper and other elements are present. For instance, the organo-mercury compounds are much more toxic than the inorganic mercury compounds whereas for arsenic the opposite is true. The detailed analysis of the forms in which elements are present is often denoted as *speciation* and asks, in general, for combinations of sophisticated techniques. For the moment it is difficult to see the role μTAS can play in such speciation studies but for field monitoring of simple compounds µTAS can be of great importance.

The solid material normally observed in fluid environmental samples may be inorganic in nature (e.g. floating sludge) but also bacteria are often present. Because many compounds preferentially adsorb on solid materials or are accumulated in bacteria to different extent, simple filtration will alter the overall composition of the sample solution. In environmental samples the presence of humic and fulvic acids deserve special attention. They not only can form complexes with many metal ions but they can also affect surface tension and even viscosity and so have influence on the performance of the analyzing system such as the dispersion of the sample plug in a flow-through analyzer.

All these effects may lead to the conclusion that some kind of sample pretreatment is often essential varying from simple dilution to decomposition and separation. For not too complex systems separation can be elegantly accomplished with capillary electrophoresis type µTAS.

In summary, the possible use of µTAS in the environment has to be sought in the field monitoring of a few single components which may be indicative for the quality of e.g. fresh waters, coastal waters, etc. The advantage of µTAS is the small size, the low energy demand to operate the systems which allows the use of batteries, and the very limited reagent consumption. All these features contribute to the portability of these micro analysis systems.

Clinical monitoring

Where in the environmental applications the reagent consumption is a major issue, in clinical monitoring it is often the available amount of sample that is a limiting factor. For *in vivo* measurements over prolonged periods of time, the aspect of biocompatibility will play a major role, but also factors like clogging and viscosity need careful consideration. Fortunately some of the detection systems that are available nowadays are sufficiently sensitive to allow dilution as a first step in the analysis. On appropriate dilution of blood and serum samples, many of these problems can be avoided or at least be reduced.

It is without saying that for direct control of patients, analyzing systems should meet the highest reliability standards. Because the analyzers will be mostly handled by personnel that is not specifically analytically trained, the instruments should also be very robust ('doctor-proof') and possibly be equipped with some kind of auto-diagnosis system that will warn the patient or the medical staff when something is wrong with the instrument.

Quality control in production processes

The required reliability and robustness of analyzers applies also to process analyzers. At this moment relatively little chemical analyzers can meet the high standards and, therefore, are seldom used as an integral part of the automatic control loop. This means that most of the chemical production processes are still automatically controlled by measuring devices for temperature, pressure, and flow rate. The analytical results from monitoring systems for the concentration of chemical components, whenever present, are only used to guide the operator who is supervising the running of the process. In addition to many points mentioned before, it has to be noticed that sample streams in most of the chemical production processes are rather "unfriendly" for any instrument with regard to temperature, pressure, corrosiveness, etc. Moreover, a maintenance-free period of at least two weeks or even one months is regarded as a minimum in process industry. Because these analyzers operate on a 24-hours per day basis for seven days a week, the low reagent consumption is in these analyzers a major point as well.

Summarizing the requirements μTAS have to fulfill in environmental, clinical and process monitoring:
low reagent consumption
low energy consumption
high reliability
good robustness also in the hands of not-analytically trained personnel
low maintenance requirements

3. Technological/scientific achievements

Possibilities of miniaturization

The great advances in this area over the last twenty years have shown already interesting results. One of the most appealing examples is the miniaturized gas chromatograph presented by Angell et al.[4]. This subject, being one of the main topics of the workshop, will be extensively discussed by scientists better qualified to speak about this subject than this author.

Data evaluation

When the analyzing system would be extremely selective, i.e. respond to one compound only, data evaluation could be very simple such as smoothing (e.g. noise filtering by means of a Sawitzky-Golay technique) and transformation of the data to a form that can be easily interpreted by its users.

However, most of the available simple analyzing systems doesn't show a sufficiently good selectivity and interference by other components in the sample has to be considered. When the presence of such other components is established and their concentration as well as their impact are known, a mathematical correction can be made, but when the concentrations of these interfering compounds are unknown other measures have to be taken. The most obvious way out is too install a series of analyzers with different selectivities for the different components and to apply so-called multivariate calibration techniques. In the domain of *chemometry*, these techniques have shown a fast development during the last two decades. Apart from the techniques that are appropriate for linear calibration systems, methods for the calibration of systems

34

where the interdependencies are non-linear are of importance. In this respect the use of, e.g. neural networks have to be mentioned. In the literature examples can be found which adequately illustrate the scope of these networks [5]. It is true that this gives a nice possibility to circumvent the limitations due to low selectivity but at the expense of the need of an enhanced stability of the sensors. Because the training of neural networks is quite time consuming, this effort pays back only when the trained network can be in operation for longer times without the need to retrain the network due to changes in the instrumental parameters of the analyzing system.

Data evaluation doesn't necessarily refer to the purely analytical results only. It is also possible to use sophisticated methods for the analysis of the shape of the signals as a tool for fault detection. This may be an aid in the development of the auto-diagnosis system mentioned before. Hardly any attention has been paid to this aspect yet.

New sensing principles

Until now most efforts in the development of miniaturized analyzers are based on the use of photometric and electrochemical detection. During the workshop many examples will be given. It can be easily shown that miniaturizing the devices with respect to diameters of channels, injected volumes, etc. asks for a more than proportional reduction in the detector volume [6]. In this respect the important development of (ultra-) microelectrodes has to be mentioned. Because electrochemical techniques, but also optical ones like surface plasmon resonance, are essentially based on surface phenomena they are critically dependent on the condition of the sensing surfaces. As these surfaces can often easily be affected in an irreversible way, extreme care in handling is required to get reproducible results. The search for other techniques related to bulk properties of the samples should therefore be continued. These may comprise optical as well as magnetic or enthalpimetric principles.

New chemical compounds selectively complexing analytes

Because hardly any compound in a sample will show one unique type of interaction with a sensing device, selectivity will remain a major issue. This selectivity can be created by the use of 'selectors', i.e. compounds that exhibit an almost exclusive complexation with the analyte. Nature is a good guide in this respect in view of the large number of enzymes that fulfill this requirement. During the last two decades also great progress is made in the field of *supramolecular chemistry*. The large number of ionophores developed all over the world and used both in ion-selective electrodes and optodes are good examples. It is not difficult to predict that these developments will continue for the next years and will certainly lead to new natural or synthetic compounds.

4. Conclusions

It can be expected that μTAS can offer several advantages that may give them a prominent place in the total arsenal of analytical tools of the future. Large scale and cheap production may be one of the major advantages. Their small size will be of particular interest for field monitoring and in many medical applications. The inherent small reagent consumption is of importance in all types of applications.

Being quite optimistic about the chances of μTAS in the (near?) future, a warning seems in place. The introduction of new analytical techniques has always been accompanied by presentations giving the impression that the ultimate technique has been developed. After a short period of such an over-estimation there has always been a backlash which was detrimental for the smooth and successful development of the technique. A striking example was the booming interest in ion-selective electrodes at the end of the sixties and begin of the seventies which was followed by an almost complete neglect of these devices in industry for the following two decades because in real practice most of the expectations could not be fulfilled. It is only recently that electrodes gradually get their legitimate place in the analytical routine laboratory.

5.References

1. P.M.E.M. van der Grinten, *Statistica Neerlandica*, 22(1968)43
2. G.Kateman, and F.W.Pijpers, *Quality control in analytical chemistry*, Wiley, New York, 1981
3. P.J.W.M.Müskens, and W.G.J.Hensgens, *Water Res.*, 11(1977)509
4. J.B.Angell, S.C.Terry, and P.W.Barth, *Sci.Am.*, 248(1983)36
5. A.Bos, *Artificial neural networks as a tool in chemometrics,* Thesis, Enschede, 1993
6. W.E.van der Linden, *Trends in Anal.Chem.*, 6(1987)37

μTAS FOR BIOCHEMICAL ANALYSIS

I. Karube

Research Center for Advanced Science and Technology (RCAST),
University of Tokyo, 4-6-1 Komaba, Meguro-ku, Tokyo 153, Japan

0. Abstract

Miniaturized enzyme columns integrated with a microelectrochemical flow cell were developed for μTAS. The devices consisted of a glucose oxidase-immobilized column and a flow cell with a microelectrode, which were fabricated and integrated on a silicon wafer by micromachining techniques such as photolithography, anisotropic etching, thermal oxidation, vacuum evaporation, and anodic bonding. A silicon wafer was photolithographically developed and anisotropically etched to form a V-shaped groove 1 m long, 100 m wide, and 70 m deep. A glass plate with thin-film electrodes was anodically bonded to the silicon before enzyme immobilization. The flow injection system with each device gave a linear relationship between peak current and glucose concentration.

1. Introduction

Recent progress in science and technology requires detection and measurement of chemical, biological, environmental parameters. A solution is the biosensor, combination of physical transducers and biomaterials. In principle, biomaterials transduce biochemical parameters to physical parameters followed by the detection with physical transducers. Typical requirements such as *in situ* or *in vivo* monitoring, real time or fast analysis, and high sensitivity, encouraged us to fabricate miniaturized biosensors. First, we fabricated miniaturized needle-type biosensors that had enzyme(s) on the top of the needle as an electrode. This had some advantages including fast analysis, high sensitivity, low-cost (due to small amount of enzyme(s)), and reduced requirement of analyte area. The sensors, in principle, require analyte only around the detector with the thickness of the efficient diffusion layer. This advantage enabled us to apply the sensors for *in situ* and *in vivo* monitoring.

Strong dependence of enzyme reaction on pH and temperature, however, made samples unsuitable, because no sensor had the ability to adjust pH and temperature. Then, in the second stage, we constructed flow injection analysis (FIA) system. A FIA system requires simple diagram, that consists of carrier reservoir, injector, reactor, and detector connected serially with a certain pipe. Unfortunately, a typical FIA-system occupies about one desk-area, and is not portable.

37

A. van den Berg and P. Bergveld (eds.), Micro Total Analysis Systems, 37–46.
© 1995 *Kluwer Academic Publishers.*

The advanced IC fabrication techniques have opened up a new window to fabricate miniaturized devices, and are called micromachining It has permitted the fabrication of miniaturized gas chromatographic systems for the use in space laboratories Subsequent efforts have extended the technique to the field of chemical measurement, allowing plate-type sensors, such as ISFET and thin-layer electrodes, which handle a sample outside of the device, to be produced by photolithographic techniques

In contrast, there has been limited progress in liquid-flow-type systems, which handle a sample inside of the device, because they require three dimensional structures with higher pressure resistance than that in gas-flow systems [1], and they require micropumping mechanisms as well Some attempts have been made to fabricate flow systems for chemical analysis by micromachining techniques, including a FIA system [2] and a capillary electrophoresis (CE) system [3]

Manz et al have discussed the advantages of miniaturized analysis systems [4] Miniaturized biosensing systems show great promise for medical, biochemical and environmental applications In these fields, it is desirable for many different substances to be analyzed rapidly using minimum amounts of sample and of expensive reagents, miniaturization offers obvious benefits in this regard In addition, miniaturization allows analytical systems to be portable, enhancing their value for field use in environmental monitoring

In this paper we describe flow type biosensing systems fabricated by micromachining techniques In these studies, we fabricated some enzyme immobilized columns integrated with an electrochemical flow cells on a silicon wafer, and demonstrated the functionality of this device for the determination of glucose by connecting it to a conventional flow injection system

2. Experimental

2 1 STRUCTURE AND GEOMETRY OF THE DEVICES

Figure 1 shows a schematic overview of the devices fabricated in this research Figure 1-(a) to (e) are an electrochemical flow cell, a long enzyme -immobilized column, an enzyme-immobilized column integrated with an electrochemical flow cell, a two-step enzyme reactor cell and a packed column, respectively The typical design is as follows the device of fig 1-(c) consists of an anisotropically etched silicon chip and a glass plate with thin film gold electrodes The silicon chip has a V-shaped groove (100 m in width, 70 m in depth, 962 mm in length) folded 66 times [5] to fit into an area 28 mm × 19 mm, and penetrated holes (1 mm × 1 mm) at both ends of the groove All patterns for the inner corners were compensated [6, 7] with squares of 120 m to prevent undercutting The surface of the silicon chip was oxidized to insulate the electrodes The electrochemical flow cell is constructed with four thin film gold electrodes (0 5, 1, 1 5, and 1 mm in width, 0 05, 0 1 0 15 and 0 1 mm^2 in effective area, respectively, 18 mm in length, spaced 0 5 mm from each other) at the end of the

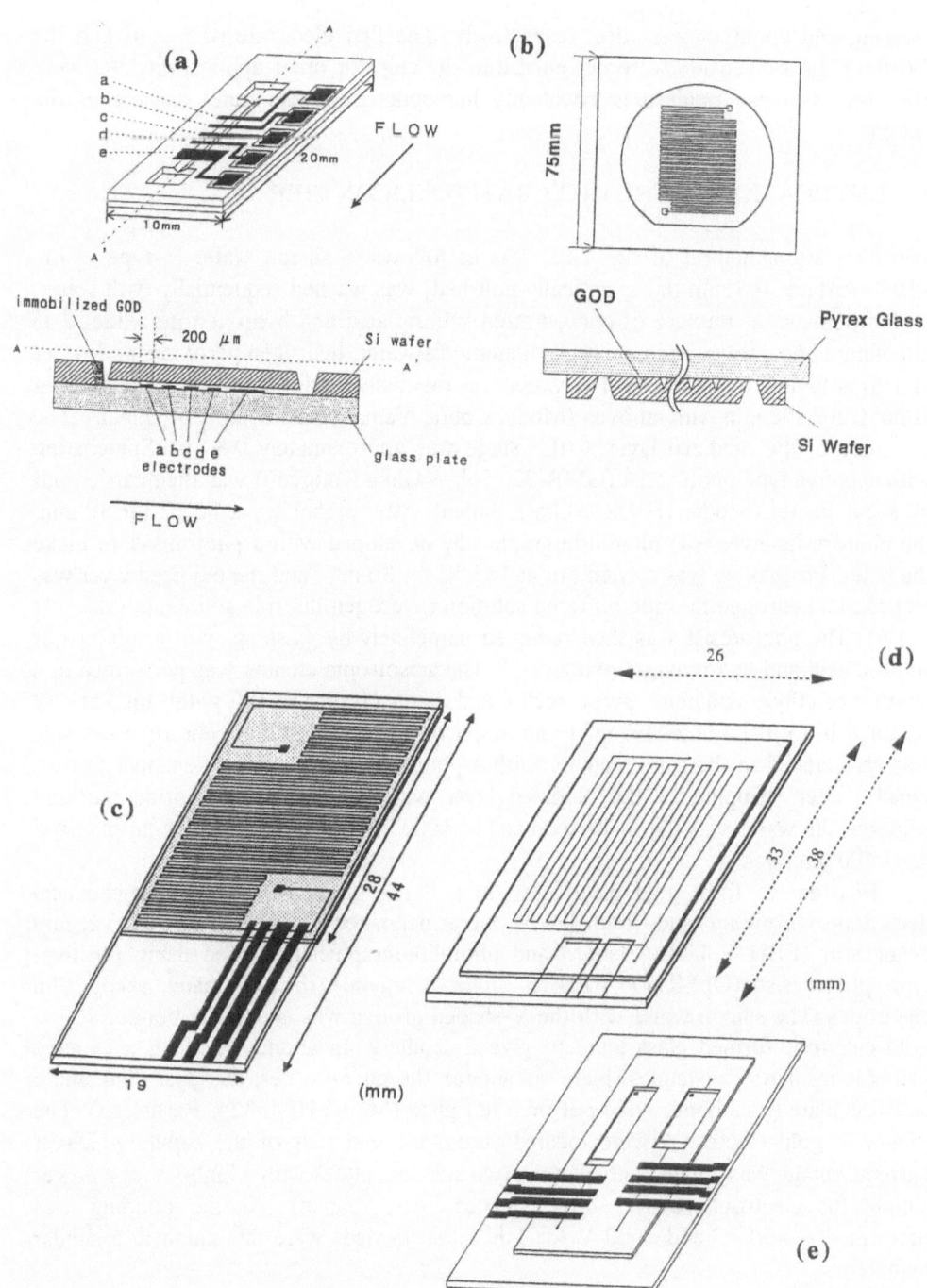

Figure 1. Schematic overview of the devices fabricated in this research.

column The second electrode of 1 mm in width and the third of 1 5 mm were used as working and counter electrodes, respectively The first electrode (0 5 mm) and the fourth (1 mm) electrode were designed into the chip for other applications and were not used Glucose oxidase is covalently immobilized on the inner surface of the column

2 2 FABRICATION OF THE INTEGRATED SILICON CHIP

The fabrication method of fig 1-(c) was as follows a silicon wafer [n-type, 2 in, <100> surface, 0 5 mm thick, optically polished] was washed sequentially with water, trichloroethane, a mixture of concentrated sulfuric acid and hydrogen peroxide (2 1) at boiling temperature, and a mixture of ammonia water, hydrogen peroxide, and water (1 1 6) at boiling temperature The wafer was thermally oxidized in pyrogenic water at 1000 C for 200 min with an oven (Model Cobra, Yamato Semiconductor, Japan) The thickness of the oxidized layer at this stage was approximately 0 8 m Spincoating with negative type photoresist (OMR-83, Tokyo Ohka Kougyou) was then carried out by a spin coater (Model 1H-D2, Mikasa, Japan) After prebaking at 80 °C for 30 min, the photoresist layer was photolithographically developed with a photomask to make the holes Postbaking was carried out at 135 °C for 30 min, and the oxidized layer was etched in a hydrogen fluoride-buffered solution (hydrogen fluoride ammonium fluoride = 1 6) The photoresist was then removed completely by washing with a mixture of sulfuric acid and hydrogen peroxide (2 1) The anisotropic etching was performed in a mixture of ethylenediamine, pyrocatechol and water (1000 mL 180 g 480 mL) at 110 °C for 8 h to make holes for inlet and outlet of a carrier solution The reversed side was also etched in the same manner with a photomask to make a V-shaped groove Finally, after stripping of the oxidized layer with the hydrogen fluoride buffered solution, the wafer was thermally oxidized at 1000 °C for 50 min to give an oxidized layer 400 nm thick

In order to form gold electrodes on a Pyrex glass plate for electrochemical detection, chromium and gold layers were deposited on the plate by vacuum evaporator (EBH-6, Ulvac, Japan) and photolithographically etched using positive-type photoresist (OFPR-800, Tokyo Ohka Kougyou) to give band shape film electrodes The silicon wafer with the V-shaped groove was anodically bonded to the gold electrode-formed glass plate to give a capillary In anodic bonding, a ceramic plate (an insulator), a stainless plate (an anode), the silicon wafer, the glass plate and a stainless plate (a cathode) were put on a hot plate (Model HP46824, Barnstead) The thin-layer gold electrodes were located across the end part of the capillary Direct current voltage was applied between the two stainless plates with a high-voltage power supply for electrophoresis (Model SJ-1065, Atto, Japan) Anodic bonding was performed at 450 °C and 1000 V for 1 h Other devices were fabricated in a similar manner

2 3 IMMOBILIZATION OF GLUCOSE OXIDASE

For the immobilization of glucose oxidase (GOx), amino groups were covalently attached inside the capillary (3-aminopropyl)triethoxysilane [γ-APTES, 10 L of 10% (v/v)] in toluene was passed through the capillary at room temperature, leaving behind a thin layer of γ-APTES on the inner surface of the capillary, followed by heating overnight at 115 °C After 1 h treatment with 2 5% (v/v) glutaraldehyde in phosphate buffer (pH 7 1), the capillary was filled with 10% (w/v) GOx (type II, Sigma) in phosphate buffer (pH 7 1) and kept overnight to immobilize GOx Finally, flange-type female unions made of polyether ether ketone (PEEK) for fingertight fitting were divided into two pieces and bonded onto the inlet and outlet of the capillary with cement (Araldite, Ciba-Geigy)

2 4 CONSTRUCTION OF THE FIA SYSTEM

The FIA system constructed consisted of the enzyme-immobilized column integrated with gold electrodes on a silicon chip, a pump for HPLC (minimum flow rate 0 1 L min^{-1}), an injector for HPLC (sample size 0 2 or 0 5 L, Rheodyne), a potentiostat (handmade), and a data processor Commercially available flange-type fittings were used for the connection of the biosensing device with other components of the FIA system

3. Results and Discussion

3 1 BATCH MODE OF THE ELECTROCHEMICAL FLOW CELL

First, we fabricated the electrochemical flow cell, and used it in a batch mode (without flow system) [8] A sample solution was directly introduced into the cell by the capillary effect In order to immobilize GOx, a crosslinking method was performed with bovine serum albumin (BSA) and glutaraldehyde (GA) onto the inlet hole The 1 5 L mixture of 100 mg of GOx, 300 mg of BSA and 500 l of phosphate buffer was put into the inlet hole to fill about half the volume of the hole It was left in a box saturated with GA vapor at room temperature for 4 hours

We used glucose oxidase (GOx, EC1 1 3 4) for the device because of its advantages Glucose oxidase is one of the most suitable enzymes to evaluate these enzymatic devices or systems because it is stable, inexpensive, and can be easily applied to clinical or industrial use The GOx also catalyzes the reaction that consumes oxygen and produces hydrogen peroxide in solution

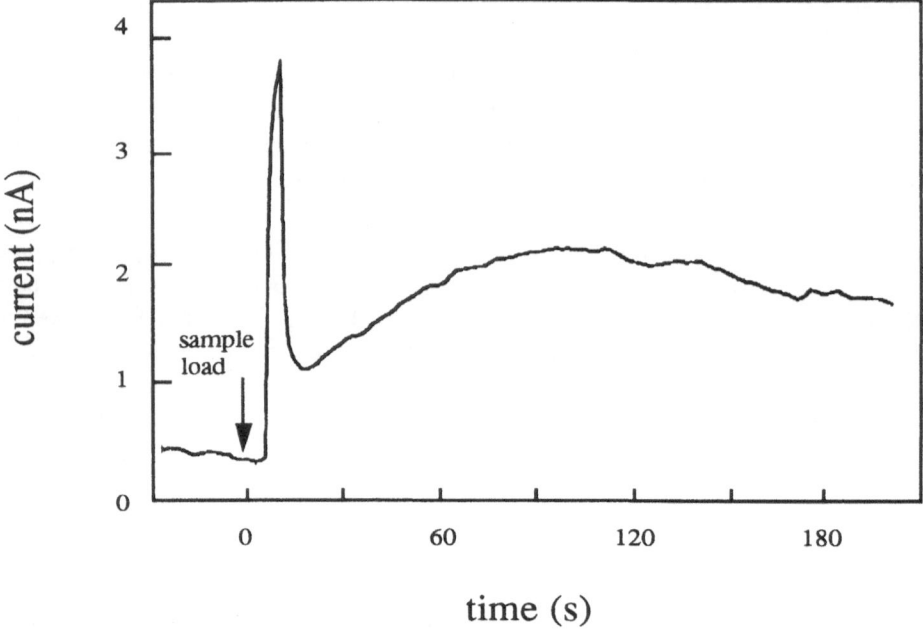

*Figure 2. Response curve of the glucose sensor to 1 mM glucose solution
with 0.5% Triton X-100 (pH 7.0, 25 °C)*

The sample filled in the experiment before was removed by nitrogen gas blowing, and the sample drop was put on the inlet hole, then it was introduced into the cell by capillary effect Hydrogen peroxide produced by GOx at the inlet hole was detected when 700 mV was applied between the counter electrode and one of the working electrodes Surfactant was added into the sample solution to enhance the capillary immediately after sample introduction Presence of the sample was observed as a broad effect Figure 2 shows a typical response curve of the device when used as a glucose sensor The charge current of the electric double layer was observed as a sharp peak peak about 2 minutes after its introduction The latter peak current was handled as the response value When 2 L of glucose solution in phosphate buffer (pH 7 0, 10 mM) with 0 5% of the surfactant (Triton X-100) was introduced to the device, the linear relationship was observed between the response and the glucose concentration in the range of 0 4 to 1 0 mM The response value depends on the flow rate and diffusion rate at the electrodes These factors vary, however, due to manual handling of the sample We did not attempt to improve the reproducibility of the response, but instead, we turned to the investigation of the FIA system

3 2 ELECTROCHEMICAL FLOW CELL IN FIA SYSTEM

The device was combined into a FIA system to evaluate its characteristics [8] Phosphate buffer (pH 7 0) at a flow rate of 50 L min^{-1} and 0 2 L of sample was pumped through the system, applying 500 mV between the electrodes First, hydrogen peroxide was detected with the system The system response was linearly proportional to the concentration of hydrogen peroxide in the range 0 2 to 50 mM with a sensitivity of $4 2 \times 10^{-7}$ A M^{-1} (r = 0 997) The linearity of the calibration means that the device acts as a kind of tubular electrodes and has enough ability to act as a hydrogen peroxide electrode Glucose was also detected by the same method The system current response was linearly proportional to the concentration of glucose in the range 2 to 115 mM with a sensitivity of $5 3 \times 10^{-8}$ A M^{-1} ($r = 0 996$) Although the enzyme reaction requires dissolved oxygen, which was less than 10 mM in our buffer, it gave a response to greater than 100 mM glucose The main reason is because of sample dispersion This is especially significant at the inlet hole that has a volume of more than 1 L which is much greater than the volume of the sample (0 2 L) Sample dispersion at the inlet hole is, therefore, the main reason for the sensor response at high glucose concentration One reason for the difference between the sensitivity to hydrogen peroxide and glucose may be due to the fact that the GOx reaction was not complete at the inlet hole A long residence time was expected to give a higher response To confirm this relation, therefore, the flow rate was varied to alter the residence time (flow rate inversely proportional to residence time) The response was proportional to the residence time $1/Q$ below 50 min mL^{-1} (greater than a 20 L min^{-1} flow rate) This result suggests that the rate determining step was the enzyme reaction and that a large amount of enzyme needs to be immobilized to achieve good sensitivity The pH dependence of the system was in agreement with the characteristics of GOx

3 3 LONG ENZYME REACTOR

To immobilize a large amount of enzyme, we fabricated a long enzyme reactor [9] This column was about 2 6 m in length, and had immobilized-enzyme on its inner-surface A commercially available electrochemical detector was used in the FIA system constructed for the evaluation of the column Phosphate buffer (pH 7 1) was used as carrier solution at a flow rate of 60 L min^{-1}, 5 L of 1 mM glucose containing sample was injected and a voltage of 500 mV (vs Ag/AgCl) was applied The oxidation current of hydrogen peroxide peaked at about 90 s after injection Coefficient of variation was 4 6% ($n = 17$) when a flow rate was 40 L min^{-1} with a volume of 5 L of 100 M glucose solution A linear relationship was observed between current and glucose in the range of 5×10^{-5} to 10^{-3} M with a volume of 5 L of glucose solution The minimum volume of commercial mechanical injector available to us is 0 2 L A linear relationship was also observed in the range of 10^{-3} to 5×10^{-2} M with a volume of 0 2 L of glucose solution, which includes typical human blood glucose values

One possible advantage of the silicon capillary is that it is fabricated by bonding of a groove with a plate as a lid This means that a part of the inner surface of the capillary can be selectively modified before bonding, avoiding the fact that modification to a completed capillary would uniformly affect the whole inner surface If certain parts of a chip can be modified before bonding, it may be possible to create new varieties of capillaries with substantial added value A set of gold electrodes for an electrochemical flow cell is a typical example

Materials suitable for use in modification are restricted by the bonding temperature Recently, Esashi reported an anodic bonding method performed near room temperature [10] This technique may allow us to modify the chip with chemical or biochemical substances before bonding, although problems due to glass strength and uniformity in bonding still remain to be solved The ability of the silicon capillary to support value-added functions may enhance its advantages as compared to traditional drawn glass tubes or injection-molded plastic tubes

3 4 INTEGRATION OF REACTOR WITH ELECTROCHEMICAL FLOW CELL

Then we integrated the enzyme reactor with the electrochemical flow cell [11] In this system, the carrier and sample solution containing glucose come from the inlet hole at the end of the column Passing through the column, glucose reacts with dissolved oxygen in the carrier solution in a reaction catalyzed by the GOx on the inside wall of the column to give gluconolactone and hydrogen peroxide Eventually the electrodes detect oxidation current from the hydrogen peroxide Phosphate buffer (pH 7 1) was used as carrier solution fed at 10 L min^{-1}, with 0 2 L of glucose contained sample injected and 700 mV applied between two integrated electrodes The oxidation current of hydrogen peroxide peaked at 1 min after injection The baseline was noisy, and we think one reason was absence of the reference electrode, such as Ag/AgCl If 0 2 L sample zone passes the flow cell without any dispersion, it takes 1 2 seconds at 10 L min^{-1} However, half-value widths were approximately 20 seconds

A linear relationship was observed between current and glucose in the range of 1 to 25 mM, which includes typical human blood glucose values The correlation coefficient, r, is 0 993 ($n = 23$) in the range of 1 to 25 mM at a flow of 10 L min^{-1} The sensing system gave no response to 100 mM of maltose Glucose solutions in the range below 1 mM were injected, but peak heights were difficult to measure against the baseline noise

The reaction rate was investigated by changing flow rate The flow rate dependence of the response to glucose or hydrogen peroxide was also examined Because the peak current of hydrogen peroxide depends upon its mass transportation rate, a faster flow gave a higher response The peak current of glucose, however, depends upon reaction time in the reactor, and a slower flow gave a higher response Comparing the response of hydrogen peroxide and glucose at the same concentration (1 mM), we found that glucose was not completely oxidized about 10% of the glucose was oxidized at 20 L min^{-1} When the flow rate was 5 L min^{-1}, glucose reacted completely

Fabricated sensors remained usable for more than 2 months after glucose oxidase immobilization In addition, we were unable to detect any decrease in sensor response after 1 month's storage at room temperature

3 5 TWO-STEP ENZYME REACTOR CELL

A silicon capillary has some advantages over a drawn-glass capillary such as the ability to construct very complex capillary configurations and integrate the structure with total microfabricated systems In addition, a glass capillary cannot have a T-intersection, while this is easily formed in a silicon capillary Manz has fabricated an intersection which possessed a several-nanoliter injection mechanism [3]

The analysis of many analytes using biosensors require more than one enzyme reaction step For example, detection of acetylcholine requires acetylcholinesterase and choline oxidase Cholesterol ester detection also requires the esterase and oxidase As a development of the miniaturized devices for FIA, we have fabricated a new device that can be applied to a two-step enzyme reaction Glucose is also detectable using a two-step enzyme system such as GOx and peroxidase with a mediator These enzymes are stable and possess high activities, and response is comparable with the result of the devices we have reported before We fabricated a micromachined biosensing device incorporating a bienzymic capillary applied to the detection of glucose by connecting it to a conventional flow injection system [12]

A calibration curve was obtained for this device for injection of buffered 0 5 l glucose solution at a potential of +150 mV (vs Ag/AgCl) and at a flow-rate of 10 l min^{-1} The response time from injection to peak was approximately 150 s The half-height peak response was approximately 70 s for 1 mM glucose, with a coefficient of variance of 9 4% (n = 10) The response to glucose solution or hydrogen peroxide solution was negative (reduction current), while the response to ascorbic acid was positive There was no response to 1 mM glucose when the carrier did not contain mediator, and no response to 100 mM sucrose solution when the carrier contained mediator

Cyclic voltammetry using a gold wire electrode was performed to determine the redox potential of the mediator The mediator in carrier solution gave an oxidation peak at +420 mV and a reduction peak at +358 mV (vs Ag/AgCl) The response had no typical dependence to the electrode potential from 0 to +300 mV (vs Ag/AgCl) Background current, however, increased when the potential was far from +100 mV It was difficult to find a peak in several times bigger background than the peak at the potential of less than 0 mV Thus, we performed following experiments at +150 mV

Low electrode potential enhances a specificity to the analyte Real samples such as blood or extracts from foods often have other electroactive interferences such as ascorbic acid or dopamine A peak current of 1 mM ascorbic acid was 0 48 or 5 8 nA at +150, or +300 mV, respectively Because the concentration of ascorbic acid in human blood is less than 0 1 mM, we see that the use of the mediator reduced any possible effect from interfering substances

3 6 PACKED COLUMN

These devices described above have immobilized-enzyme on itself [13] The standard method to give an enzyme-immobilized column, however, is a packed-column method

We also fabricated a packed column with integrated electrodes GOx was immobilized on glass beads, approximately 0 1 mm in diameter, with standard method using γ-APTES and GA The beads suspended in the enzyme solution were directly injected with micropipet into the column Microscope observation made it clear that a dense packing of the beads was achieved Dilution rate calculated from the difference between the peak current of hydrogen peroxide through the column and that of without column also suggested this A linear relationship was observed between current and glucose in the range of 1 to 10 mM The correlation coefficient, r, is 0 955 (n = 14) in the range of 1 to 25 mM at a flow rate of 10 L min^{-1} The response time was 170 s when phosphate buffer was fed at 10 L min^{-1}, whereas without phosphate buffer this was 140 s

References

1 S C Terry, J H Jerman and J B Angell, Gas chromatographic air analyzer fabricated on a silicon wafer, *IEEE Trans Electron Devices, ED-26* (1979) 1880-1886
2 S Shoji and M Esashi, Micro flow cell for blood gas analysis realizing very small sample volume, *Sens Actuators B8* (1992) 205-208
3 D J Harrison, A Manz, Z Fan, H Ludi and H M Widmer, Capillary Electrophoresis and Sample injection systems integrated on a planar glass chip, *Anal Chem* (1992) 1926-1932
4 A Manz, N Graber and H M Widmer, Miniaturized total chemical analysis system a novel concept for chemical sensing, *Sens Actuators B1* (1990) 244-248
5 M G Guvenc V-Groove capillary for low flow control and measurement, *Micromachining and Micropackaging of Transducers,* Elsevier Amsterdam(1985) 215-223
6 X P Wu and W H Ko, Compensating corner undercutting in anisotropic etching of (100) silicon, *Sens Actuators 18* (1989) 207-215
7 B Puers and W Sansen, Compensation Structures for convex corner micromachining in silicon, *Sens Actuators A21-A23* (1990), 1036-1041
8 Y Murakami, T Uchida, T Takeuchi, E Tamiya, I Karube and M Suda, Micromachined electrochemical flow cell for biosensing, *Electroanalysis,* in printing
9 Y Murakami, M Suda, K Yokoyama, T Takeuchi, E Tamiya and I Karube, Micromachined enzyme reactor for FIA system, *Microchem J , 49* (1994) 319-325
10 M Esashi, A Nakano, S Shoji and H Hebiguchi, Low-temperature silicon-to-silicon anodic bonding with intermediate low melting point glass, *Sens Actuators A21-A23* (1990) 931-934
11 Y Murakami, T Takeuchi, K Yokoyama, E Tamiya, I Karube and M Suda, Integration of enzyme-immobilized column with electrochemical flow cell using micromachining techniques for a glucose detection system, *Anal Chem , 65* (1993) 2731-2735
12 Y Murakami, T Takeuchi, K Yokoyama and I Karube, in preparation

DETECTION PRINCIPLES FOR µ-TAS

Hans-Joachim Ache
Kernforschungszentrum Karlsruhe, Institut fur Radiochemie,
Postfach 3640, 76021 Karlsruhe, Germany

Abstract

Detectors used in Micro Total Analysis Systems (µ-TAS) have to comply with
high standard of reliability and low production costs in order to compete with
conventional analytical equipment
In particular besides meeting the technical specifications of selectivity,
sensitivity, suitable response and recovery time, long term stability in a quite
frequently aggressive environment will be the most important feature In
view of these requirements two types of chemical sensors, optochemical and
mass-sensitive sensors (SAW's) and the concept of the evanescent field
absorbance microprobe (EFAS) will be discussed by referring to ongoing
microanalyzer system development at the KfK (Kernforschungszentrum
Karlsruhe)

Introduction

Detection principles for Total Chemical Microanalysis Systems (µ-TAS) have to
be discussed in view of the demands made on the performance of the µ-TAS
and the type of application for which it is intended to
As in the case of any analytical instrument, a µ-TAS will have to be designed
as flexible as possible but still with a special application in mind
That is to say that it will have to provide a satisfactory selectivity for the
analyte of interest, with tolerable interferences caused by potential
contaminants, efficient sensitivity in the concentration range under
investigation, coupled with the desired limit of detection, a suitable response

47

A van den Berg and P Bergveld (eds) Micro Total Analysis Systems 47–70
© 1995 *Kluwer Academic Publishers*

and recovery time as well as stability and longevity under realistic operational conditions

In this connection it should be emphasized that the main purpose of miniaturization of an analytical system is not only reducing the physical dimensions of the instrument Certainly a reduction in size may result in some important advantages, e g in in-vivo medical diagnostics, where the sensor may have to be introduced into small cavities etc or it may lead because of its small size and light weight to a greater mobility, to a smaller demand for space, less chemical consumption and perhaps a shorter analysis time

But the major justification for miniaturization by using microsystem technological solutions has to be seen in gaining a much higher reliability of the analysis system as compared to macrosystems This is achieved by integrating the various functions of the microsystem - electronic, optical chemical fluiddynamical and acoustical - within a small volume using compatible process technology Short connections, no complicated wiring or contacts enhance the reliability of the product Inherent to this technique are technical improvements, a better performance which otherwise would require in macrosystems an immense and costly technical effort if it can be achieved at all The second major reason for turning to microanalysis systems conceived by microsystems technology is the expectation that in this case the availability of advanced software can be efficiently introduced to provide greater flexibility and adaptivity as far as the application of the system to various analytical problems is concerned, as well as it can provide the basis for a costefficient production technology, even in the case of the projected smaller a medium sized number of units produced

Since nearly all analytical problems can be handled by the conventional (macro) analysis systems their widespread replacement would depend on the availability of microanalysis systems with the same or better technical specifications which are clearly more reliable (i e widely maintenance free, user friendly etc , thus reducing the operating costs) as well as being available at considerable lower purchase costs than the conventional systems

Generally a complete chemical analytical microsystem exists of several subsystems including those for sampling, conditioning measurement, connected by a microfluidic handling system, data processing and control functions For each one of these components the same high requirements of reliability and low production costs apply

In the following only the subsystem which performs the actual measurement, identification and/or quantification of the analyte, shall be discussed at greater length

It consists of chemical sensors or analytical microprobes which interact with the analyte compounds in the environment, e g atmosphere, process solution etc , to which they are exposed Analytical microprobes utilize the physical interaction with the analyte with e g photons (spectroscopy) and are obtained by downscaling conventional analysis equipment, while chemical sensors depend on a reversible chemical interaction for the identification and quantification of the analyte This interaction will be transformed into a signal, electronic or optical, which will be processed by means of microelectronic data processing systems The resulting data can then be used to determine the required actions, e g closing or opening valves, initiate heating or cooling etc to be carried out by the so-called actuators

Considering the highly advanced state of the microelectronics available and the associated data processing, chemical sensors or analytical microprobes to be included into such a microsystem will have to be compatible with these sophisticated microelectronic components That means only miniaturized microelectronic compatible chemical sensors or analytical microprobes produced by utilizing modern cost efficient miniaturization techniques, and in this way reaching the standards of the associated microelectronic data processing, as far as geometric dimensions, reliability, stability and production costs are concerned, can be considered for inclusion in such microsystems

While physical microsensors, i e sensors which measure physical parameters, such as temperature, pressure, acceleration etc , have reached already a high state of art, the development of chemical microsensors or analytical microprobes for the identification and quantification of chemical compounds has so far shown much less success The reason for that can be seen in the fact that in contrast to most physical sensors such as an acceleration sensor which are carefully packaged and protected from the outside, the very nature of the chemical sensor requires the exposure of the sensor element to the quite often very aggressive surroundings which contain the analytes to be measured.

All chemical sensors presently in use or in development are based on a two-step detection mechanism [1]. In the first step, chemical selectivity is achieved by a chemical reaction or chemi- or physisorption on a chemically selective surface as listed in table 1. In the second step, a physical change, which is the result of the first selective chemical step, is transformed by a suitable transducer into an electrical signal. This physical change can be a variation of the chemical potential, caused by the reaction or sorption process, a change in optical properties, a change in mass, conductivity, surface resistance or conductance. The physical effects listed in table 2 in column 1 can now be combined with the various transducers listed in column 2 in a suitable way. They will yield the total spectrum of chemical sensor types presently under investigation.

Table 1

Chemical interaction between surface and analyte.

Complex formation, catalysis, electrochemical reaction, chemi- or physisorption, ion exchange, membrane transport, antigen- or antibody binding, binding on immobilized receptor

Table 2

Physical changes resulting from chemical interactions:	Transducers:
Chemical potential	Field effect transistor
Optical properties	Optrode
Heat	Thermistor
Mass	Piezocrystal SAW
Conductivity	Dielectrometer
Surface resistance	Chemiresistor
Current flow	Amperometer

From this large variety of chemical sensor concepts four types have found the greatest attention:

- Optical sensors (optrodes)
- Mass sensitive sensors (surface acoustic wave devices, SAW's)
- ChemFET's (chemically modified field effect transistors)
- Metaloxidefilms (MOF; conductivity sensors)

All four types are well suited for the incorporation into microsystem technological solutions, e. g. FET's or MOF's are available via the usual semi-conductor mass fabrication techniques, the same is true for the piezocrystals equipped with interdigital structures (IDT's) for the SAW sensors, as well as for integrated waveguide structures in the case of the optochemical sensors (optrodes)..

In the following optochemical and SAW sensors will be discussed in connection with some ongoing development on microanalyzer systems at KfK. Electrochemical sensors are the subject of two KfK poster contributions at this conference to which reference is made.

As a third example the development of an analytical micro probe by "down-scaling" of a macro evanescent field absorbance sensor" (EFAS) previously produced at KfK shall be demonstrated.

OPTOCHEMICAL MICROSENSORS AND SENSOR SYSTEMS FOR DETECTION OF POLLUTANTS IN GASES AND AQUEOUS SOLUTIONS

In order to compete with existing (macro)analyzer systems the µ-TAS's have to show a high reliability and longevity, which are to a great extent determined by the quality of the chemical detection device (chemical micro sensors or microanalyzer).

As far as the opto chemical microsensors are concerned it is a fact, that in spite of the evident and very promising advantages which they may offer very few such devices with the desired specifications are available on the market, perhaps with a few exception, which are operated as one way sensors. This is mainly due to the fact that the long term stability of these microsensors cannot generally be guaranteed for an extended period of time nor is it presently possible to set up correlations between signal efficiency and aging of these sensors, which can be used for self-correction or self-calibration as it is commonly done with sensors measuring physical parameters. On the other hand the detection of chemical compounds by optical sensors has some unique advantages with respect to other (e. g. potentiometric) sensors, in that no reference element is needed, influences of electrical interferences are minimal and baseline drifting can be eliminated by using multi-wavelength calibration. In principle, the optochemical sensors are based on receptor or recognition molecules such as complexing agents or dyes, immobilized in a suitable matrix, polymer or sol-gel glasses, which show spectral changes when interacting with the analytes of interest. In its simplest form it can be a quartz glass fiber, to which mostly at the end of the fiber the matrix with the immobilized recognition agent has been attached (optrode) (fig. 1).

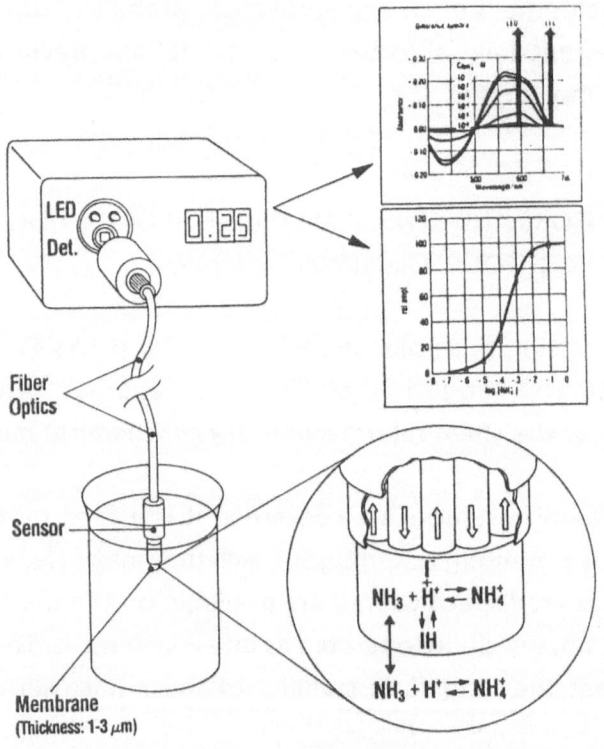

Fig. 1.
Ammonium-
Optrode

If this reagent forms reversible complexes with an analyte, either in the gaseous state or if the optrode is immersed in a solution, the optical properties of the complex will differ from that of the uncomplexed reagent, i. e. the maximum of light absorption has shifted to a different wave length, or fluorescence light of a different wave length is emitted. Light either from a broadband source, such as a xenon or quartz-halogen lamp of if suitable from a line source, laser or light-emitting diode is transported via the fiber optics to the reactive site of the optrode and back to a detector which recognizes the resulting changes of the optical properties of the reagent and correlates them with the analyte concentration.

This is demonstrated in fig. 1, where an optrode designed to operate as an ammonium detector for drinking water is shown. A chemical reaction in which ammonia (NH_3) which is in equilibrium with ammonium ion (NH_4^+) in the solution leads to a change of the colour of an indicator molecule immobilised at the end of a quartz fibre in a membrane. This colour change, which is a function of the concentration of ammonium ion present in the solution, causes enhanced absorption of the mono-chromatic light which is transported by fibre optics to the membrane and back to a detector unit.

The most serious defiency encountered with optochemical sensors of this type is the leaching out of the recognition molecules, a process which leads to uncontrollable signal variations. Therefore every effort has to be made to securely anchor the recognition molecules in the matrix without affecting their reactivity towards the analytes. The latter usually requires a certain mobility of the molecule in the matrix.

This point shall be demonstrated by the immobilization of porphyrine derivatives in a porous glass matrix.

Porphyrine derivatives are a promising class of chelating agents, being well established as highly sensitive indicators for the spectrophotometric determination of metal ions. The high complexation constants of porphyrins allow very sensitive detection. Furthermore, this class of dyes has extremely high extinction coefficients which can be used for a very sensitive optical detection. In addition, the different metal complexes of porphyrins show very specific absorption bands. This class of dyes makes a multicomponent analysis feasible, with sensors which can detect several metal ions by their specific reaction with one receptor molecule. For example, a simultaneous and accurate determination of Hg, Cd and Pb ions is possible after deconvolution of the complex spectra using standard chemometric procedures, e. g. partial least square regression (PLS) [2,3].

Fig. 2 shows the detection principle for detection of metal ions together with some representative spectra of metal complexes of a porphyrine derivateve (5, 10, 15, 20-tetra(4-N-methylpypidyl)porphyrin, TMPyP).

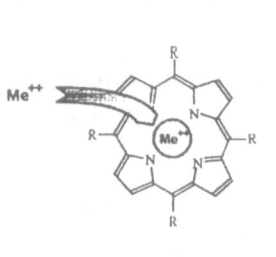 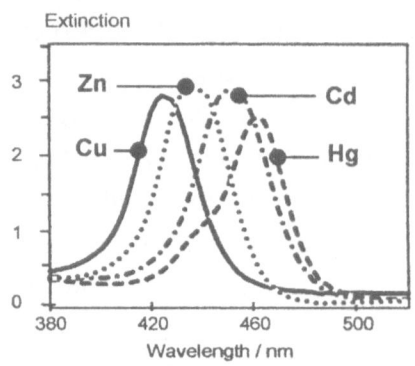

Fig. 2.
Porphyrines as receptor molecules for heavy metal ions

Immobilization of the porphyrines by simply adding them to a mixture of tetramethoxysilane (TMSO) methanol, H_2O and a catalyst (HCl or NaOH) which forms after sol-gel formation and upon drying at slightly elevated temperatures a porous glass was not found satisfactory. As shown in fig. 3 within an operating time of a few days the sensor signal deteriorated significantly. Drastic improvements could be achieved by covalently binding the porphyrines first to a macromolecule (dextran) before adding them in this

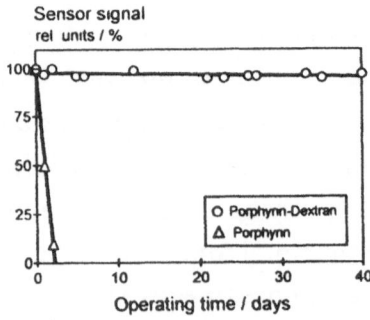

Fig. 3.
Longterm stability of heavy metal ion sensor

form to the sol-gel mixtures [4]. Probably because of the large size of the polymer the leaching out process has slowed down to a point where it became negligible and the sensor signal remained unchanged even after an operating time of 40 days or more. Detection limits of 5 µg/L for Cd(II) and 30 µg for Hg(II) were obtained. The response time was typically 5 - 10 minutes. Another successful variation of this techniques was the covalent binding of the porphyrin receptor to the Si-OH groups with spacer molecules.

Optochemical multisensor arrays with individual sensors of different functionality which will extend the possibilities of a multicomponent analysis can be obtained via processes similar to those used in microelectronics (thick film printing) or as shown in fig. 4 by applying a modified lift-off technique to form porous glass structures [5].

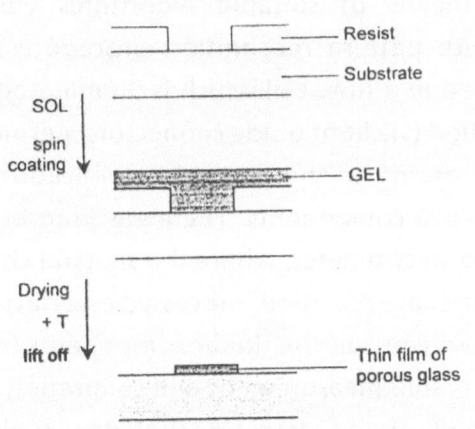

Fig. 4.
Schematic of Lift-off Process
(Structuring of sol-gel-glass films)

The improvements discussed here, as important as they are with regard to increasing the longterm stability of the chemical sensor, however, may not suffice in cases where extreme demands on reliable data acquisition exist. Therefore the chemical microsensors are incorporated into a microsystem as shown in fig. 5, which allows a self-testing and self-calibrating of the sensor by alternatingly recording signals from the sample and from calibrating solutions [6].

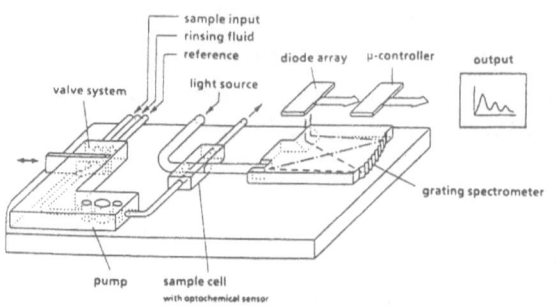

Fig. 5.
Layout of
optical
pollutant
analyser

The key component of the microsystem is an optochemical sensor either for toxic gases or for heavy metal ions. As an integrated system, it will be able to perform on-line sampling with the help of actor components (pumps, valves), to carry out cycles of calibration and measurement, and to evaluate and display the data acquired by means of suitable algorithms (spectral deconvolution and evaluation with pattern recognition procedures). The optochemical sensors are integrated in a flow cell which is illuminated with white light. The modular construction (via light guide connectors) permits the use of various sensor elements according to the particular application, and allows rapid replacement of defective components. The measuring beam is then transmitted to the LIGA microspectrometer, where the spectral changes arising from contact with the analytes are recorded, processed and output via a serial port. The microprocessor also controls the fluidic components (valves, pumps), so that variable, programmable measurement and calibration cycles can be carried out. The evaluation of the data (spectral deconvolution, calculation of the current concentration) and presentation of the results can be performed by a processor module.

Provisions for conditioning the sample and cleaning of the sensor from surface contaminations are features to be added to the system in the future.

SURFACE ACOUSTIC WAVE MICROSENSORS AND MICROSENSORSYSTEMS

Mass sensitive devices especially of the surface acoustic wave type have [7] recently attracted new interest because of the wide range of potential applications and since by combining them with advanced electronic signal processing drastic improvements in the limit of detection of analytes could be achieved.

Fig. 6 demonstrates the principle of a SAW sensor. By applying radio-frequency up to 433 MHz to the piezoelectric substrate, usually quartz, a surface wave is generated (Raleigh wave). The frequency of this wave is influenced by the mass of the substance deposited on the surface, e. g. by sorption or complex formation. The quartz surface is covered with a chemically selective layer in the simplest case a polymer, which interacts preferentially e. g. with one particular gaseous compound of the surrounding atmosphere via a reversible sorption a process leading to an equilibrium determined by the laws of distribution between the two phases polymer and gas. The analyte concentration can be determined by comparing the resulting frequency with the frequency displayed by a quartz crystal without a chemically selective layer. The observed frequency shift can be correlated with the analyte concentration. In this way the limit of detection could be extended to the sub-ppm range with response times of less than a minute.

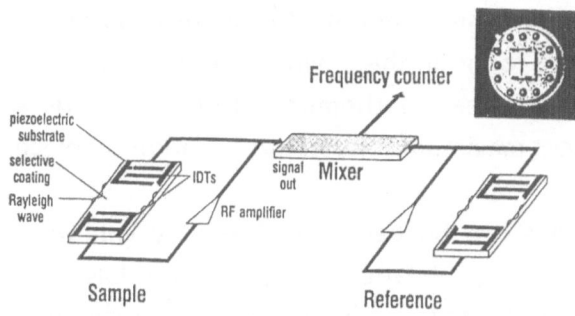

Fig. 6.
Surface acoustic
wave sensor:
IDT = interdigital
transducer

A high degree of selectivity for one particular analyte compound is normally not achieved with one sensor material, it can however be obtained if one is willing to spend enough research effort on such a development. On the other hand it seems more advantageous to achieve the desired selectivity for one or in this case for a limited number of analytes by employing several SAW's prepared with various chemically active layers showing different responses toward the compounds under investigation as determined by the different equilibria between the analyte and the various phases. It is then possible by using chemometric data evaluation or neuronal networks to identify and quantify them even in complex gas mixtures [8]. A typical response pattern displayed by eight differently prepared SAW's toward the components of a model mixture containing compounds known to cause multiple chemical sensitivity allergies is shown in fig. 7.

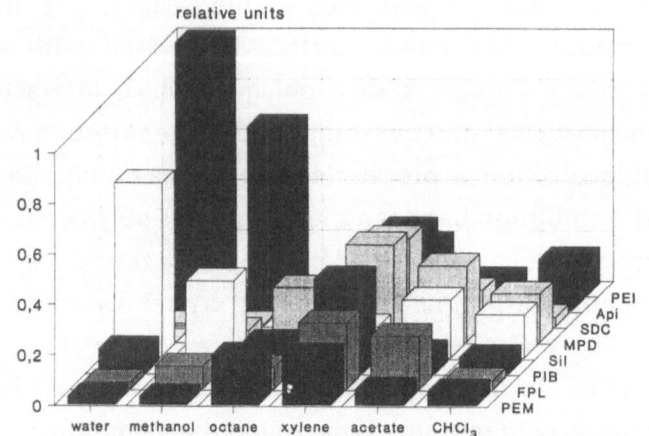

Fig. 7.
Sensitivity
Pattern of
various organic
vapors

Other applications of this technique, which can detect gaseous compounds down to the sub-ppm range can be found in the environmental monitoring of laboratories, workshops, storage areas of chemicals, refineries, service stations etc. (survey of the maximum tolerable working place concentrations).

Suitable sensor array combinations can be easily predicted by computer simulation based on the Linear Energy Relationship (LSER) Models. This model correlates the solubility parameters with known physicochemical properties of the materials such as polarizability, polarity, acidity, basicity and geometrical factor, placed as sorption materials on the SAW's. Thus such a computer simulation leads to an optimization of the sensor array combination for a given task and replaces costly and time-consuming experimental work in the laboratory.

SAW sensor array systems conceived in this manner compensate the lack of specificity shown by the individual sensors towards the analytes by employing a sophisticated data processing software. In this sense it will be more a complex system, however, it is more universally applicable, it can therefore be produced in greater numbers leading to a reduction of the production costs per unit. The individual SAW sensors can be coated very flexibly by spin-coating or spray-on techniques with materials found optimal for a given purpose without significant changes in the production process. Essential for a good performance of the system with regard to good limits of detection, is an excellent signal processing electronic. Typical noise levels obtained with the unit consisting of eight SAW's sensors developed at KfK (fig. 8) are 1 - 2 Hz

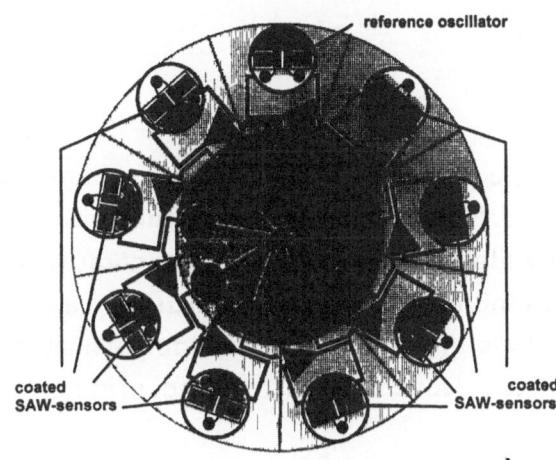

Fig. 8.
SAW-Sensor-Array
(radial configuration)

- Eight polymer coated SAW sensors and only one common reference SAW sensor

- Radial setup

- High frequency distributed by a central impedance adapted knot

- Short wires
  → minimal cross talk
  → minimal attenuation

- Monolithic housing with 9 chambers

- Ideal radial gas routing

from which limits of detection for a number of compounds were determined as shown in table 3 [9].

Furthermore since the distribution of the analytes between two or more phases depends strongly on the temperature, the analysis system is operated under carefully controlled thermal conditions.It is also imperative that no

60

Table 3   Detection Limits

Solvent	ppm [ml/m³]	g/m³	MAK-Values * [g/m³]
Xylene	6	0,026	0,44
Ethyl acetate	17	0,063	1,40
Chloroform	24	0,120	0,05
Water	14	0,010	-
Methyl alcohol	28	0,039	0,26
n-Hexane	59	0,216	0,18
n-Heptane	30	1,125	2,00
n-Octane	17	0,080	2,35
n-Nonane	7	0,035	-
n-Decane	5	0,027	-

* MAK - Values maximal permitted
workshop place concentration

dust particles can become deposited on the selective layers, therefore appropiate filters have been installed. Finally it is necessary to regenerate the system quickly. Thus air purified by filters incorporated in the apparatus can be used to purge the system. It was found quite useful to switch from a two-minute measuring cycle to an equally long purge cycle, a procedure which can be accomplished by a micropump and valves arrangement as shown in fig. 9.

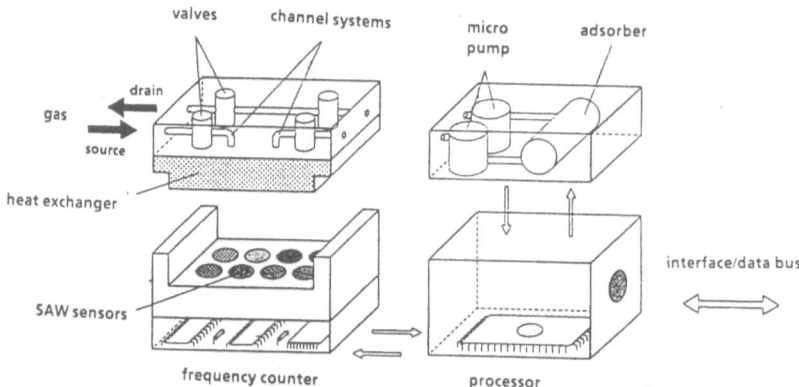

Fig. 9.   Analytical Microsystem for the Detection of Organic Gases

The system, as shown in the photograph (fig. 10), also contains a micro data processing and handling facility. As a stand-alone instrument careful consideration has been given to its energy consumption, which has been kept to a minimum.

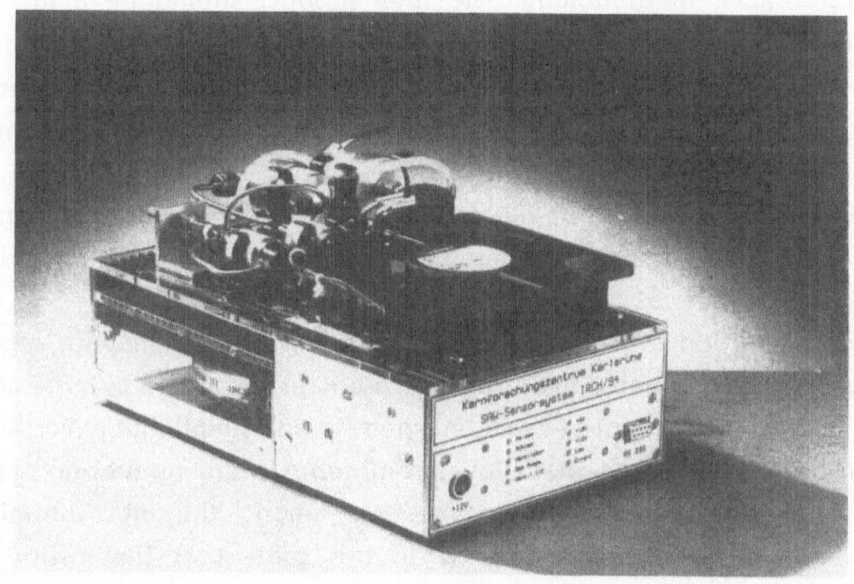

Fig. 10. Photograph of SAW Sensor array Prototype

CONCEPTS FOR MINIATURIZED ANALYTICAL MICROPROBES
INTEGRATED-OPTICAL NIR EVANESCENT WAVE SENSOR SYSTEM FOR CHEMICAL ANALYSIS

Analytical microprobes are obtained by down-scaling conventional laboratory analysis equipment In this case, the advantage is the availability of long proven methods, e g based on physical interaction between the analyte and photons, as in optical spectroscopy Although in principle the analytical systems for laboratory and in-line use can be very similar, major differences arise from the on-line need to maintain continuous flow gauging, real time measurement and signal processing, and the demands for the instrument to withstand the industrial or environmental conditions Major challenges can be seen in ensuring the representativeness of the considerably reduced material volume encountered in the measurement, selecting the sensing head principle and construction, and in developing proper calibration and disturbance methodology. The final product should be a low-cost integrated compact sensor-like device, a chemical analyzer performing the analyzing function in-line, with minimum maintenance, produced by modern miniaturizing techniques and compatible with the associated components of the complete microsystem.

The kind of approach can be examplified by the ongoing development of chemical microanalyzer systems based on evanescent field sensing principles at the KfK.

The contamination of drainage and industrial waste waters with organic substances like chlorinated hydrocarbons, aromatic substances or mineral oils requires the development of rapid, inexpensive and simple but nevertheless powerful chemical sensors that allow a continuous in situ monitoring of such pollutants. Opto-chemical microprobes are among the most promising approaches to fulfill such demands. In this context at the Institut für Radiochemie of the Karlsruhe Nuclear Research Center (KfK) during the last years an evanescent field absorbance sensor (EFAS) built from a quartz glass fiber with polysiloxane classing has been developed [10]. In combination with a conventional NIR spectrometer it allows to monitor nonpolar organic compounds in aqueous and gaseous media down to the low mg·l-1 range. Its sensing principle is based on the enrichment of these substances in the hydrophobic polysiloxane membrane and the measurement of their evanescent wave near-infrared (NIR) absorption spectra without spectral interferences from broad water O-H absorption bands (fig. 11).

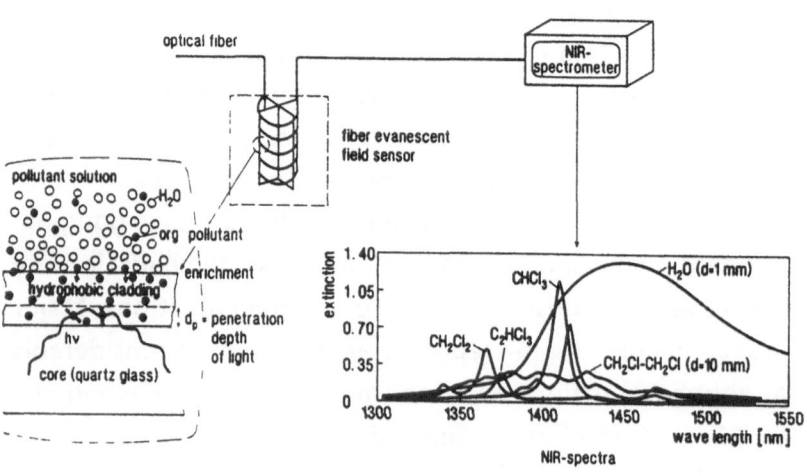

Fig. 11. Evanescent field absorbance sensor operating in near infrared (NIR) region

Current research is now directed towards realizing a miniaturized sensor by using integrated optics technology and planar waveguide structures. The construction of planar evanescent wave absorbance sensors offers some advantages compared to the existing fiber-optic sensor:

- in contrast to a cylindrical fiber geometry a planar substrate allows easy deposition of tailormade polysiloxane superstrates, which act as sensing layer.

- the planar structure provides a much higher mechanical stability compared to a sensing fiber coiled up close to its minimum bend radius on a support.

- due to the smaller waveguide dimensions a higher number of reflections and thus an increased sensitivity per unit length and a higher fraction of light intensity in the evanescent field are obtained.

The basic set-up of the integrated-optical evanescent field absorbance sensor (IO-EFAS) is shown in fig. 12. One part of the sensor is formed by a surface waveguide structure, produced by $Ag+/Na+$ ion-exchange in a BGG31 borosilicate glass substrate. Since the maximum refractive index (RI) of the waveguide is directly at the glass surface and due to the smaller waveguide dimensions, compared to an optical fiber a higher fraction of the modal field of the light wave reaches out of the surface. The sensing membrane that covers this waveguide structure is built by a suitable silicone polymer with lower refractive index, that acts as a hydrophobic matrix for the reversible enrichment of nonpolar organic contaminants from water or air. If light from the near-infrared range is coupled into the waveguide the evanescent wave part of the light field penetrating into the silicone layer can interact with the extracted organic species and light intensity in the spectral region from 1600 - 2000 nm will be absorbed at thecharacteristic frequencies of the corresponding C-H overtone and combination band vibrations.

The waveguide structure in the BGG31 borosilicate glass substrate has been realized in cooperation with IOT (Integrierte Optik Technologie, Waghäusel), a company that mainly produces integrated-optical components for tele-communication and sensor applications on a pre-production scale. The present surface waveguide structure was generated by the silver-sodium ion-

64

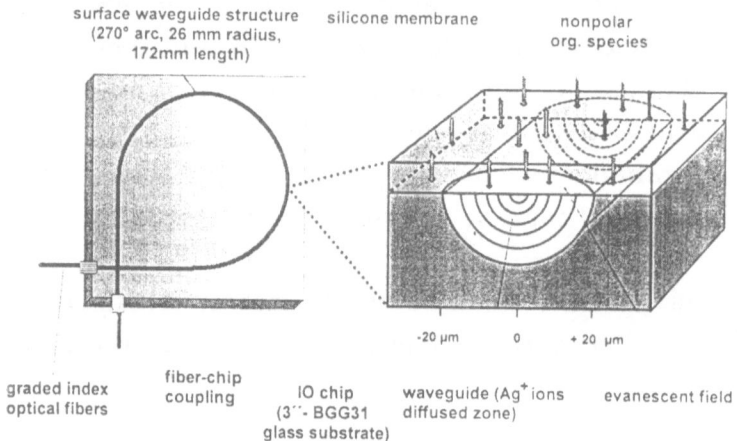

surface waveguide structure (270° arc, 26 mm radius, 172mm length)   silicone membrane   nonpolar org. species

-20 µm   0   + 20 µm

graded index optical fibers   fiber-chip coupling   IO chip (3''- BGG31 glass substrate)   waveguide (Ag⁺ ions diffused zone)   evanescent field

Fig. 12.   Basic-set up of integrated-optical evanescent field absorbance sensor

exchange process developed by IOT [11] The procedure starts by writing of the mask set for photolithography by an e-beam machine. The next steps of waveguide production in glass are carried out under clean room conditions. After a careful cleaning the glass wafer is evaporated with a thin titanium or chromium layer. In the following photolitthographic process (resist spinning, baking, exposure, developing, etching) the structure of the mask set for the later formed strip waveguide is copied into the titanium layer. After removal of the resist the wafer with the titanium mask is dipped into a $AgNO_3$ or $AgNO_3/NaNO_3$ melt The thermal Ag + /Na + ion-exchange process takes place at temperatures between 220 and 500°C The refractive index increase in the glass which is necessary to form a waveguide is caused by the higher polarizability of silver ions compared to sodium ions. After cooling to room temperature and removal of the titanium mask the waveguide structure can be cut out from the wafer.

The present multimode surface waveguide structure design for the IO-EFA sensor consists of a 270° arc with 26 mm radius terminating in two straight lines that intersect in a 90° angle The total length of the strip waveguide is 172 mm and the dimensions of the light guiding zone are around 35 x 20 µm.

To increase the penetration depth of the evanescent wave and hence the sensitivity it is essential that the RI of the superstrate approaches the RI of the waveguide as close as possible On the other hand with a smaller RI difference between waveguiding zone and superstrate the light losses increase Hence, one has to find a compromise between good sensitivity and bad signal-to-noise ratio

At the moment the preferred sensing materials for nonpolar organic substances, which are synthesized at the IRCH are polysiloxanes with methyl and phenyl side chains, which allow to change the refractive index of the superstrate in a rather broad range from 1 41 (pure polydimethylsiloxane) - 1.55 (methyl-phenyl-polysiloxane with 50 % phenyl groups) Due to this dependence of the silicone RI on the degree of phenylation, it is possible to approach the RI of the sensing membrane to the higher value of waveguiding zone, which has a graded index profile The RI value in the waveguide zone of BGG31 glass can be adjusted in the range from 1 459 - 1 519, depending on the ion-exchange conditions The "tailor-made" polysiloxanes for the sensitive coating of the waveguide structure are synthesized from the corresponding dichlorosilanes by hydrolysis [12] The reaction products are α,ω-dihydroxy-alkyl-aryl-polysiloxanes, that are obtained as viscous liquids. After purification by rectification, they can be used as precursor for the polymers After adding a curing catalyst the α,ω-dihydroxy-polyalkyl-(aryl)siloxanes are dissolved in toluene and poured onto the IO-glass substrate fixed in a mould Due to their low surface tension they spread easily over the wafer The solid silicone membrane is formed within 24 h of curing at room temperature

EVALUATION OF THE SENSOR PERFORMANCE

To optimize the sensor design and for testing the sensor response to organic pollutants the IO-EFA sensors are adapted over a lens system to a conventional incandescent light source (100 -W tungsten halogen lamp) and a Czerny-Turner spectrograph with a 300 lines/mm planar grating. A 256-element InGaAs diode array that can be applied at wavelengths up to 1680 nm is used as light detection element The measuring light is conducted from the tungsten halogen lamp to the sensor and back to the spectrograph via 50-µm all-silica graded index fiber of 2 m length, that are glued to the

66

waveguide structure by silicon fiber-waveguide couplers The experimental setup allows to measure NIR absorbance spectra with the IO-EFA sensor over a 120 nm wavelength range.

In this way trichloroethene (TCE) has been determined in the aqueous as well as in the gas phase.

For gas phase measurements the absorbance signal of the IO sensor with a waveguide length of only 172 mm is already higher than the signal of a fiber-optic sensor with a length of 11920 mm, thus showing the increased sensitivity of the IO component compared to the fiber-optic sensor. The increase in sensitivity per unit length of the waveguide is on the order of a factor of 60 (fig. 13).

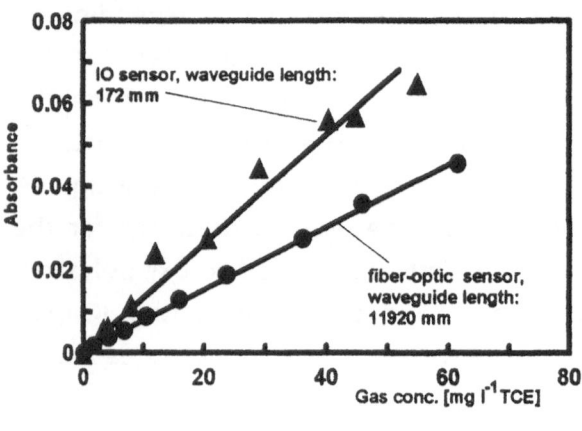

Fig. 13.
Comparison of calibration plots for trichloroethene gas measurements with IO-EFA sensor and fiber-optic EFA sensor

SPECTROSCOPIC MICRO SYSTEM FOR CHEMICAL ANALYSIS

An optimized evanescent wave micro sensor allows the integration into a spectroscopic micro system for chemical analysis, by combining it with a miniaturized incandescent light source, an NIR micro spectrometer and a diode array detector. A concept of this set-up is shown in fig. 14 [13]. A miniaturized incandescent light source, which provides a continuous emission spectrum in the near-infrared spectral range could be formed by a thin silicon filament, which is heated to 1100°C by resistance heating. Such a microlamp has been developed by Mastrangelo et al. In a first approach also already available miniaturized tungsten halogen lamps, which are used in mini pocket lamps seem to be applicable.

For the light dispersion modular element a planar grating spectrograph realized in LiGA technique, as previously discussed, could be used. For the fabrication of the spectrograph with the LIGA process polymethyl methacrylate is used (3-layer resist), which has only a good transparency at wavelengths up to 1100 nm. Therefore, for measurements in the spectral range from 1600 - 2000 nm the spectrograph has to be fabricated in deuterated PMMA, which is wellsuited at higher wavelengths.

For the light detection in the NIR range either a InGaAs photodiode array detector or a PbSarray detector could be used. In GaAs arrays, whose diodes are light sensitive up to wavelengths of 1700 nm, are applicable only to a restricted number of analytes because the first overtone C-H absorptions of most organic compounds are found at wavelengths >1700 nm. Therefore, the use of photo conductive lead sulphide detectors, which can be applied up to wavelengths of 3 µm seems to be advantageous.

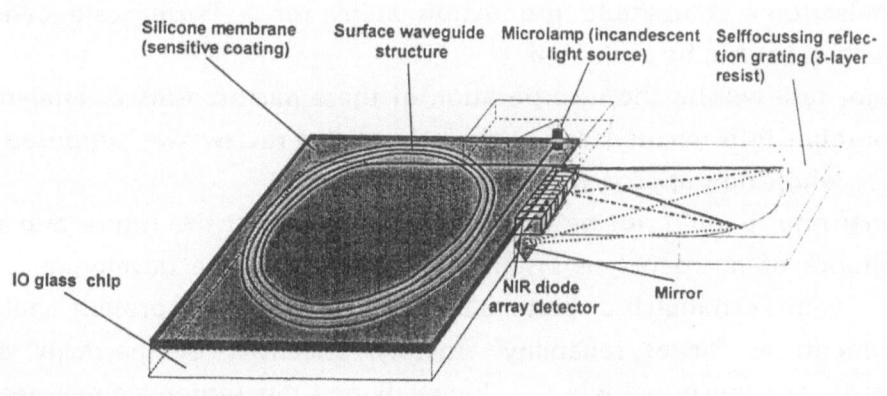

.Fig. 14.   Concept of Integrated EFAS-IOT-Spectroscopy
            System

For the realization of such a micro system the combination of the optimized single elements by suitable adhesion and bonding techniques has to be developed Moreover, miniaturized data acquisition and signal processing electronics and a corresponding intelligent software, for calibration and evaluation purposes have to be generated If these problems can be solved, the described concept compared to the fiber-optic approach with a conventional spectrometer allows to build up a cheaper but nevertheless powerful system for in-situ chemical analysis of organic pollutants

Summary

By evaluating the present development directed towards complete analyzer systems it should be recognized that the very fact that chemical microsensors or analytical microprobe at least in part have to be exposed to an aggressive environment containing the analytes to be measured requires new technological solutions in many areas For the fabrication of the chemically selective layer, new materials containing the recognition molecules have to be synthesized and new techniques producing thin layers, PVD, laser ablation, sol-gel, photo-lithographic structuring etc have to be explored The adhesion of these layers to the transducers has to be improved Microelectronic compatible production steps for a large scale sensor production have to be evaluated

A major task will be the incorporation of these microsensors or analytical microprobes into microsystems for which, as this review was supposed to show, a whole new approach has to be developed

Summarizing it could not be emphasized enough that the future and the acceptance of microanalysis systems will depend on the development of devices which can match or better surpass the properties of present analysis instruments as far as reliability, stability, selectivity or specificity and sensitivity are concerned whereby longevity and fast response times are of equal importance Last not least new flexible production techniques will have to be developed to produce these systems at low cost even in relatively small or mediumsized numbers.

References

[1.] For a general review of chemical sensors, see e. g. Sensors, W. Göpel, J. Hesse, J. N. Zemel ed., Vol. II and III.
Verlag Chemie, Weinheim, 1991

[2.] A. Morales-Bahnik, R. Czolk, J. Reichert, H. J. Ache: An optochemical sensor for Cd(II) and Hg(II) based on a porphyrin immobilized on Nafion membranes, Sensors and Actuators B, 13, 424 (1993)

[3.] H. J. Ache, Optochemische Mikrosensoren, KfK Nachrichten 26, 28 (1994)

[4.] M. Plaschke, J. Reichert, H. J. Ache, unpublished results

[5.] S. C. Kraus, R. Czolk, J. Reichert, H. J. Ache
Optimization of the sol-gel process for the development of optochemical sensors, Sensors and Actuators B, 15 - 16, 199 (1993)

[6.] W. K. Schomburg, R. Rapp, B. Büstgens, J. Reichert, O. Fromhein
Mikromembranpumpen als Elemente eines optochemischen Mikroanalysesystems
KfK Report 5238, 76 (Sept. 1993)

[7.] H. Wohltjen, Mechanism of Operation; Considerations for Surface Acoustic Wave Device Vapour Sensors; Sensors and Actuators 5, 305 (1984)

[8.] M. Rapp, M. Balzer, W. Coerdt, O. Fromheim, T. Kühner, J. Reichert, A. Voigt
Analytisches Mikrosystem auf der Basis von Surface Acoustic Wave Bauelementen
KfK Report 5238, 83 (Sept. 1993)

[9.] M. Rapp, unpublished results

[10.] J.-P. Conzen, J. Bürck, H. J. Ache, "Characterisation of a fiber optic evanescent wave sensor for non-polar organic compounds", Appl. Spectrosc. 47(6) (1993) 753 - 763.

[11.] L. Roß, "Integrated optical components in substrate glasses", Glastech. Ber. 62 Nr. 8 (1989) 285 - 297.

[12.] B. Zimmermann, J. Bürck, H. J. Ache "Verteilungsverhalten organischer Substanzen in Wasser/Siliconpolymer-Systemen", KfK-4967 (1991)

[13.] H. J. Ache, "Wege aus der Kostenklemme", Chemische Industrie 1 (1993) 40 - 43

[14.] C. H. Mastrangelo, R. S. Muller and S. Kumar, "Microfabricated incandescent lamps", Appl. Optics Vol. 30 No. 7 (1991) 868 - 873.

MICROFABRICATED
LIQUID HANDLING ELEMENTS

H. Sandmaier, R. Zengerle and A. Richter
Fraunhofer-Institut fur Festkorpertechnologie, Hansastr 27d,
D-80686 Munich, Germany

Microfabricated liquid handling elements can be regarded as key devices in very different application areas. They will play a major role in the development of innovative systems for chemical process control like μTAS, but also in medical drug delivery systems, environmental control, consumer goods or industrial equipment. Micropumps and active microvalves as the main parts of these systems are connected by closed channels together with other "plumbing" elements, like check valves, branchings, mixers or reservoirs. The progress in silicon micromachining and related micro structuring technologies like LIGA has lead to the development and successful demonstration of a multitude of devices. The current effort is to combine the appropriate liquid handling elements with chemical and biochemical transducers for the development of μTAS. This will be a hybrid system to allow the integration of components from different sources and fabricated with different technologies. Only this will lead to efficient production processes and the necessary economy of scale for successful commercial exploitation.

Little is known about the complex dynamic interaction in microfluid systems, consisting of several pumps, valves and narrow flow channels and the requirements on the liquids to be pumped, preconditioning and priming of the system. The peripheric fluid system has a pronounced effect on the pumping performance of micropumps. Due to the inertia and friction of the liquid in the system, the flow rate can be drastically reduced beyond the limit of the static pressure drop. Preliminary guidelines for the system design can be now given and the necessity for the methodic design for microengineered fluid devices will be made clear.

For an electrostatically activated micromembrane pump fabricated by bulk micromachining [1] transient measurements of the pump chamber pressure and the inlet and outlet pressures were made. From these values also the time dependent flow rate can be deduced. A complex dynamic behaviour can be observed, with a low frequency oscillation of the maximum pressure. Measurements can be compared and predicted with the results of an simulation tool PUSI [2] for micropumps connected to a peripheral fluid system.

71

A van den Berg and P Bergveld (eds), Micro Total Analysis Systems, 71–72
© 1995 *Kluwer Academic Publishers*

By the use of the simulation tool, the performance of an electrostatic actuated diaphragm pump with the outer dimension of 7x7x2 mm^3 was optimized. A maximum hydrostatic pressure of 310 hPa (3 m H$_2$O) and maximum pump rates of 360 µl/min have been achieved at a supply voltage of 200 V   The power consumption is less than 1 mW.

1. R. Zengerle; A. Richter; H.Sandmaier; "A Micro Membrane Pump with Electrostatic Actuation"; Proc. IEEE MEMS-92 ; (1992) pp.19-24; 4.2.-7.2.1992; Travemünde; Germany

2. R. Zengerle, W. Geiger, M. Richter, J. Ulrich, S. Kluge, A. Richter; "Application of Micro Diaphragm Pumps in Microfluid Systems"; Proc. Actuator '94; (1994) pp. 25 - 29; 15. - 17.6.1994; Bremen, Germany

MICROMECHANICAL COMPONENTS FOR µTAS

Jan H. Fluitman, Albert van den Berg and Theo S. Lammerink
MESA Research Institute, University of Twente
P.O. Box 217, 7500 AE Enschede, The Netherlands

0. Abstract

The Modular Fluid System (MFS) concept as base system for the realization of Micro-TAS, as well as a number of different micromechanical components for use in Micro-TAS are presented. The correspondence of MFS to electronic breadboards is discussed, and an example of a possible "mixed" fluidic/electronic board is given. The consequences of downscaling for the operation of sensors in Micro-TAS are discussed, and a number of components, sensors, sieves, mixers, valves and pumps are presented. Finally, the importance of the development of design tools and rules, especially bondgraph modelling, for MFS is emphasized.

1. System approach

In this contribution we introduce the Modular Fluid System concept [1] as a generic Micro-TAS breadboard, comparable and additional to its electronic counterpart. We will treat the components or functions, necessary to build such a system. Systems may greatly differ in their specifications. This means that components must be available in a range to meet these specifications. Hopefully a set of components can be designed to meet this range, while on the other hand they fit into a system with standard interconnection rules.

If such a system is developed the most important function is the interconnection, which must be reliable, easy to reach and leakfree. One can think of a floorplan with a set of buried channels for fluids on one hand and a set of components that can be fixed between channel in- and outputs on the other hand. An impression of such a system with a floorplan and building stones is given in figure 1.

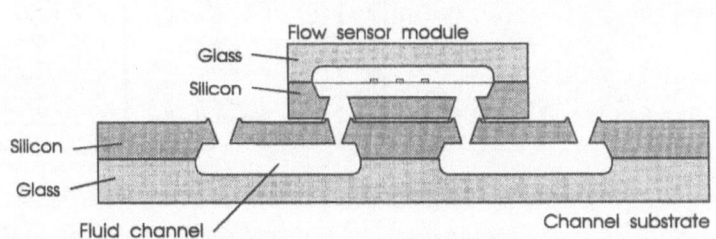

Figure 1 Flow sensor module on channel substrate.

A van den Berg and P Bergveld (eds), Micro Total Analysis Systems, 73–83
© 1995 Kluwer Academic Publishers

The most simple component is the two-point link and its function is just to connect channels. Of course the floorplan will contain channels of varying lengths and directions, but it will sometimes be impossible to have all the desired channels in the floorplan. This will certainly be the case for many-point links. Think of a splitter, that connects the fluid flow from one channel into two or more other channels, or the reverse thereof. In principle these components are not difficult to fabricate.

More complicated components are supply buffers, (active) valves, (active) switches, pumps, sieves, fluid mixers, reaction chambers, sensor units (flow, pressure, temperature, fluid parameters like viscosity and density, chemical composition, etc.), separators, waste buffers, etc. The fluid in- and output ports generally are in the floor plane, but it might be necessary to have injection units and/or outflow units as components as well.

A general design rule for every system is, that is must be able to be filled up in order to reach the starting conditions. Undesired blocking of fluids as a consequence of gas bubbles in a liquid or of pieces of dirt, must be prevented. So, the total design might contain parts with a function to remove blocking objects and with a function to check the overall status of the system. This includes an alarm system to become active, when the system is not functioning according to its purposes. This means that, like in microelectronics, testable design is necessary. Additional components will be added to check, in the case of failure, what the cause of the failure is, and if redundancy is possible to switch the flow to another part of the system.

All components that have an interface to electronics, must have electrical connections. Factually, besides the fluid flow systems there must be a flow system for electric currents to connect the components to an output connector or to the bonding pads of IC's that are built on the system. So apart from the fluid flow channels the floorplan should contain a set of electrical connections and bonding pads. In Figure 2 an idea of such a mixed system is given.

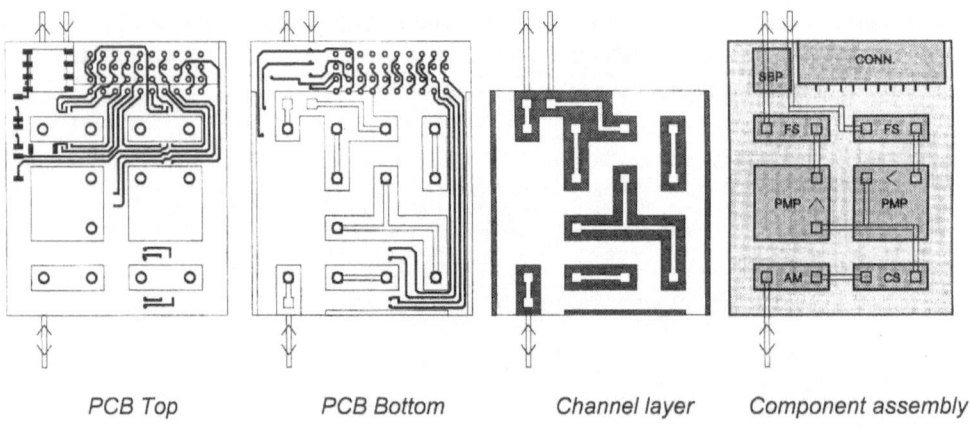

PCB Top   PCB Bottom   Channel layer   Component assembly

Figure 2 Schematics of floorplan with mixed electrical and fluid connections.

It is impossible to treat all possible elements in this contribution, therefore we have made a choice. The choice is based on relevance of the components, combined with the "in house" experience in our Institute. We will treat: sensors, sieves, mixers/reaction chambers, valves and pumps.

2. Sensors

2.1 CHEMICAL SENSORS

Although the chemical sensors are the crucial parts in a Micro-TAS, and are therefore extensively investigated in our Institute, it is impossible to spend much attention to them, because the chemical sensors form a separate and very extended field of research . Many conferences or conference-sessions are devoted to this subject and we have to refer to the proceedings and journals treating this subject.

However, the choice of the chemical sensors may have a large influence on the system as a whole. For instance we can have sensors with mainly a surface interaction with the fluid to be analysed, e.g. amperometric electrodes, or with a bulk interaction, e.g. in some optical sensors. In the case of surface-type sensors, the question is how much time T it takes to reach a stationary situation. Consider the reagent flowing with velocity V along the sensor. For V=0 the sheet thickness of the fluid S must be such that there is room for full development of static equilibrium in the desired time T. If V≠0 the supply of fresh fluid over the sensor leads to a stationary situation after a time T with a reduced sheet thickness required for a good measurement. So an increase of V, leads to a decrease of the required value of S, so that the product F = VSB, being the volume flow F for a channel with width B is less than proportional to V. If V is the determining factor for good sensing indeed, this means that a desired increase of V does not require a proportional increase of F.

The pressure difference Δp needed to reach a certain velocity in a thin slit is proportional with the reciprocal of S^2, while the volume flow F is proportional with the reciprocal of S^3B. So, in miniaturisation, the effect of V on Δp is two orders of magnitude less than the effect of F on Δp. This means that the requirements concerning pressures in a system, based on V-values are modest compared to mass injection systems based on F-values.

In the case of optical sensing this issue must be analysed for the case of "bulk" measurements, e.g. focusing a light beam through a fluid reservoir. Obviously, optical sensing methods based on the influence on the evanescent "light fields" along planar integrated optical waveguides, brings us again in the region of surface sensing techniques and the advantages thereof.

Of course there may be reasons, to pay attention to the minimum values of volume flows. It is conceivable that the flow velocities in low resistive connection elements becomes so small that other system requirements cannot be fulfilled, e.g. the number of analyses per hour, the procedures to clean the system and the suppression of sedimentation in dead corners.

2.2 PRESSURE SENSORS AND FLOW SENSORS

Pressure and flow sensors cannot be missed in a fluid flow system. Of course one can make use of well defined pressure and flow sources, but such sources relate their reliability on built-in pressure and flow sensors. Any pressure or flow source must be monitored for long term drift. (The same is true for the sensors, which must be calibrated by well defined sources: in the ideal case each actuator should have a calibrating sensor, each sensor should have a calibrating actuator.)

Pressure sensors mainly have monitoring and control functions in a Micro-TAS. But it is conceivable, that the pressure in a reaction chamber (e.g. as a function of time and/or concentration of the known reagent) can be used as an additional source of information about the chemical process and the ratio of chemical constituents. Silicon membrane pressure sensors of the type, which are widely used for automotive and

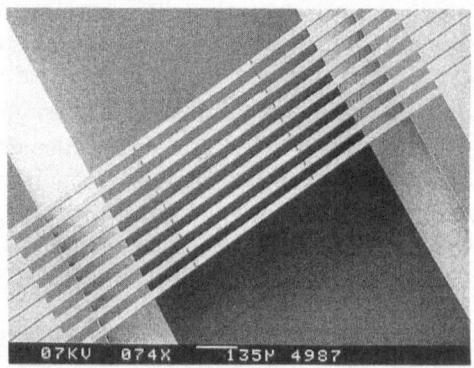

Figure 3 One dimensional multifunctional sensor/actuator array with 9 metal film resistors above the etched flow channel

medical applications are the most important candidates. We refer to the abundance of literature about these sensors. An example of the use of commercial pressure sensors of this type is given by the set up of the proposed fluid control system of Redwood Company [2].

For monitoring purposes, use of pressure and flow sensors can be made to check for sedimentation in components, blocking of channels, and so forth. Accurate flow sensors are needed to check for the amounts of fluid involved. A very interesting type of sensor is the array sensor described by Lammerink *et al.* [3], see figure 3.

The array consists of a number of Si_3N_4 carrier bridges, 1 micron thick, 40 microns wide with a spacing of 80 microns. The bridges are covered with 200 nm metal films acting as resistive heaters or thermoresistive sensors, whatever the choice. Three of them are used in a standard anemometric sensor and two of them as a thermal time of flight sensor. Since these two types act in a different way on gas parameters it is possible two measure the mass flow as well as the composition of, say (used in our experiment), a He/N_2 mixture. An (artificial) neural network is used for the determination of the output values (figure 4).

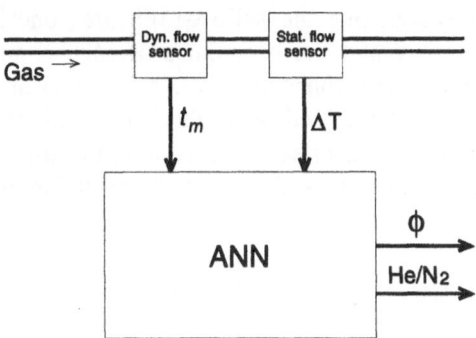

Figure 4. Artificial Neural Network (ANN) to determine flow rate and gas mixture indirectly from two flow-meters.

After covering the bridges with a thin noble metal film, they can also be used to measure heat developed by (bio)chemical reactions either in liquid phase or in the gas phase (catalytic converters) [4]. In the latter case, selectivity can be obtained by using different catalytic metals and by defining the working temperature of the resistors. In this way one can get accurate results from an array of sensors, where the elements themselves are not very selective (figure 5).

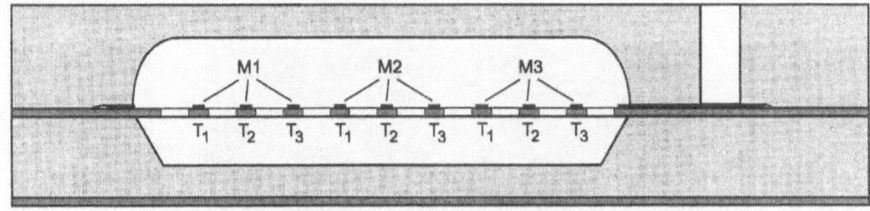

Figure 5. Cross-section of an array gas sensor with three different catalytic metals (M1-M3) and three different temperatures (T1-T3).

A matter of great concern is the durability of the sensors, in fact of any component in the system. Therefore it might be necessary to shield the sensors. It has been shown that Teflon coatings can be made, which are free of pinholes [5]. Although, this technology is in its infancy, the prospects are good.

2.3 TEMPERATURE SENSORS (AND ACTUATORS)

Temperature detection can easily be performed with the help of thermo-resistive elements. We do not restrict ourselves to materials that are compatible with the silicon IC-process. The fluid and electronic systems should be combined in a "hybrid" system. Smart fluid systems in the sense of, say, integrated sensors are out of reach. Smart components, used in a system like the MESA-MFS are conceivable. However, one should consider the necessity and the economics of such components in an overall system, where mounting, bonding, connecting, etc. is necessary anyway.

Temperature can also be sensed using thermo-electric effects in doped silicon structures. Since this transduction effect is reversible, the use of Peltier actuators is worth to be considered in processes that must be well controlled with respect to temperature.

3. Sieves

A sieve (as depicted in fig.6) is necessary as an obstruction for dirt and possibly for gas bubbles in a liquid. The latter function might be useful in filling up a system. A gas bubble separator is well conceivable, and it seems possible to design bubble outflow parts. Sieves can be made rather easily in silicon technology. The pore holes can be designed in diameter and depth, using dry chemical etching techniques. In a system such sieves can be designed to function at several places, combined with a channel system to remove dirt or bubbles. One has to look to such components as in electronics. If the production and mounting of such parts is cheap, one can even design with some "overkill" with a very limited increase in cost, while the system integrity increases significantly.

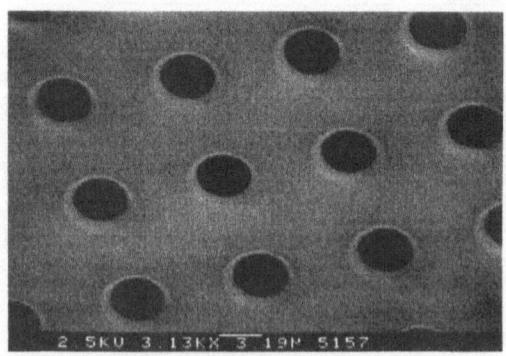

Figure 6. Electron micrograph of a part of a microsieve with holes of 5 μm [6].

4. Mixers/reaction chambers

The problem with mixers/reaction chambers is, that the down scaling of the fluidic process lead to Reynolds numbers which are much smaller than the values that are characteristic for the onset of turbulence [7]. Therefore it is not possible to increase the contact area of fluids to be mixed by "stirring". One is forced to look for solutions that increase the contact area in laminar flow situation. An elegant proposal for such a mixer/reaction chamber is from Miyake *et al.* [8], developed at our Institute. One of the fluids is pressed through a matrix of tiny holes perpendicular to the flow direction of the second fluid. The bulbs of the first fluid as they arise out of the holes into the second fluid leads to a large increase of the contact area and consequently of the diffusion process (see figure 7). As far as we know there are no other miniaturised mixers/reaction chambers described in the literature.

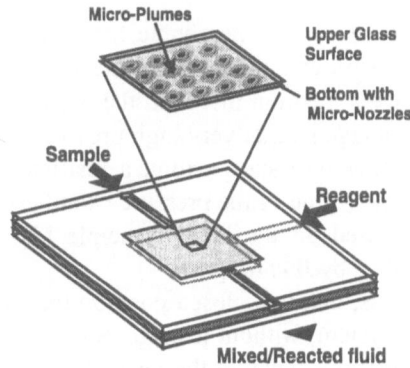

Figure 7. Basic principle of the micro-mixer

If one thinks of a straight forward mixer, being a channel with two sheet flows at the entrance, diffusion takes about 1 second to mix two 10 micron thick sheets [8]. So one can reduce the channel length of such a straight forward mixer by having a low velocity V. Here again the relevance of the velocity V over the volume flow F shows up.

5. Valves

Passive valves open and close as a consequence of the pressure distribution of the fluid at both side of the valve. They function as safety guards or as checker valves in a pump (next section). Active valves are controlled by signals from the system and therefore need an actuator to deliver the work for opening or closing the valve or put it in a correct position in a flow controller. Apart from the action, the actuators must be designed such, that they can maintain the valve position in a steady state. There are many valve designs, most of them are of the type that moves a membrane onto a valve seat (figure 8).

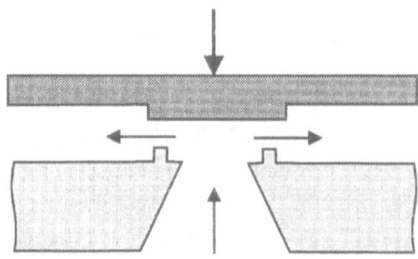

Figure 8. Valve with restriction perpendicular to fluid flow.

The favourite actuators to move the membrane are piezo stacks [9], thermal bimorphs [10], and thermal expansion chambers. The latter may comprise a liquid or gas, but the best solution is to have a medium that is in the phase change region, e.g. a liquid gas mixture. The great advantage is, that very high pressures can be reached with moderate power and that the temperature stays within a restricted range. A possible disadvantage may be the wafer scale filling of the pressure chambers. Anyway, the most successful commercial valve is based on the latter principle [7]. An analysis of its working principle has been published by Ji *et al.* [11].

Very important for a valve is, that in a closed position the valve must withstand the possible fluid pressure in the system, without leakage. Since a normally open valve must be active in this situation, one must analyse the open/closed ratio under operation in order to make a choice for a normally open or normally closed valve. It is also possible to design a valve with pressure compensation, such that the over pressure to be controlled is reduced [12] or even be used to enhance the valve position [10].

Controlled valves can be used in analog mode, e.g. as pressure reducers. In any flow analysis system, controlled pressures are needed to keep things moving. It fully depends on the application whether compressors, pumps and/or injectors are in the system or outside. In the latter case, on board controlled valves can do the job to create the desired flow pattern in the system. Gas bubbles must be prevented in a micro liquid flow system, because of their blocking effect. However one can also make use of gas bubbles in an open/close valve, and even as an actuator principle for pumping, as shown by Lin et al. [13].

6.Pumps

Many proposals and feasibility studies for micro pumps can be found in the literature. The most important micro pump type is the miniaturized checker valve pump (figure 9).

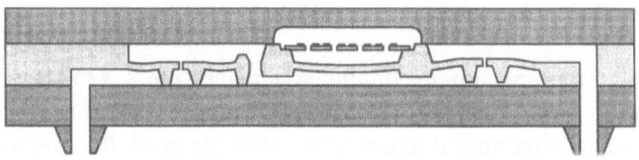

Figure 9. Cross section of a thermopneumatic checker valve pump [14].

It depends mainly on the actuator, which is the central element in the pump, what counter pressures can be reached. In the literature "low pressure" pumps are presented [15], but the desired pressures can be upgraded by using the phase change type of actuator, the thermal bubble jet principle [16], or even by making use of a pneumatic actuator driven by pressures generated in macro systems outside the micro system. The latter solution might seem odd, but it completely depends on the requirements of the application, how the system is designed. If, for instance, the micro system is designed for minimal use of liquids, minimal production of waste and a high throughput, but can remain on the "bench", there is no reason to exclude pneumatic actuation generated in a macro device on the same bench.

So, this again is an example of an application, which shows how difficult it is to talk in general terms. One should never overlook available solutions from the macro world, or the world of precision engineering, to design the best performing system for the application one has in mind.

A point with regard to the checker valve pumps (and also other pump types [17]) is the "heart beat". Indeed the pump in the system can be seen as the heart, but it may be required that other core elements must receive a flow with only a restricted ripple. Two pumps, side by side, pumping out of phase partly solve this, but it might be necessary to have a material buffer and a constriction to level of the ripple analogous as is done in electronics. Of course the introduction of a buffer suffers from the disadvantage that in a single analysis action the whole buffer must be filled with the liquid, which might be far more than necessary for the analysis itself. Again a consideration of possibilities starting with the application requirements is necessary.

A non checker valve type worth mentioning is the diffuser/nozzle pump [17] based on asymmetry of the pump structure. This pump is not free from backwards leakage, which can be solved by putting an open/close valve in series. Finally we want to mention the electro hydrostatic pump mechanism [18], which can produce a steady flow because it generates "body forces" on the liquid. However this pump requires high voltage and can be used only for a restricted kind of liquids.

It is a matter of course that a pump must always be accompanied by a flow sensor to have a well described component: an active micro flow controller or dosing system.

7. Design of Micro Fluidic Systems

An important design tool for micro fluidic systems is the CAMAS- or the older TUTSIM-design tool based on the use of bondgraphs. Unfortunately bondgraph modelling is not very popular and ill understood. Zengerle *et al.* for example, state that "more general design rules are needed", while bond graph modelling factually reflects physical systems in the most general way thinkable. It is based on energetic and entropic principles and cover physical processes in any physical domain. Introduced in 1961 by Paynter [19] a thorough background in physical systems theory has been worked out by Breedveld [20]. However, there is a lack of modern introductions, because the use of bondgraphs require a mental adaptation that is experienced by many persons as "user unfriendly". Nevertheless there is the review paper of Van Dixhoorn [21] and the CAMAS-introduction of Broenink [22], that might be of any help.

The elegance of bondgraph modelling is that it connects physical phenomena from whatever physical domain and that it can be used in a hierarchical way: supersystems, systems, subsystems. To give an idea of the latter, see figure 10, taken from [23] .

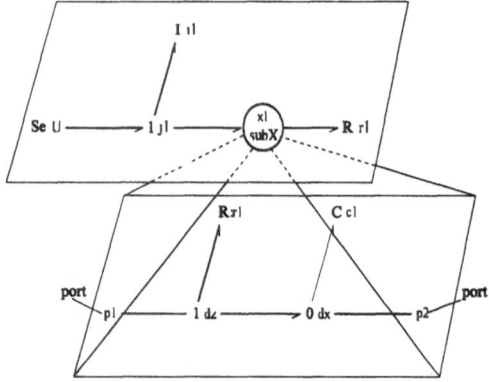

Figure 10. Bondgraph modelling of system with subsystem

For example, the subsystem might be a pump, which can be modelled with the help of bondgraphs and easily be introduced in the system as a whole, also modelled with bondgraphs. Subsystems, modelled by means of other computational techniques, can also be introduced by means of and adaptation of the subsystem's behaviour in terms of the input/output relations at the ports in the total system's bondgraph.

Van der Pol, in his thesis work [14] has shown, how the thermal, the pneumatic and the hydraulic subsystems, present in a thermally driven pump can be modelled. So a subsystem does not need to be a part that can be separated geometrically, but also a "part" that is contains the relations in a physical domain, e.g. mechanical or thermal.

Future work will show the advantage of bondgraph modelling and simulation and we expect CAMAS to be an important tool in designing micro fluid systems.

References

1 Theo S J Lammerink, Albert van den Berg and Jan H J Fluitman, *Brochure MESA-Modular Fluid Systems*, available on request

2 *Brochure Redwood Microsystems Inc*, 959 Hamilton Avenue, Menlo Park, CA 94025, USA

3 Theo S J Lammerink, Albert van den Berg and Jan H J Fluitman, "Micro System Array Sensors",*Proc Sensor Technology 1994*, National Conference, ISBN 90-73461-06-5, pp 173-177

4 P T Walsh and T A Jones, "Calorimetric Chemical Sensors", in W Gopel *et al*, "Sensors A comprehensive review", Vol 2, VCH Verlag, Weinheim, Germany, (1991), 531

5 H V Jansen, J G E Gardeniers, J Elders, H A C Tilmans and M Elwenspoek, "Applications of Fluorcarbon in Micromechanics and Micromachining", *Sensors and Actuators*, A41-42, 1994, pp 136-140

6 C van Rijn, personal communication

7 Peter Gravesen, Jens Branebjerg and Ole Soendergard Jensen, "Microfluidics - A Review", *Proc Micro Mechanics Europe 1993*, pp 143 - 164

8 Ryo Miyake, Theo S J Lammerink, Miko Elwenspoek and Jan H J Fluitman, "Micro Mixer with Fast Diffusion", *Proc MEMS-Workshop 1993*, pp 248-253

9 Masayoshi Esashi, "Integrated Micro Flow Systems", *Sensors and Actuators*, A21-A23, 1990, pp 161-167

10 Hal Jerman, "Electrically Activated, Micromachined Diaphragm Valves", *Digest Workshop Hilton Head*, 1990, pp 65-69

11 J Ji, L J Chaney, M Kaviany, P L Bergstrom and K D Wise, "Microactuaton Based on Thermally-Driven Phase-Change", *Proc Transducers 1991*, pp 1037-1040

12 Michael Huff and Martin A Schmidt, "Fabricaton, Packaging and Testing of a Wafer-Bonded Microvalve", *Digest Workshop Hilton Head 1992*, pp 194-197

13 L Lin, A P Pisano and R S Muller, "Silicon Processed Microneedles", Proc Transducers '93, Yokohama, Japan, (1993), 237

14 Frans C M van de Pol, "A Pump based on Micro-Engineering Techniques", *Ph D -Thesis*, University of Twente, 1989

15 R Zengerle, W Geiger, M Richter, S Klugge and A Richter, "Application of Micro Diaphragm Pumps in Microfluid Systems", *Proc Actuator 94*, Bremen 1994 pp 25-29

16 Theo Lammerink, personal communication

17 Eric Stemme and Goran Stemme, "A Valveless Diffuser/Nozzle-Based Fluid Pump", *Sensors and Actuators* A39, 1993, pp 159-167

18 A Richter, A Plettner, K A Hofmann and H Sandmayer, "Electrohydrodynamic Pumping and Flow Measurement" *Proc MEMS Workshop*, 1991, pp 271-276

19 H M Paynter, "Analysis and Design of Engineering Systems", MIT Press, Cambridge, Mass 1961

20 P C Breedveld, "Physical Systems Theory in Terms of Bondgraphs", *Ph D -Thesis*, University of Twente, 1984

21 J J van Dixhoorn, "Bondgraphs and the Challenge of an Unified Modelling theory of Physical Systems",*Progress in Modelling and Simulation*, F E Cellier (ed), Academic Press, New York, 1982, pp 207-245

22 *Internal Report* Introduction to PC-CAMAS, modeling and simulation" Available on request Control Lab, Faculty of Electrical Engineering, University Twente, POB 217, 7500AE Enschede, Netherlands

23 Jan F Broenink, "Computer-Aided Physical-Systems Modeling and Simulation A Bond-Graph Approach", *Ph D-Thesis*, University of Twente, 1990

MATERIAL SCIENCE FOR FUTURE (BIO-)CHEMICAL MICROSYSTEMS: THE KEY ROLE OF TAILORING INTERFACES

W. Göpel

Institute of Physical and Theoretical Chemistry and "Center of Interface Analysis and Chemical Sensors", University of Tübingen, 72076 Tübingen, Germany

0. Abstract

Modular components of future chemical sensor systems are introduced briefly Their development involves, in particular, new or fine-tuned (well-known) sensor-active materials and transducers A molecular understanding of the sensing mechanisms is shown to be a prerequisite for the development of a new generation of sensor systems with particular emphasis on their miniaturization and integration This understanding requires comparative microscopic, spectroscopic and sensor test studies on prototype materials to be performed, which requires to use experimental setups that combine the usual techniques of interface analysis with sensor preparation and sensor test chambers A few selected case studies on prototype materials are chosen to illustrate recent trends in the development of new materials and transducers for integrated chemical and biochemical sensor systems

1. Introduction

The general task of a molecular or elemental analysis to identify chemical or biochemical species in different environments is usually solved by applying the common instruments of analytical chemistry Considering the present marking share, chemical and biochemical sensor systems are still of minor importance but for well-known reasons, such as their low costs, small size, microelectronic compatibility etc their share is increasing steadily Progress in this trend is basically determined by progress in the understanding of sensor-active materials and their preparation technologies [1]

New structures obtained by microstructurization ("top-down" approaches) and chemical synthesis ("bottom-up" approaches) will on principle lead to a new generation of physical and chemical sensors with dimensions down to the nanometer range provided that the *stabilities* of interfacial structures, *sensitivities*, and *selectivities* of output signals ("sss") can be controlled sufficiently [2-5] The atomistic understanding of these phenomena is therefore essential in the research and development (R+D) of

85

A van den Berg and P Bergveld (eds), Micro Total Analysis Systems, 85–93
© 1995 *Kluwer Academic Publishers*

these sensors This includes in particular the interaction mechanisms between the different particles, waves, or fields with the different molecular, supramolecular or biological units The same understanding is required to apply the different scanning tunneling microscopy (STM) and related ("SXM") techniques for investigating and controlling the sensor structures or to reach the ultimate limits in the fabrication of reliable sensors for measuring physical or chemical parameters

In the following, only chemical and biochemical sensors are considered They make use of specific "key-lock" interactions which convert chemical to electronic information Three different tasks are usually fulfilled by chemical sensors, i e the quantitative and selective determination of individual particles (such as molecules or ions in gases or liquids), the determination of gross parameters (such as toxicity), or the quantitative characterization of odors (such as smells monitored qualitatively by the human nose) These requirements can only be achieved with *sensor systems* which in the most general case contain ten components for analyzing gases or liquids [4]

Like in a chain, the component with the weakest performance determines the overall performance of the sensor system For most systems the weakest component is the individual sensor element [4] Current sensor research therefore concentrates at three approaches to reduce this bottleneck by

- empirical optimization of sensor materials and of transducers in systematic tests under realistic measuring conditions,

- systematic studies of elementary steps of chemical sensing under thermodynamically or kinetically controlled conditions by means of microscopes and spectroscopies, and

- theoretical calculations of elementary steps of selective "key-lock" interactions

Commonly used transducers monitor chemical compositions by monitoring the phenomenological properties have been described extensively in earlier papers [1] Current microsystem technologies lead to new miniaturized transducers such as interdigital structures (for complex impedance measurements), thermopiles (for temperature measurements), piezoelectric oscillators based upon bulk, surface or plate waves (for mass measurements), integrated optics components (for optical measurements), multi-electrode arrays (for electrochemical measurements including recent developments for electrical connections to nerve systems), or arrays of sensor elements with integrated transducers as well as data processing (for integrated "electronic noses")

The trend towards improved performance and lower size and price per sensor element requires the systematic optimization of interface properties by means of the common tools of microscopy and spectroscopy of interfaces This requires in particular to realize *completely reversible bonding* of the detectable particle to the sensor-active coating and *completely irreversible chemical bonding* of this sensor-active coating to the transducer

This will be illustrated now briefly for typical sensor materials which show characteristic "key-lock" interactions to detect and identify particles Particular emphasis will be put on the discussion of ultimate limits in the miniaturization of sensor elements as a prerequisite to develop systems of nanosensors which will operate at the physical limits of miniaturization

2. Selected Case Studies

2 1 ELECTRON CONDUCTORS SnO_2

Electron conductors such as SnO_2 are commonly used as chemical sensors to monitor oxidizing and reducing gases, such as O_2, H_2, CO, NO_2, hydrocarbons etc   Current R&D of electron-conducting sensors aims at optimizing contact geometries, film thicknesses, particle size distributions, doping, operation temperatures and operation frequencies of thin film structures   Usually, film thicknesses are chosen in the micrometer range with a recent trend to stabilize nanosized SnO_2 particles with identical diameters in these films   On principle, the charge transfer even of individual electrons to or from the chemisorption complexes should be detectable in such sensor-active materials by, e g , monitoring steps in scanning tunneling spectroscopy (STS) spectra of nanosized SnO_2 particles in an arrangement similar to the one chosen in recent electron tunneling experiments which occur in the presence of specific oxidizing or reducing gases [6]

2 2 MIXED CONDUCTORS

Electron as well as ion-conduction of mixed conductors is typically monitored at elevated temperatures T>700 K in gas sensors based on oxides such as TiO_2 or a variety of perowskites all of which may be modified by different noble metal dopants to improve their selectivities towards specific gas detections by improved selectivities towards specific gas detections as a result of an improved catalytic activity of the sensor-active surface

Recent investigations aim at miniaturizing Ohmic contact and Schottky barrier devices with a control of metal atom depositions down to the atomic scale   The two different sensor operation modes may be adjusted intentionally at medium temperatures by externally applying a positive or negative DC-voltage to noble metal electrodes (such as platinum) and thereby switching the effective charges of the metal from neutral to ionic species and vice versa   The next step is to perform the switching with individual Pt atoms in an STS setup and to modify the current/voltage curves taken at individual surface or subsurface Pt sites   Chemical sensing occurs through charge transfer reactions as they occur during chemical sensing of donor or acceptor type of molecules in thin film devices [7]

2 3 ION CONDUCTORS

As a typical example, ZrO_2 is widely used as chemical sensor to monitor O_2 by utilizing its high temperature ion (O^{2-}) conduction in the bulk   In potentiometric devices, the cell voltage is determined by the pressure difference between two Pt contacts of ZrO_2 samples and in amperometric devices the current is monitored at a given cell voltage   Of key importance for both applications is the stability of the three phase boundary at which oxygen (O_2) molecules from the gas phase are converted to ions (O^{2-}) in the solid By externally applying a voltage, the value of the Fermi level at the electrode and thereby the concentrations and chemical reactivities of different oxygen-derived species at the surface may be shifted intentionally and thereby completely different selectivities may be adjusted

There is a huge potential in controlling surface and interface properties of new chemical sensors by controlling electronic, ionic as well as mixed conduction across interfaces of new chemically active sensor materials Similar to electronic contacts which may be Ohmic or Schottky barriers, ionic contacts may be created and three phase boundaries may be optimized towards their selective transport of electrons and/or certain ions Because of the experimental problems to prepare ideal epitaxial thin films of new materials and sandwich structures with a structural control in the monolayer range, the different transport phenomena of charged or neutral particles in these films and across their interfaces are not yet understood on the atomic scale Consequently, ultimate limits for the miniaturization of this new generation of chemical sensors cannot be estimated precisely yet Particular problems arise from the fact that long-term stability and reversibility are the most important properties for any application of future nanosensors [4, 8] and reasonable operation voltages of such devices are in the order of mV or above This often leads to high electric fields, electromigration and hence drift problems

2 4 SUPRAMOLECULAR AND POLYMERIC STRUCTURES

Molecular, supramolecular and polymeric compounds are particularly suitable for tailoring chemically sensitive coatings for thin film gas sensors to, e g , monitor volatile organic compounds (VOC s) Their molecular recognition occurs either by specific key-lock interactions in the first monolayer or by incorporation in the bulk The separation of surface reactions from bulk dissolution effects is deduced from the thickness dependence of sensor effects of prototype materials which show different mechanisms of molecular recognition The latter may be interpreted either in the framework of a rigid lattice (solid state) approach to describe key-lock interactions or in the flexible lattice (liquid state) approach to model the sensor response behavior and to systematically tailor new materials for this type of chemical sensing

The choice of new supramolecular structures is stimulated by theoretical calculations of high and selective partition coefficients, by new concepts which follow from an analysis of experimentally observed outstanding transducer patterns of sensor coatings with well-known supramolecular structures, by mimicking synthetically the local order of biological receptors with their well-known recognition functions, and by controlling the geometric arrangement of those gas specific recognition sites which have so far been optimized only empirically (e g , by modifying the polymeric coatings of columns in gas chromatographs or by modifying individual supramolecular structures in the liquid state towards their controlled attachment at solid surfaces) It appears to be of increasing importance in the synthesis of new supramolecular units to also take into consideration their subsequent molecular assembly in the thin film formation, e g , by providing these units with functional groups for their subsequent covalent attachment to gold or silicon substrates or for their subsequent two- or three-dimensional cross-linking by utilizing reactions which are triggered by self-assembly, thermal treatment, photon- or electron-beam bombardments etc [9]

2 5 BIOMOLECULAR FUNCTION UNITS

Hybrid structures with inorganic, organic and biological function units that make use of biospecific recognition may be divided into four categories with an increasing complexity of structures and functions, i e , systems based on bioaffinity (receptor-based interactions), systems based on catalytic activity (enzyme-based interactions), transmembrane channels, carriers or pumps and membrane receptors, cells and tissues The atomistic understanding and control of information transfer across interfaces is evidently of primary importance for a successful miniaturization and integration of components in any biochemical sensor system This has been illustrated for a variety of model systems in the past [10]

2 6 PATTERN RECOGNITION WITH (BIO-)CHEMICAL SENSOR SYSTEMS

A huge variety of data evaluation schemes and pattern recognition approaches is available to analyze the individual components of complex chemical systems of even of odors by an analysis of static or dynamic sensor response properties A clear trend is to be seen towards integration of data preprocessing on the chip with the latter containing a variety of sensor and calibration elements [11]

3. Microscopic and spectroscopic analysis of sensor transducer interfaces

Any optimization of above-mentioned sensor-active materials towards selectivity, sensitivity and stability during their use in integrated sensor systems requires a structural control down to the molecular scale [1-5, 12]

Because of the existing huge market of *standard materials for chemical and biochemical sensing* and because of an extensive know-how concerning their synthesis, film preparation, structure, and chemical reactivity , their use as chemically active sensor coatings has been so far and will be for the near future of major practical importance However, sensors of the next generation will require better sensitivities, selectivities and stabilities This will require a better control of the elemental and molecular structure and of the thin film formation This will also require new chemical syntheses

In the latter context, a clear trend is to be seen towards chemical sensing with molecular materials in general This trend is similar in other potential fields of future applications of the same molecular materials as, for instance in molecular electronics or integrated optics It concerns the use of *ultra-pure oligomers* instead of polymers as starting materials and the *growth of well-ordered monolayer or multilayer films* with such molecular or supramolecular units

The systematic improvement of sensor performances, failure analysis, and unequivocal understanding of chemical sensing in this next R+D step requires a complete control of chemical compositions, geometrical arrangement of atoms and of electronic and dynamic structures with particular emphasis on the structure of the interfaces to the gas phase and to the transducer substrate

3.1 SAMPLE PREPARATION

The *sample preparation* includes
- the elaboration of reproducible experimental conditions which lead to atomically flat and chemically homogeneous substrates with particular emphasis on Si or Au,
- the extraction, preparation or synthesis of well-characterized molecules for chemical sensing and their subsequent purification by carrier gas preparation, UHV sublimation in a temperature gradient, zone refining, chromatographic cleaning techniques etc.,
- the subsequent preparation of layer structures with these molecular or supramolecular units for which a large amount of different technologies is available such as Langmuir-Blodgett film formation, Knudsen-evaporation from the gas phase, self-assembly from solution, electrochemical deposition, spray or spin-coating deposition, any deposition and an additional subsequent crosslinking by thermal, optical or other treatments etc., and
- application of a suitable periphery such as contacts for electrical transducers.

The latter requires to identify experimental conditions for thermodynamically controlled or for kinetically hindered interactions of thin films with metal atoms, electronically conducting metal-oxides or conducting organic molecules.

3.2 MICROSCOPIC AND SPECTROSCOPIC SAMPLE CHARACTERIZATION

The *sample characterization* and *failure analysis* after different preparation steps, or before and after its practical use as complete chemical sensor requires to apply a variety of different tools of surface and interface analysis. The structures have to be investigated microscopically, spectroscopically and with respect to these phenomenological properties which are monitored by appropriate transducers. The most important spectroscopic techniques used on the laboratory scale are based on electrons, photons, ions, and atoms as probes to monitor the geometric, electronic, and dynamic structure as well as the chemical composition. Others require to apply synchrotron radiation (e.g. near edge x-ray absorption fine structure (NEXAFS) or x-ray diffraction at surfaces under grazing incidence conditions). Of increasing interest for characterizing and also structuring thin films are scanning tunneling microscopy ("STM") techniques and a variety of different modifications thereof ("SXM"). These techniques monitor various spatial variations, e.g., in electron densities attributed to different energetic states, in attractive or repulsive forces perpendicular or parallel to the surface, in ion currents, in heats, in optical or in magnetic properties for further details on SMX approaches. Unfortunately, some commonly used and powerful techniques for the detailed structural analysis of macromolecules (including in particular 2D-FT-NMR or X-ray crystallography) are not (yet) sensitive enough to analyze thin film structures at the interface with the required resolution down to the molecular scale.

The critical film preparation steps should be analyzed with the techniques mentioned above to control the correlations between structures and functions of the thin film samples in the different stages of their synthesis. In addition, the same systematic studies are also required to then investigate the influence of all those molecules or ions from the gas or liquid phase which are present under the various experimental conditions of the sample preparation and of the subsequent device application These studies require again the combined use of both, the spectroscopic or microscopic techniques of interface analysis and those techniques monitoring phenomenological layer properties with appropriate transducers

Of particular importance in this context are comparative investigations performed under clean room or even ultrahigh-vacuum conditions on the one side (characterization of "model systems") and under controlled atmospheric pressure or electrochemical conditions on the other side (characterization of "real systems") If at all possible, in-situ techniques should be applied which monitor structures under both, ideal conditions as well as under those natural environment conditions in which the final device is used These include in particular various optical techniques such as absorption and reflection spectroscopy between the IR and UV range, measurements of dichroic ratios, Raman spectroscopy, surface plasmon resonance spectroscopy, spectroelectrochemistry, or SXM-techniques [12]

3 3 EXAMPLES

The following examples may illustrate briefly the usefulness of spectroscopic techniques for designing chemical sensors based upon just one specific class of materials, i e supramolecular compounds [9] For further details, see an extensive literature on this topic [1,2,3,12 etc]

- The first example concerns the characterization of the elemental composition of resorcinarene monolayers to monitor perchloroethylen with particular emphasis on preparation conditions for a perfect orientation of the oxygen-containing head groups facing to the gas phase as compared to the sulfur-containing tail groups facing to the transducer This has been analyzed by angle-dependent X-ray photoemission (XPS) measurements
- The second example concerns the characterization of binding energies of organic molecules within molecular cages as compared to their binding energies in suitably chosen "non-sensing" spacer groups to align these cages This has been analyzed by mass-spectrometric thermal desorption (TDS) experiments
- The last example concerns the optimization of Knudsen-evaporated thin films of molecular cages (calixarenes) towards their reproducible interaction energies at the surface and in the bulk with minimum activation energies of bulk diffusion This has been analyzed by TDS experiments

4. Conclusions and outlook

The few case examples and studies discussed in this paper illustrate the importance of a controlled preparation and, in particular, a controlled signal transduction across interfaces of carefully chosen sandwich systems for today's or future integrated chemical sensor systems

The molecular understanding of chemical sensing ipso facto requires to understand "key-lock" structures on the nanometer scale One might therefore consider all chemical sensors as nanosensors The geometric arrangement of chemically sensitive structures may be done in the form of an organized monolayer, which shows

92

nanometer dimensions only in the direction of the film normal but macroscopic dimensions in the two other directions perpendicular to the film normal The next step is to arrange chemically sensitive structures with nanometer sizes in two dimensions or even three dimensions with an increasingly difficult task to address individual chemically active sites with controlled signal transduction

The "man-made" inorganic materials and, in particular, the electron conducting sensor materials will make it possible to develop "bottom-up" concepts in which individual active sites of chemical sensors are addressed and controlled Of particular interest in this context is the scanning tunneling spectroscopy (STS) combined with STM and AFM imaging in different contact modes

The use of "natural" biological materials is particularly promising if a combination of biological and synthetic structures, i e , a hybrid approach is chosen to build chemical sensor systems Again the controlled signal transduction across interfaces plays the central role in current research

By systematically replacing, adapting or exchanging step by step the different components of integrated chemical sensor systems in the two different worlds (i e , the world of the "electronic nose" and of the "human nose"), a systematic improvement in the understanding of recognition and signal generation, signal processing, and data storage is obtained This may finally lead to experiments, in which even individual reversible reactions are controlled on the molecular level in a miniaturized chemical analysis system

References

1 W Gopel, J Hesse, J N Zemel (Eds), "Sensors A Comprehensive Survey", VCH, Weinheim (FRG) 1992, 8 Volumes
2 W Gopel, Ch Ziegler (Eds), "Nanostructures Based on Molecular Materials", VCH, Weinheim (FRG) 1992, ISBN 3-527-28416-8
3 W Gopel, "State and Perspectives of Research on Surfaces and Interfaces", Report for DG XII, Commission of the European Community, Luxemburg, Report EUR 13108 EN, 1990, ISBN 92-826-1795-5
4 W Gopel, "New Materials and Transducers for Chemical Sensors", Conf Proc Eurosensors VII, Budapest (H) 9/1993, and Sensors and Actuators, B 18-19 (1994) 1
5 W Gopel, "Nanostructured Sensors for Molecular Recognition", Conf Proc NATO ARW, Cambridge (GB) 4/1994 and Far Proc Roy Soc , in press
6 See, e g , W Gopel and K D Schierbaum, "SnO$_2$ Sensors Current Status and Future Prospects", Conf Proc Eurosensors VIII, Toulouse (F) 9/1994, and Sensors and Actuators, in press, and references given there
7 K D Schierbaum, X Wei-Xing, and W Gopel, "Schottky Barriers and Ohmic Contacts with Pt/TiO$_2$(110) Implications to Control Gas Sensor and Catalytic Properties", WEH Symposium "Oxides", Bad Honnef (FRG) 2/1993, and H -J Freund, E Umbach (Eds), "Adsorption on Ordered Surfaces of Ionic Solids and Thin Films", p 268, Springer, Berlin 1993, ISBN 3-540-57416-6, and references given there
8 H -D Wiemhofer, U Vohrer, and W Gopel, "Interface Analysis for Solid State Electrochemistry", Conf Proc Third Int Symp on Systems with Fast Ionic Transport, Holzhau (FRG), and Mat Sci Forum 76 (1991) 265
9 K D Schierbaum, A Gerlach, W Gopel, W M Muller, F Vogtle, A Dominik, and H J Roth, "Surface and Bulk Interactions of Organic Molecules with Calixarene Layers", Fres Z Analyt Chem 349 (1994) 372

K D Schierbaum, and W Gopel, "Functional Polymers and Supramolecular Compounds for Chemical Sensors", Conf Proc E-MRS Spring Meeting 1993, Straßburg (F), and Synth Met 61 (1993) 37

K D Schierbaum and W Gopel, "Selective Chemical Sensing Molecular Recognition with Cage Compounds and Polymeric Permselective Layers", in G Harsányi (Ed), "Polymer Films and Sensor Technologies", Technomic, Lancester (USA) 1994, in press

W Gopel, "Supramolecular and Polymeric Structures for Gas Sensors", Conf Proc 5th Int Meeting on Chemical Sensors, Rome (I) 7/1994, and Sensors and Actuators, in press

10 See, e g , W Gopel, "Controlled Signal Transduction Across Interfaces of "Intelligent" Molecular Systems", Conf Proc 2nd European Workshop on Bioelectronics, Frankfurt (FRG) 11/1993, and Biosensors and Bioelectronics, in press, and references given there

11 A Hierlemann, U Weimar, G Kraus, G Gauglitz, and W Gopel, "Environmental Chemical Sensing Using Quartz Microbalance Sensor Arrays Application of Multicomponent Analysis Techniques", Sensors and Materials, in press

F Davide, C Di Natale, A D'Amico, A Hierlemann, U Weimar, and W Gopel, "Feedback Calibration of Sensor Arrays Based on QMB Polymer Coated Gas Sensors Sensor Physics, Non-Linearities and Cooperation in the Array", Conf Proc Eurosensors VIII, Toulouse (F) 9/1994, and Sensors and Actuators, in press

J W Gardner, A Pike N F de Rooij, M Koudelka-Hep, A Hierlemann, and W Gopel, "Integrated Polymer Array Sensor for Detecting Organic Solvents", Conf Proc Eurosensors VIII, Toulouse (F) 9/1994, and Sensors and Actuators, in press

F Davide, C Di Natale, A D'Amico, A Hierlemann, J Mitrovics, M Schweizer, U Weimar, and W Gopel, "A Novel Neural Network System for the Recognition of Gas Mixtures", Conf Proc 5th Int Meeting on Chemical Sensors, Rome (I) 7/1994, and Sensors and Actuators, in press

F Davide, C Di Natale, A D'Amico, A Hierlemann, J Mitrovics, M Schweizer U Weimar, and W Gopel, "Autoregressive Techniques for Dynamical Calibration of Sensor Arrays Based on QMB Polymer Coated Sensors" Conf Proc 5th Int Meeting on Chemical Sensors, Rome (I) 7/1994, and Sensors and Actuators in press

M Schweizer-Berberich, J Goppert, A Hierlemann, J Mitrovics, U Weimar, W Rosenstiel and W Gopel, "Application of Neural Network Systems to the Dynamic Response of Polymer-Based Sensor Arrays", Conf Proc Eurosensors VIII, Toulouse (F) 9/1994, and Sensors and Actuators, in press

A Hierlemann U Weimar, G Kraus M Schweizer-Berberich, and W Gopel, "Polymer-Based Sensor Arrays and Multicomponent Analysis fo the Detection of Hazardous Organic Vapours in the Environment" Conf Proc Eurosensors VIII, Toulouse (F), and Sensors and Actuators, in press

C Di Natale, F Davide, A D'Amico, W Gopel, and U Weimar, "Sensor Arrays Calibration with Enhanced Neural Networks", Conf Proc Eurosensors VII, Budapest (H) 9/1993, and Sensors and Actuators, B 18-19 (1994) 654

12 W Gopel, "Chemical Sensing, Molecular Electronics, and Nanotechnology Interface Technologies Down to the Molecular Scale", Conf Proc Eurosensors IV, Karlsruhe (FRG), and Sensors and Actuators B 4 (1991) 7

OPTICAL MICROSYSTEMS FOR
(BIO)CHEMICAL ANALYSIS

Otto S. Wolfbeis
Karl Franzens University, Institute for Organic Chemistry,
Analytical Division, Heinrich Str 28, A-8010 Graz, Austria

0. Abstract

The bulkiness of optical sensor instrumentation has prevented - so far - the use of optical chemical sensors ("optodes") in micro total analytical systems (μ-TASs) This is going to change in the next future   Three kinds of approaches toward optode miniaturization are presented which appear promising in context with μ-TAS schemes   The following are considered to hold particular promise
    (a) miniaturization of non-fiber systems in small flow cells,
    (b) the use of optical waveguide systems integrated into micro-capillaries,
    (c) miniaturization of fiber optic systems using extremely small fibers
These schemes, along with the use of solid-state electronics (light emitting diodes, pin diodes) may contribute to pave the way for a hyphenation between the optode and μ-TAS technologies

1. Introduction

Various approaches have been made to miniaturize the components of analytical systems, and to totally integrate them into one single unit in order to end up with a micro total analytical system (μ-TAS) Optical spectroscopy is a preferred method of detection, but - to the best of our knowledge - no systems have been described so far that make use of *optical chemical sensors* (optodes) In contrast to direct spectroscopy which detects any species absorbing at a specific wavelength, chemical sensors possess a certain (or total) specificity for a single analyte   Typical analytes for which specific optical sensors are known include oxygen, ammonia, and pH   The main reason why optodes have not been integrated yet into μ-TAS is the bulkiness of the opto-electronic periphery (light sources, photodetectors) of optical sensor instrumentation which makes its implementation into the whole system difficult

Optical (fiber) chemical sensors [1] are capable of measuring a single species in an untreated sample by simply bringing into contact sample and sensor Separation steps or addition of chemical reagents are not required This is the preferred method in case of samples where the matrix does not vary to a large extent, e g   blood However, optodes also represent useful detectors for use in chromatography, flow injection analysis

A van den Berg and P Bergveld (eds), Micro Total Analysis Systems, 95–103
© 1995 *Kluwer Academic Publishers*

(FIA), or capillary zone electrophoresis (CZE) Such methods are frequently employed in case of samples having a strongly varying composition The combination of optode technology with FIA [2] turned out particularly useful because the optode can provide a high degree of selectivity when detecting a single species in a plethora of potential interferents

2. Optical Chemical Sensors

In order to obtain an optical chemical sensor, a planar support - or the tip of an optical fiber - is covered with a coating whose optical properties vary in accord with the concentration of an analyte with which it is in contact and for which it is specific Coatings have been reported for most fundamental analytes including pH, oxygen, ammonia, ammonium, carbon dioxide, alkali and earth alkali metals, several heavy metals, along with a variety of biomolecules such as glucose, urea, and penicillin A number of optical detection schemes including reflectometry, fluorescence, and (N)IR techniques may be employed to measure the spectral changes of the sensing chemistry

Two kinds of optical chemical sensors may be differentiated, namely the fiber optic and the non-fiber optic sensors In each case, it is the selectivity and sensitivity (along with the stability) of the *chemical coating* which usually requires the most effort in sensor development Once such a sensor material is obtained, it may be deposited on fibers by dip-coating [3,4] or photo-polymerization [5], and on planar supports by spin-coating [6] or covalent immobilization [7,8] A most common technique involves the spreading of a chemically sensitive layer onto polyester foils [1,9] which paves the way for mass manufacturing of sensor spots Such spots may form one wall of a micro flow cell or, alternatively, be placed at the tip of an optical fiber Coated polyester membranes may also be obtained by spin coating [1,10-11] Fig 1 shows a typical cross section of a planar sensing membrane Occasionally, a light-impermeable so-called optical isolation may be placed on top of the sensing layer to prevent ambient light and sample fluorescence to interfere

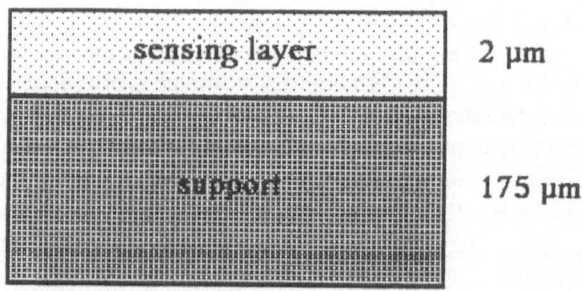

sensing layer   2 µm

support   175 µm

Fig. 1. Cross-section through a planar optical sensor membrane composed of an inert solid support, and an optically responsive sensor chemistry. A light source is placed underneath the sensor along with a photodetector which measures the intensity of the light emitted by the sensing layer. Sensor membranes of that kind are used in flow cells as shown in Fig. 3.

The sample is passed through a cell where it changes the optical properties of the sensor membrane Such cells, frequently made from steel or durable plastic, are robust, bulky, expensive, and have relatively large internal volumes The sensor membrane is attached to - or forms one side of - the cell This is a critical step and, in practice, the cell-membrane interface represents a notorious source of system failure due to gas permeation, leakage, swelling, scratching, or delamination

An interesting feature of optodes results from the fact that they can be operated simultaneously at several wavelengths This not only enables the design of sensors with a measuring wavelength and a reference wavelength, but also of sensors that specifically respond to more than one analyte simultaneously Thus, a planar sensor of the kind shown in Fig 1, but with a sensing layer composed of two indicators having different optical properties, has ben used to sense oxygen and carbon dioxide simultaneously in respiratory air The sensor spot contains two emitters, one possessing a green fluorescence, the other a red luminescence When illuminated with blue light, the green fluorescence is specific for carbon dioxide, and the red luminescence is specific for oxygen Hence, one sensor spot was capable of detecting two species at the same time

In addition to their use in planar sensing membranes, chemically sensitive optical materials also have been deposited at the tip of optical fibers Such devices, in principle, can be placed directly in the sample (e g , the artery) Hence, both sampling and transportation of the sample become unnecessary However, this approach suffers from other shortcomings including the fiber bending effect, mechanical rupture, interferences by false light because of inadequate optical isolation from both ambient light and sample fluorescence, and poor reproducibility of the thickness of the coatings on fiber tips Unless components with integrated fiber pigtail are used, the reproducibility in coupling light into and out of the tip coating (or a tip membrane) is rather poor Fig 2 shows a cross-section of a tip of an optical fiber sensor

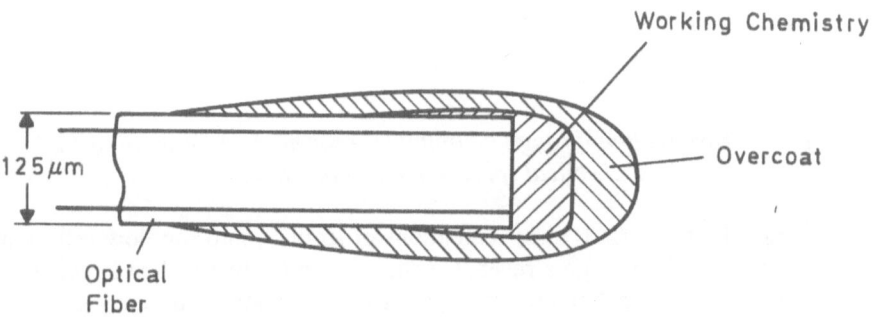

Fig. 2. Cross-section through a optical fiber chemical sensor composed of a fiber waveguide, an optically responsive "working" chemistry. The outer overcoat may be the optical isolation, or a layer of immobilized enzyme.

Both planar sensor membranes (Fig 1) and optical fiber sensors (Fig 2) have their respective merits and may be applied in various kinds of microsystems.

98

3. Micro Cells

Given the ease of making and handling planar optical sensors, and because they exhibit a more intense optical signal than do fiber optics, they were the first to be used in miniaturized systems. An example is provided by a triple sensor for measuring oxygen, pH and carbon dioxide in a bioreactor broth [12]. The sensor comprises a main electronic unit and a flow-through cell. The electronic unit consists of three independent opto-electronic units (one for each chemical species) controlled by one micro-controller and built into one housing. The flow-through cell unit comprises three stainless steel sensor modules mounted together on a steel support. A cross section through a typical sensor module is shown in Fig. 3. It has a sample volume of about 25 μL.

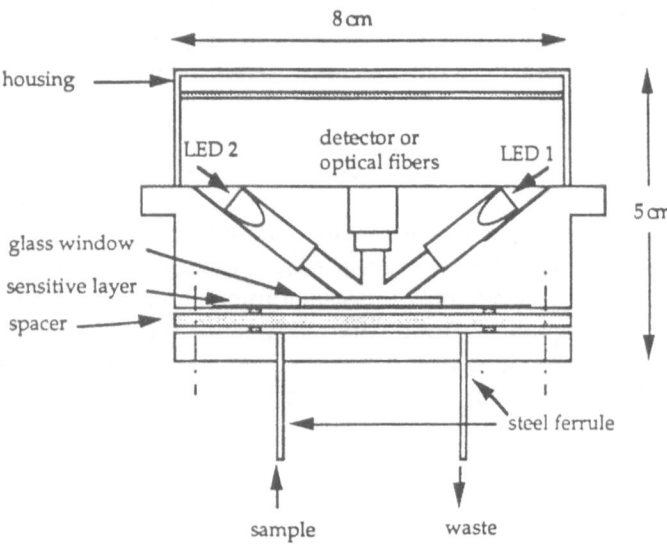

Fig. 3. Cross-section through an optical sensor module for monitoring a chemical species in a sample flow.

The liquid sample is transferred, under conditions of sterility, into the flow cell. Optical sensor membranes which undergo a reversible change in reflectivity when in contact with varying concentrations of analyte are placed on a glass window. A measuring LED (#1) and a reference LED of different wavelength (#2) are used, and the light reflected at different wavelengths is collected by the photodetector (top) or conveyed to a photodetector via optical fibers. The top of the housing contains a complete data processing unit and gives, via a standard port, a digital signal of the actual concentration of the analyte.

Another microcell has been proposed that may be used as a microcell for pseudo-continuous sensing, or as a disposable blood gas analyzer [13]. Three sensor spots, responsive to, respectively, pH, oxygen and carbon dioxide, are placed in a 30-μL microcell and filled with blood   On contact with the sample, the spots change fluorescence accordingly, and fluorescence intensity is read by a small meter   A schematic of such a "blood gas kit" is shown in Fig 4

A small syringe pumps blood from the blood stream into the microcell where readings are performed   After typically 2 min, the sample is pumped back into the blood vessel   This enables a pseudo-continuous monitoring, with data obtained every 2 min   In an alternative application, the kit is filled with blood by direct sampling at either the artery or the vein, and then placed in a stand-alone instrument which analyzes the three 3-mm i d   sensor spots which form part of the microcell

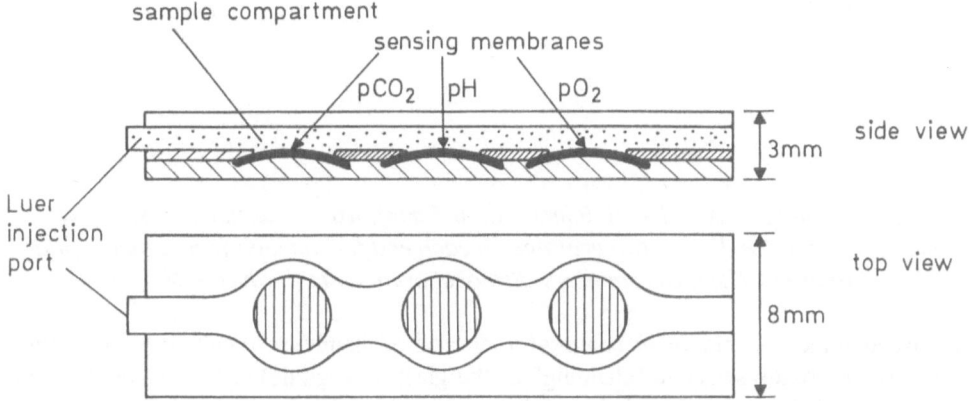

Fig. 4. Schematic of a disposable kit for monitoring blood gases and pH. The disposable containing three fluorescent sensor spots is filled with blood and placed in an instrument which reads the fluorescence intensity of the sensor spots and converts them into appropriate units (pH, Torr).

4. Capillary Flow Sensors

Conventional glass capillaries with an inner diameter of typically 700 µm, and coated - on the inner wall - with a sensor chemistry, recently have been shown [14] to be excellent "integrated micro-sensors" (Fig 5)   In contrast to former capillary type devices [15,16], such capillary sensors have attractive novel features including (a) a minute sample volume (typically 8 µL), and (b), a sample cavity which also acts as an optical waveguide so they can easily be subjected to optical interrogation   Most noteworthy, such capillaries may be used as devices for direct sampling

Sensor "chemistries" were deposited inside glass capillaries by pumping a cocktail solution through the capillaries and evaporating the solvent   Disposable glass micro-pipettes with a inner diameter of 0 7 mm were used in the particular case   The optical arrangement comprises an input coupler and an output coupler, respectively, at either end of the capillary (Fig 5)   In essence, a 0 5 x 0 8 mm coupler spot was ground onto the outer surface using conventional optical grinding paper   The resulting, highly scattering surface acts as a very efficient coupler, and the glass capillary with its index of refraction of 1.54 acts as the waveguide

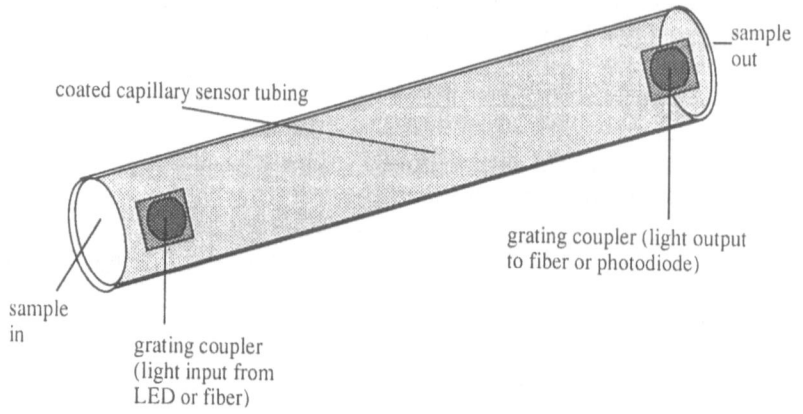

Fig. 5. Capillary sensor (l = 20.0 mm, i.d. 0.7 mm), with an chemically sensitive coating on the inner wall, and two gratings on each end for incoupling and outcoupling of light. The glass capillary also acts as an optical waveguide.

The refractive index of the inner coating (composed of mainly ethyl cellulose and acting as the carbon dioxide-sensitive "cladding" of the glass waveguide) is 1 35   The 488-nm line of an Ar ion laser was coupled into the capillary waveguide at one end   Light was attenuated by the blue carbon dioxide-sensitive inner coating on its way to the outcoupler.   A standard lens was used to focus the outcoming light onto a silicon photodiode.   In order to make use of such a capillary cell in a μ-TAS, an LED may be placed at one end, and a photodiode at the other end.

In order to demonstrate the utility of such a capillary microsensor, a chemistry sensitive to carbon dioxide was deposited on the inner wall of the capillary.   It is of blue color in the absence of carbon dioxide, and turns to yellow (via green) when exposed to increasing levels of carbon dioxide   The process is fully reversible.

Expired air contains 2 to 5% carbon dioxide   Although most carbon dioxide optodes are known to be particularly slow in response, the response time of this kind of sensor is less than 0 3 s (!) for 90% of the total signal change to occur.   This is obviously due to the very thin coating.   Such a fast response makes it possible to instrumentally monitor the concentration of carbon dioxide in respiratory air, but also to visually observe the continuous color change (from blue to yellow and back) while breathing through the capillary

Fig. 6 shows the response to varying concentrations of carbon dioxide.   It is obvious that not only the signal change is large, but also the response time is much faster than that of a conventional sensor chemistry deposited in much thicker layer

Because the capillary tube can act as a "tubular waveguide", changes in the absorption or fluorescence of the sensor layer within the evanescent field of the waveguide can be monitored [17], while interferences by color and turbidity of the sample (e g , blood) are widely eliminated, since the coating is thicker than the penetration depth of the evanescent wave which is in the order of 1 µm

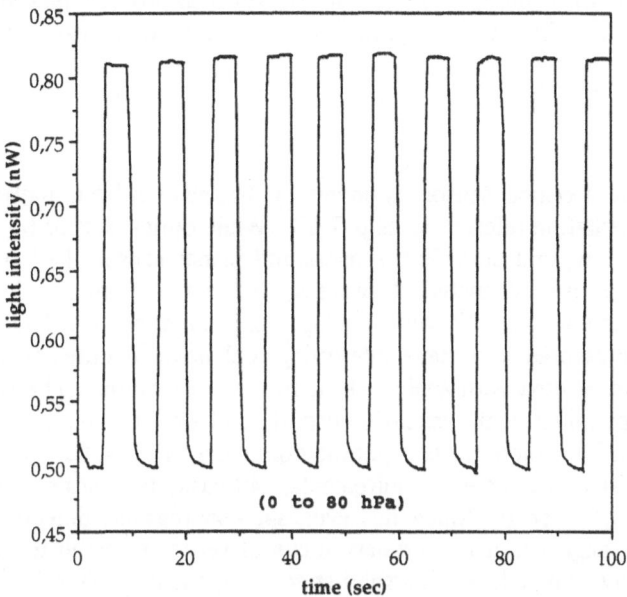

Fig. 6. Response time, reversibility, and relative signal change of a capillary sensor for carbon dioxide on switching from 0 to 80 hPa (0 to 61 Torr) pCO₂.

5. Microfiber Detectors

More recently, another configuration has been reported that hold promise in context with µ-TAS   A chemically sensitive material was deposited at the tip of a micropipette using near-field photo-polymerization [5,18]   The method lends itself to fabrication of miniature fiber optic pH sensors which require minutest sample volumes (e g the volume of a living cell) and have millisecond response times   Sensors were fabricated by drawing a conventional 100-µm fiber to diameters of a few microns (but also as small as 0 1 µm)   Then, the clad at the end of the fiber was coated with aluminium and the tip provided with a pH-sensitive chemistry by polymerizing, onto the fiber tip, a polymer-to-dye (fluorescein-acrylamide) conjugate   Light from a 488-nm laser was coupled into the fiber, and the fluorescence of the fluorescein-acrylamide conjugate observed through a microscope lens (rather than through the fiber)

Fibers of such a small diameter are capable of monitoring the pH of fractions of volumes of a cell which, typically, has a volume of 4 femtoliter (assuming a cell diameter of 2 µm)   It is noted that such a small volume contains only around 250 protons at pH 7 0   Hence, the number of dye molecules in the sensing layer (which act as a buffer) must be distinctly smaller

The most attractive feature of such as sensor is its small size   Conceivably, it lends itself to monitoring of femtoliter-size samples in an autoanalyzer system   At present, such sensors are excellent research tools, but fragile and unlikely ever to be used in an adverse environment or by unskilled personell   However, if properly placed in the benign environment inside a μ-TAS, they seem to represent a unique micro-tool for continuous chemical sensing

6. Conclusions

The miniaturization of optical sensors is underway, but has not been implemented yet in the kind of instrumentation referred to as μ-TAS   While optical sensor tips can be made in very small size now, and both fiber sensors and planar sensors have their merits in conjunction with micro-total analytical systems, they suffer from the disadvantage of bulky peripheral components   Notwithstanding these limitation, we think that some of the sensor types presented here have interesting and novel features that makes them potentially useful in context with analytical micro-instrumentation   The solution to the problem of bulky periphery is the exclusive use of solid-state components such as light-emitting diodes (LEDs), diode lasers, photodiodes, and other solid-state components   They have minute size, low power requirements, can easily be mass manufactured, and are available at very low costs   Since, however, such sources and detectors do not fully cover the spectral range yet, it is a primary focus of sensor research to make chemical optical sensors fully compatible with solid-state components   This is a considerable challenge to organic chemists but finally will enable the complete integration of optical detection schemes into miniaturized analytical instrumentation

7. References

[1]   O S Wolfbeis (ed), *Fiber Optic Chemical Sensors and Biosensors*, CRC Press, Boca Raton, Florida, vols 1 and 2, 1991
[2]   O S Wolfbeis, *J. Mol. Struct. 292* (1993) 133
[3]   C Munkholm and D R Walt, *Talanta 35* (1988) 109
[4]   J L Gehrich, D W Lubbers, N Opitz, D R Hansmann, W W Miller, J K Tusa and M Yafuso, *IEEE Trans. Biomed. Engg. BME33* (1986) 117
[5]   W Tan, S Zhong-You, S Smith, D Birnbaum, and R Kopelman, *Science 258* (1992) 778
[6]   K Seiler and W Simon, *Anal. Chim. Acta 266* (1992) 73
[7]   B G Harper, *Anal. Chem. 47* (1975) 348
[8]   O S Wolfbeis, H Offenbacher, H Kroneis, and H Marsoner, *Mikrochim. Acta (Vienna) 1984 I*, 153
[9]   O S Wolfbeis, L Weis, M J P Leiner and W E Ziegler, *Anal. Chem 60* (1988) 2028
[10] T P Jones and M D Porter, *Anal. Chem. 60* (1988) 404
[11] B H Weigl, A Holobar, N V Rodriguez and O S Wolfbeis, *Anal. Chim. Acta 282* (1993) 335

[12] B H Weigl, A Holobar, W Trettnak, I Klimant, H Kraus, P O'Leary and O S Wolfbeis, *J. Biotechnol. 32* (1994) 127

[13] M J Leiner, K Harnoncourt, G Kirchmayer, E Kleinhappel, H List, H Marsoner, O S Wolfbeis and W E Ziegler, US Pat 5,080,865 (1992)

[14] B H Weigl and O S Wolfbeis, *Anal. Chem.,* in press (1994)

[15] V Chernyak, R Reisfeld, R Gvishi, and D Venezky, *Sens. & Materials 2* (1990) 117

[16] I Kuselman and O Lev, *Talanta 40* (1993) 749

[17] B D MacCraith, *Sens. Actuat. B11* (1993) 29

INTEGRATION OF ANALYTICAL SYSTEMS INCORPORATING CHEMICAL REACTIONS AND ELECTROPHORETIC SEPARATION

D.Jed Harrison, Karl Fluri, Zhongui Fan, and Kurt Seiler
Department of Chemistry, University of Alberta, Edmonton, Canada

0. Abstract

Microfluidic systems have been micromachined in glass chips for systems intended for chemical analysis or sensing Using electroosmotic pumping, applied voltages control the direction of fluid flow without the need for valves Mixing of reagent solutions, chemical reactions and separation of compounds in mixures can be achieved. Demonstration of pre-separation mixing of fluorescent compounds on chip, and post-separation fluorescent labelling on chip is presented

1. Introduction

Microflow systems can be prepared in glass substrates using silicon micromachining techniques These systems can offer advantages in performance relative to stand-alone sensors, as has been demonstrated for ISFET-based analyzers [1,2] We have etched systems composed of complex networks of intersecting capillaries, in which chemical reactions, sample injection and separation of individual components of a sample can be achieved [3,4] The systems use electroosmotic effects for fluid pumping at velocities up to about 1 cm/s in 20 μm capillaries Electroosmotic pumping also allows control of the direction of fluid flow at the intersection of capillaries, without a need for valves or other moving parts Separations are achieved using electrophoretic effects, i e the differing mobilities of ions within an electric field result in different migration rates and this leads to separation   The separations are as efficient as in conventional capillary electrophoresis systems, and can be an order of magnitude faster, owing to the short distances involved We have shown we can separate and detect mixtures of amino acids within 3 to 14 sec [4] and even more rapid separations have been reported by Effenhauser et al [5] The potential applications of such systems include miniaturized analytical systems that could compete with bench-top instruments or chemical sensors in terms of performance, analysis time or durability [3-7]

A van den Berg and P Bergveld (eds), Micro Total Analysis Systems, 105–115
© 1995 *Kluwer Academic Publishers*

To practically realize the use of electroosmotic pumping in a complex manifold of channels it is necessary to develop an understanding of the factors that control flow within such systems, particularly at the intersection of two channels containing different solutions One of the best ways to achieve this is to visually image the flow process within the channels [4] In addition, the quantitative study of flow rates and study of the ability to control the direction of solvent flow using applied fields is required In this report we present both images and quantitative studies of diffusional and convective mixing of solutions at channel intersections

The applications of micromachined devices could be quite varied if the chips can be made versatile Performing chemical reactions on-chip is a key aspect of extending the versatility of these devices To this end we have demonstrated that a variety of device layouts can be used for mixing reagents on-chip, the first step in inducing a chemical reaction, and reaction of amino acids with a fluorescent label is presented

2. Experimental results and discussion

2 1 ELECTRICAL CHARACTERISTICS

A device referred to as COCE was prepared, consisting of three channels intersecting at a T-shaped junction, and this was used to study the application of potential to all three solvent reservoirs simultaneously The device geometry and the labeling scheme we have used for the channel reservoirs, lengths and resistances are shown in Figure 1 As a first step the impedance characteristics of the network were determined The dc electrical impedance of the intersecting capillaries was modelled as a network of resistors, as is shown in Figure 1b, and the validity of the model was examined

Table 1 gives the measured lengths and resistances between the reservoirs and the intersection point, and the ratios of these values to those for the channel length l_a Channel length l_b consisted of both a 23 mm length that was 220 µm wide and a longer 30 µm wide segment The wider segment length was expressed in terms of an equivalent length of 30 µm wide channel, specifically 3 1 mm, for the calculations The data show that the resistances R_a, R_b, and R_c are indeed proportional to the channel lengths within experimental error, as would be expected providing there is no defect in the bonding of the glass cover plate to the etched piece

With two voltage sources and a near ground potential connected to the channels, as illustrated in Figure 1b, the current in each channel and the potential at the intersection are readily expressed using Kirchhoff's Rules where the assumed directions of the currents I_1, I_2 and I_3 are shown in Figure 1b The 10 kΩ resistor has been omitted, as it is much smaller than R_3 Solving for V_J and I_3 gives equations 1 and 2,

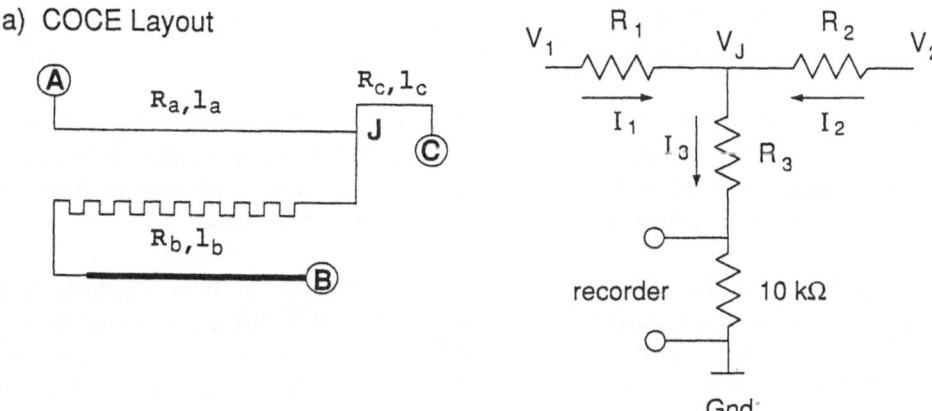

Figure 1.   a) Device COCE layout with reservoir labels and channel lengths, l_x, indicated. Overall dimensions were 2.8 by 7.0 cm.
b) Equivalent circuit for device COCE. Subscripts refer generally to the resistances of the 3 channels of the device in a).

Table I. Lengths and Resistances for COCE Device.

	A	B	C
Length (mm)	45	93.3	8
Resistance G	0.32 ± 0.02	0.72 ± 0.02	0.079 ± 0.008
L/L_a	1	2.07 ± 0.03	0.18 ± 0.01
R/R_a	1	2.3 ± 0.2	0.25 ± 0.08

A,B,C refer to the lengths shown in Figure 1. Error in the lengths is about 0.05 mm. Ratios of lengths relative to L_a, and resistances relative to R_a are given.

$$V_J = (V_1 R_2 R_3 + V_2 R_1 R_3) / (R_1 R_2 + R_2 R_3 + R_1 R_3) \qquad (1)$$
$$I_3 = [V_1 R_2 + V_2 R_1] / (R_1 R_2 + R_2 R_3 + R_1 R_3) \qquad (2)$$

While in general the potentials V_1 and V_2 can have any polarity, in this study V_1 was always positive and V_2 was always negative. Equation 2 shows that a plot of I_3 versus V_1 will be linear when V_2 is held constant, however I_3 may be positive or negative depending on the potentials applied and the resistance of each channel.

The behavior of the device with potentials applied to all three channels was evaluated according to the scheme indicated in the inset of Figure 2. The current was measured from the potential drop across a 10 kΩ resistor between ground and reservoir C, with V_1 and V_2 applied to reservoirs A and B, respectively. In this configuration the resistances for eq 1and 2 were $R_1 = R_a$, $R_2 = R_b$, and $R_3 = R_c$. The current was indeed

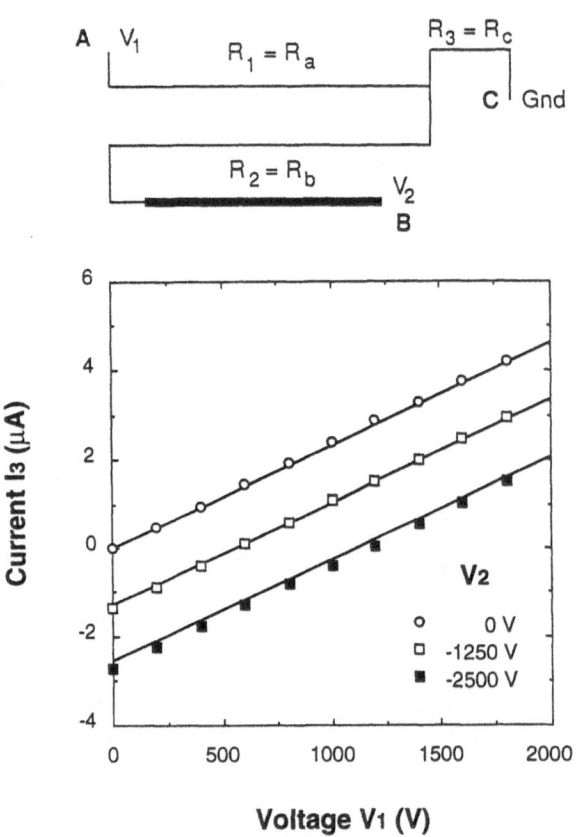

Figure 2.   Current as a function of V_1 applied to reservoir A, with the indicated values of V_2 applied to reservoir B. Symbols are for experimental data, solid lines are for current calculated from equation 2 using the data in Table 1. The inset shows the applied potential scheme.

linear in V_1 when V_2 was held constant. Figure 2 shows the data obtained for three different values of V_2. It was possible to calculate the current using eq 2 and the measured channel resistances given in Table 1, and the calculated response is shown as the solid lines in Figure 2. The very good agreement illustrated between theory and experiment shows that the model is accurate. It can be used to determine potentials and currents within a complex network of capillaries when several voltage sources are applied simultaneously.

2.2. LEAKAGE CONTROL

We have previously discussed leakage effects at the intersection of channels [3-6]. In particular, if a side channel is left floating while a potential is applied to cause flow in the main channel there will be leakage of solution from the side channel, contaminating the main channel. We have indirectly shown that both convective and diffusion effects contribute to this mixing, or leakage effect at the intersection [3,4]. In previously studied devices with various layouts, the concentration of sample leaking into the main channel was about 3% of that in the sample channel. We show here that these effects can be controlled by applying voltages to several channels simultaneously.

Figure 3 shows data obtained in another device, Jet-1, with the layout indicated in the figure. The resistances of each of the channels of this device were determined as described above. The separation of fluorescein isothiocyanate (FITC) labelled arginine (arg) and tyrosine (tyr) is shown. For these experiments sample solution was present in reservoir 2, which was held at ground. As shown in the diagram at the top, samples were injected across a "double T" injector towards reservoir 1, which was at -3 kV. This injector creates a sample plug in the separation channel about 150 μm in length.

To evaluate the ability to control leakage during a separation we applied potentials to reservoirs 2, 3 and 4 simultaneously, as shown in the insets of Figure 3. In the first electropherogram of Figure 3, 284 V was applied to reservoir 3 during the separation. The background fluorescence level was high compared to the background seen when the sample channel was instead left floating (dashed line). From the measured resistances of the channels the calculated potential at the intersection of the sample and separation channels should be -17 ± 12 V with 284 V on reservoir 3 and - 3 kV on reservoir 4. With the sample channel at ground there should be a net flow out of the sample channel resulting in an increased background, as was observed. When the potential at reservoir 3 increases, the potential at the injection point will eventually become positive, which should electroosmotically push sample solution back towards reservoir 2. As shown in Figure 3, 360 V at reservoir 3 lowered the background fluorescence below the level observed when the sample channel was left floating. The decrease in background showed that the flow in the sample channel was reversed, and establishes that leakage effects can be controlled using applied potentials.

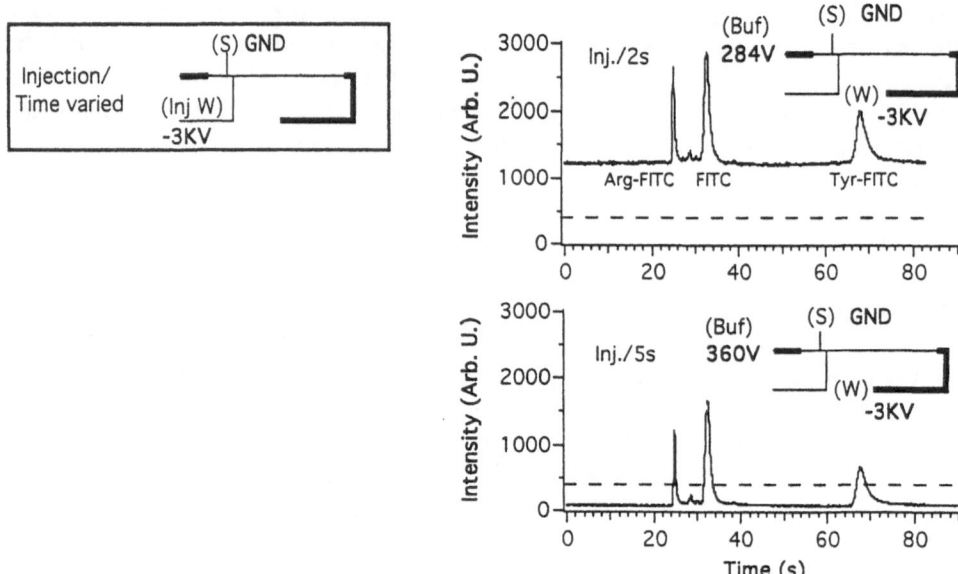

Figure 3. *Separation of 20 μM Arg-FITC and 40 μM Tyr-FITC in 0.064 M carbonate buffer, pH 9.1, with potentials applied to three reservoirs of device Jet-1. The dashed line indicates the background level when the sample channel was instead left floating during separation.*

2.3. MIXING OF SOLUTIONS USING VOLTAGE CONTROL

When electroosmotic flow is present in a capillary the linear flow velocity, v, is given by eq 3,

$$v = (\mu_{eo} + \mu_{ep})E \tag{3}$$

where μ_{eo} and μ_{ep} are the electroosmotic and electrophoretic mobilities, respectively, and E is the electric field applied.[14] The overall mobility, μ, is the vector sum of these two mobilities, and μ_{eo} is generally larger than μ_{ep} in absolute terms. The velocity of species i in each of the channels is given by equations 4-6,

$$v_{i,1} = \mu_i (V_1 - V_J) / l_1 \tag{4}$$
$$v_{i,2} = \mu_i (V_2 - V_J) / l_2 \tag{5}$$
$$v_{i,3} = \mu_i V_J / l_3 \tag{6}$$

where V_J is given by eq 2, and the various v_i refer to the velocity of species i in each channel of length l. The subscripts 1, 2 and 3 refer to the velocity and channel length associated with the potential sources V_1, V_2 and ground, respectively. It is assumed the ionic strength (resistivity) and pH in each channel is the same. These expressions indicate that controlled mixing of solutions should be possible at the intersection of the three channels by adjustment of the potentials V_1 and V_2.

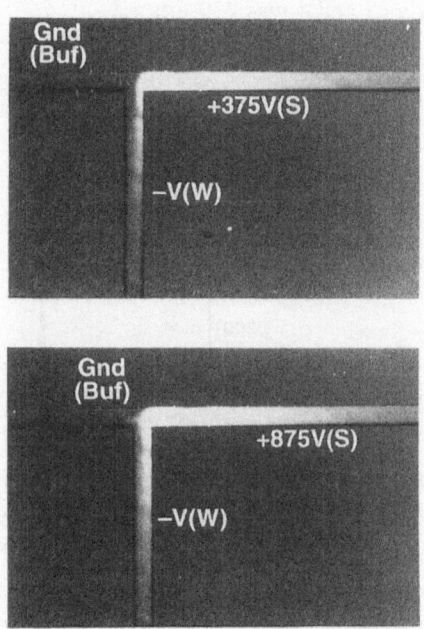

Figure 4.   Photomicrographs showing controlled mixing of 10 mM carbonate buffer, pH 9.1, and 100 μM fluorescein.   Channels are 30 μm wide. Flow direction was from the two horizontal channels into the vertical channel.

The COCE device was mounted under a microscope equipped with a camera and the region near the intersection was illuminated with 488 nm light. Any striations visible in the fluorescent dye stream seen in Figure 4 and later photos are due to non-uniformity in the illumination or collection efficiencies, as is the tendency of the intensity to fade at the edges of the photos. Potentials were applied to the three reservoirs with the polarities indicated in Figure 4. Sample (S) was driven towards the waste reservoir (W), while buffer (Buf) was also driven towards waste by the applied potentials. The two solutions mixed downstream of the intersection. The photos illustrate that increasing the potential on the sample reservoir increases the amount of dye relative to buffer downstream of the intersection.

112

2.4. MIXING CHAMBER

Figure 5 shows the layout of a device called Jet-3, which has a mixing chamber incorporated All channels of single line width are 30 μm wide, whereas the blackened bulky lines are 300 μm wide The large box between in the figure was used as a mixing chamber. The chamber was 4 5 mm long and 170 μm wide The volume of the chamber, excluding the islands, was about $3\ 6 \times 10^{-3}$ mm^3 The offset between the side channels connected to reservoirs 3 and 4 was 220 μm

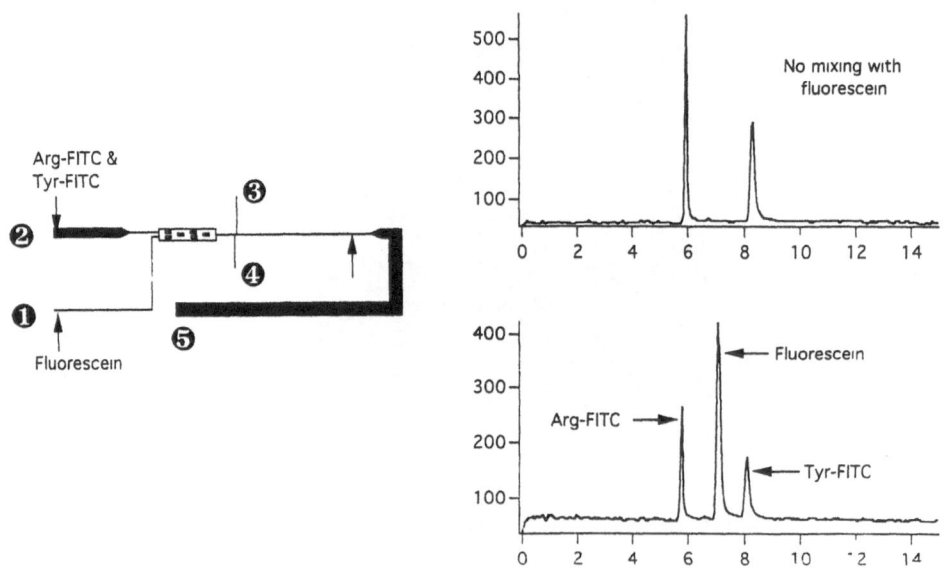

Figure 5.   *Layout of mixing chamber device. Top shows separation of amino acids, bottom shows separation of fluorescien from amino acids after mixing in chamber.*

Samples were introduced through reservoirs 1 and 2, while the other reservoirs and channels were filled with a buffer In order to drive two different sample solutions through the chamber for mixing, a positive potential was applied to reservoir 1, with another applied to 2 Reservoir 3 was at ground It is likely that diffusion served as the main driving force for mixing once the solutions were within the chamber, but this was not studied in detail. After a fixed time the potentials were switched off. Then the mixture in the chamber was injected through the double T injector. Application of a voltage between reservoirs 4 and 5 then resulted in separation.

An example of mixing two solutions followed by their separation is shown in Figure 5. One solution contained fluorescein, while the other contained two labelled amino acids arginine and tyrosine. The bottom of the figure shows that all three components were observed when the sample plug was then separated. As a comparison, an electrophrogram without mixing is also shown in the top of Figure 5. In this case only the amino acid mixture was driven into the mixing chamber and the injector, so the final separation of the sample solution shows there is no fluorescein present.

2.5. POST-SEPARATION REACTIONS

Many chemicals are not detected using a fluorescence detection scheme unless they are chemically derivatized with a fluorescent label. A reaction may be done before a separation is performed, but there are advantages if it is done after the separation. In chromatography this is known as post-column reaction. This is a chemical reaction performed "on-the-fly" in the sample stream as it moves towards a detector, and it must create a fluorescent product only when sample is present. One of the most common reagent used for post-column derivatization is o-phthaldialdehyde (OPA), which reacts with primary amines to create a fluorescent product. In electrophoresis this process might be called an in-capillary reaction, since due to the electric field the separation

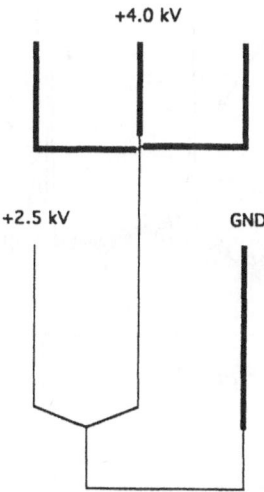

Figure 6. Layout of post-separation reaction device with typical applied voltages indicated.

would continue to occur both during and after the reaction, unlike the case in chromatography. However, for convenience we will refer to this as a post-separation reaction, making the approximation that little further separation will occur between the mixing point and the detector.

We have performed some preliminary experiments with post-separation reaction within a chip. The OPA reagent stream was delivered into the main separation channel from a side channel under potential control. Figure 6 shows the layout of the device used for this study. The reaction was monitored a short distance downstream with a fluorescence detector, fluorescence being excited with a 325 nm laser. Pyrex glass has a fluorescent background at this excitation wavelength, so that the signal to noise performance is limited. Quartz or fused silica will be needed as the device substrate to improve this. Figure 7 shows the detection of several amino acid following reaction with OPA. The efficiency of the electrophoretic separation was not significantly degraded by the following OPA reaction (22,000 (±10%) plates for a pre-labelled compound versus 19,000 (±10%) plates for a post-separation labelled amino acid). One of the amino acids was bifunctional and so reaction could have resulted in three possible products, however,

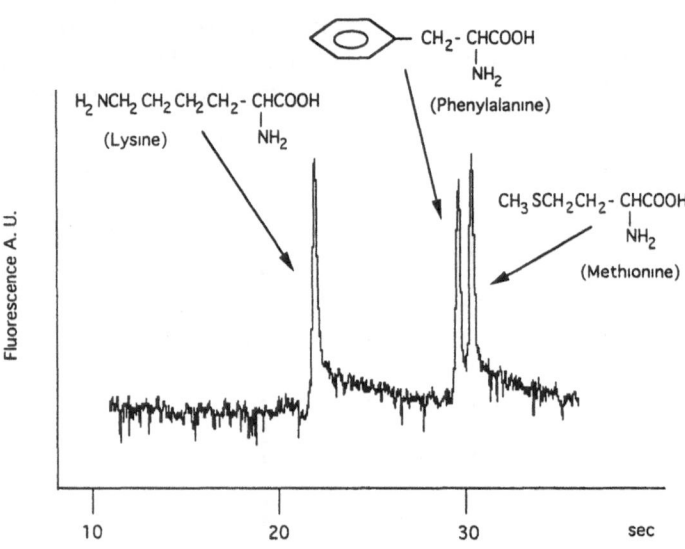

Figure 7. *Post-separation reaction of OPA with three amino acids, evidenced by fluorescence detection.*

there was no appreciable broadening of the peak for this amino acid relative to the other two This indicates the post-separation reaction does not have to lead to significant peak broadening, despite the fact that the product and starting reagents will have differing mobilities While these results are preliminary they do demonstrate that it is possible to perform chemical reactions within a chip which can be used to faciliate a chemical analysis

3. Conclusions

This study has demonstrated the feasibility of using electroosmotic pumping to control flow in a manifold of flow channels without the use of valves The fluidic control that can be achieved within these valveless devices allows for control of solution mixing, and enables us to effect reactions within the flowing streams Biologically important molecules such as amino acids can be separated, allowed to react, and then detected, all within the confines of the chip, and within a few seconds A number of different mixers can designed to perform these reactions, providing increased flexibility in terms of decreased problems from leakage at intersections and increased reaction times

Acknowledgments

We thank Ciba-Geigy and the Natural Sciences and Engineering Resarch Council of Canada for support Z F thanks the Alberta Microelectronic Centre for a graduate fellowship and use of their facilities

References

1   W Olthus, B H van der Schoot, P Bergveld, A dipstick sensor for coulometric acid-base titrations, *Sens. Actuat 17* (1989) 279-283
2   S Shoji, M Esashi, T Matsuo, Prototype miniature blood gas analyzer fabricated on a silicon wafer, *Sens Actuat 14* (1988) 101-107
3   K Seiler, D J Harrison, A Manz, Planar glass chips for capillary electrophoresis repetitive sample injection, quantitation and separation efficiency, *Anal. Chem. 65* (1993) 1481-1488
4   D J Harrison, K Fluri, K Seiler, Z Fan, C S Effenhauser, A Manz, Micromachining a miniaturized capillary electrophoresis-based chemical analysis system on a chip, *Science 261* (1993) 895-897
5   C S Effenhauser, A Manz, H M Widmer, Glass chips for high speed capillary electrophoresis separations with sub-micron plate heights, *Anal Chem 65* (1993) 2637-2642
6   D J Harrison, A Manz, Z Fan, H Ludi, H M Widmer, Capillary electrophoresis and sample injection systems integrated on a planar glass chip, *Anal Chem 64* (1992) 1926-1932
7   R J Gale, K Ghowsi, in *Biosensor Technology, Fundamentals and Applications*, Editors R P Buck, W E Hatfield, M Umana, E F Bowder, Marcel Dekker, New York, (1990), pp 55-62

FLOW INJECTION MICROSYSTEMS - THERE IS A PAST BUT WHERE IS THE FUTURE ?

Jaromir Ruzicka
Department of Chemistry, University of Washington
Seattle WA 98115, USA

0. Abstract

The paper will review past and present efforts to miniaturize flow injection systems, will discuss the principles of sequential injection and its advantages for microminiaturization New type of detector - the jet ring cell will be introduced and its use in cytoanalysis and for flow injection on renewable surfaces will be demonstrated Merits of novel class of chemical sensors with renewable surfaces will be discussed

1. Introduction

Since the advent of the silicon chip and the tremendous success of micro fabrication in the electronic industry countless efforts have been made to apply these technologies to solution handling systems of analytical instruments In some way it is a logical extension of the concept of microfabrication to create a link between computer and solution chemistries via a sensing interface which could be mass produced as easily as printed circuit boards Yet there is a vast difference between manipulating currents and charges however small and the handling of solutions molecules and suspensions This is perhaps the reason why advances on this field were not as significant as initially hoped for Even the smallest sub unit of such a system - the chemical sensor has yet to become a widely accepted practical tool in spite of a vast intellectual and material investment in its development Therefore the significance of the μTAS concept is that it accepts this challenge and that it gathers, for the first time an international group which aims at taking a broader view of the problems connected with the design of μsystems which would be capable of handling a wide range of fluid based techniques including chromatographies and reagent based chemistries This contribution will focus on the past and present efforts to miniaturize flow injection and will discuss merits and limitations of making things small and smaller yet

A van den Berg and P Bergveld (eds) Micro Total Analysis Systems 117–125
© 1995 *Kluwer Academic Publishers*

2. Flow based analyses - and their micro miniaturization.

Flow injection analysis (FIA) is an impulse - response technique, whereby the initial square wave input provided by sample injection, is transformed into a response function by means of a *modulator*. Within this modulator two processes take place simultaneously physical dispersion of the sample zone within the carrier stream and the chemical reaction of the analyte with the surrounding medium [1] Since the detector is tuned to detect the thus processed species, the readout has the form of a peak, which reflects both of the above processes It is correct to observe that this description is valid for chromatographic as well as flow injection techniques, and therefore the qualifying difference is to be found in the mechanism of the processes which are taking place in the *modulator* In chromatography it is the *column* which provides a vast amplification of the differences in migration velocities of individual analytes Therefore sample components emerge from the column serially and are sequentially detected, to yield a chromatogram In flow injection it is the *reactor* which functions to *transform analytes* through chemical reactions into species which are selectively quantified by a detector This is why chromatography and FIA share similar components (pumps, valves and detectors) though they are inherently quite different With the advent of sequential injection [1], which relies on zone stacking, flow reversal and stopped flow, (while chromatographies are based on continuous unidirectional flow) the differences between these two flow based methodologies became even more apparent

The process of transferring reagent chemistries from a test tube into integrated microconduits began in early eighties, when flow injection systems were fabricated by imprinting channels into rigid PVC plates (Fig 1) Such μFIA systems were comprised of injection valves, while detection was carried out by means of fiber optic flow trough cells, ion-selective electrodes as well as gas diffusion and dialysis units integrated with appropriate detectors [2,3] At that time, the first, and so far the most significant theoretical paper, dealing with miniaturization of FIA was published by van der Linden [4], who has shown that the limiting factor is the detector volume He also concluded that the diameter of a FIA channel should not be less then about 200 microns, unless submicroliter detectors will become available Since a decrease of channel cross-section increases the impact of diffusion on radial mass transfer, the use of narrow tubes for FIA was seriously considered several times It was Tijssen [5], who was first to point out the consequences of decreasing the radius of a FIA channel to micrometer range Only later it was realized that the greatest benefit could be obtained by simultaneous down scaling of the length of the conduit and the flow rate It is Hungerford [6] who has shown that the optimum flow velocity is

$$F = 6\ 93\ D_m/R$$

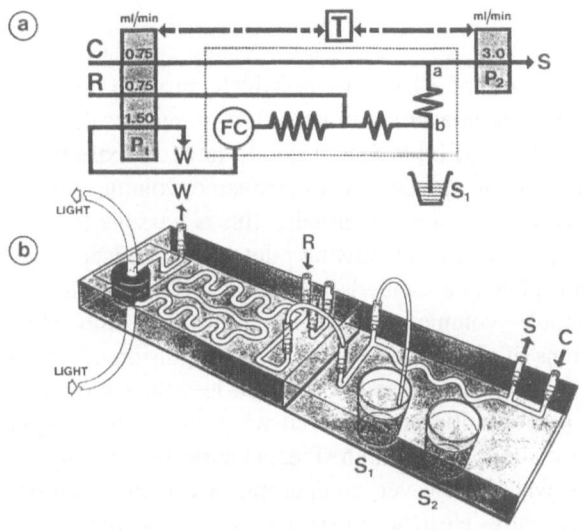

Figure 1. **(a)** *Flow scheme, and* **(b)** *microconduit integrated with fiber optic flow cell (FC), where hydrodynamic injection of a sample, the volume (17mL) of which is defined by distance between a and b has been applied. Initially both pumps are operated to fill the sample from S₁ and to wash the system. By stopping the pump on the right the sample zone is swept by carrier stream (C) towards the confluence point with reagent ® and carried into detector (FC) for monitoring. For further details see[3], where other types of microconduits and valveless injection modes are discussed. T is the timing device.*

where D_m is the molecular diffusion coefficient and R is the tube radius Thus at a flow velocity of 0 1 mm/s in 125 micron tube a narrow square wave zone will acquire a Gaussian shape traveling through only 1mm tubing, while the flow rate will be about 1nL/s, provided that the D_m value of the analyte is 1.10^{-2} cm^2 s^{-1} This high efficiency of radial mixing in conduits of sub millimeter dimensions combined with outstanding economy of material consumption is undeniably so attractive that a question must arise as to why the concept of μFIA has not so far attracted a wider attention while presently the number of publications on conventional FIA exceeds five thousand?

3. Technological obstacles.

The answer is that there is a lack of reliable tools, which provide essential operations, such as liquid propulsion, analyte injection and detection in microscale. As concluded by van der Linden, the detector for a μFIA system needs to have an extremely small internal volume. Indeed at a flow rate of nL/sec, the interrogated volume within the detector needs to be well into submicrolitre range. Admittedly, this is possible to achieve by means of fluorescence microscopy or by borrowing detection concepts from capillary electrophoresis, where both fiber optic and electrochemical detectors have been utilized. However, injection of nanolitre volumes with RSD of a few percent by mechanical means is not quite reliable, yet a tens of nL can be handled well. Propulsion of carrier stream(s) at a submicrolitre level, has in spite of intense efforts been less then successful. Traditional propulsion systems, such as peristaltic pumps, even when driven by stepper motors, are not suitable at levels below about 500mL/min. Piezoelectric micropumps seem to work well in the mL/min range - where, however, stepper motor driven piston pumps are also quite reliable. Only most recently electroosmosis has been employed and shown to be a reliable propulsion system for FIA [7]. Electroosmotic pumping is extremely attractive, since the magnitude and direction of the flow can readily be changed, being proportional to the current intensity and influenced by polarity. Starting with configurations borrowed from capillary electrophoresis, Dasgupta and his coworkers developed a novel elegant approach, whereby the propulsion and reaction systems are physically separated by a column of liquid situated in a holding coil. Therefore, as in sequential injection analysis (SIA), the reagent and sample solutions never come into direct contact with the components of the propulsion system [1]. The propulsion capillary has an I.D. of 75 mm, the holding coil has an I.D. of 250 mm and the reactor coil has an I.D. of 150 mm. Typical injected volumes were in the range of 50 to 100 nL while the flow rates were in submicrolitre range. A detailed study confirmed that flow rates ranging from 1 nL/min to 100 mL/min can be generated and maintained. The system was shown to handle typical reagent based chemistries and there is no doubt that this concept has much to contribute to future efforts toward micro miniaturization of FIA and other flow based systems.

4. Successful compromise - Biospecific Interaction Analysis.

Integrated microfluidics comprising a system of valves (Fig. 2), channels and a detector chip are the core of Biospecific Interaction Analysis (BIA) systems developed and manufactured by Pharmacia Sweden [8,9,10]. From the viewpoint of total micro-miniaturization, BIA is a compromise, since it comprises micro- as well as macro- components. Stepper motor driven syringes (volume 500mL) are delivering the carrier stream of a

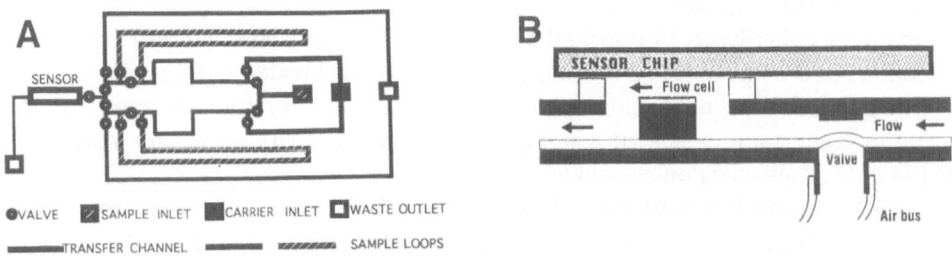

Figure 2. Flow injection scheme designed for BIA as manufactured by Pharmacia. A The sensor is integrated with two identical sets of injection loops (5 and 45 mL each), designed in such a way that the dispersion of the sample zone is minimized. By using one set of loops for measurement, while the other set is being washed and filled, the sample throughput is increased. B. Detailed side view of the flow cell shows the demountable sensor chip docked on top of the flow cell and a single air activated pinch valve.

buffer at rates of 1 to 100 mL/min. The injection loops (5 or 45 mL) are, together with a system of channels, pneumatically driven pinch valves and a flow cell (volume 60 nL) part of a planar structure. The sensor chip which interfaces optics with a flow through cell is a glass slide coated on one side with a gold film, which is in turn coated with a layer of carbomethaxylated dextran. This thin polymer film is probed by evanescent field created by illuminating the chip at a suitable angle with a 760 nm light source. Since biomolecules with a high protein content cause a change of refractive index that is proportional to their protein load, the change of the surface plasmon resonance angle will reflect the surface concentration of protein molecules adsorbed on the polymer layer. In this way binding and dissociation   of two or more molecules, such as protein-protein, nucleic acid-protein, receptor ligand, drug-protein or nucleic acid-nucleic acid can be monitored, provided that one type of the interacting molecules are immobilized on the dextran layer. Also, following the measurement, the sensing layer must be regenerated by breaking the bond (with an acid or base), or the chip must be replaced.

It might seem incidental that BIA rhymes with FIA except that the prospectus and initial publication from Pharmacia [8] state that FIA is the backbone of the fluidic system, and point out how the "sample plug dispersion has been minimized through the design of an integrated microfluidic system that places sample loops and valves near the sensor chip surface" By combining macro components (pumps, optical systems) with micro fluidics, a flow injection system has been developed that allows real time kinetic analysis of biomolecules to be carried out under conditions which are well controlled in terms of contact time and minimized dispersion of the analyte. By selecting the injected sample volume to be either large (45 or 5 mL) or decreased to very short pulses (below 1 mL) the contact time and dispersion is changed so that a gradual buildup of a binding curve can be distinguished from the disturbing effect of sample solution refractive index [10]. Such kinetic discrimination, yields an additional degree of selectivity achievable only through the use of the flow injection technique.

There is much to learn from the case of the "first biosensor based instrument", not the least of which is the cost of its development (reported to be "close to 100 million dollars over ten years "[10]), from its advantages, drawbacks and selected area of application. Targeting towards biotechnology and life sciences is opportune as there capabilities of BIA will be appreciated and development costs might eventually be recovered. Its advantage is a generally applicable detection system that does not require labeling of the target molecules - their ability to change refractive index being sufficient for their detection. The drawback of the BIA system is not so much its high cost, but the need to regenerate the sensing layer after each measurement, and the lesser sensitivity of measurement of molecules with low molecular weight.

5. Future opportunities.

It appears from the foregoing that the research in the area of sensor technology and of μTAS should be application driven rather than device driven, since its successful implementation will require considerable resources ("Anything can be made smaller, never mind physics, everything will be more expensive, never mind common sense." T. Hirshfeld). The key device in the miniaturized system will always be the flow through detector, which besides having a small volume must posses unique properties yielding novel valuable information. It is out of scope of this contribution to review the many electrochemical and optical detectors that fulfill such a requirement. Instead let us consider as a model case a novel detector recently developed in our laboratory - the jet ring cell [11].

Flow injection on renewable surfaces techniques (FI-RST [12,13]) are suitable for micro miniaturization, since their key component, the jet ring cell (Fig. 3), has in its present form a circular detection area of 800 microns and a depth of 50 to 1000 microns. This device, originally designed for FI microscopy of live adherent cells [13], has found its application in fluorescence based immunoassays and reflectance based UV-VIS spectroscopy . The sensing layer in the JR cell comprises several thousand 35 micron polymer beads, which serve as reactive surfaces for protein and reagent adsorption. The cell is a part of a sequential injection system, which allows, besides conventional FI operations (sample injection, carrier pumping, reagent addition) also injection of a well

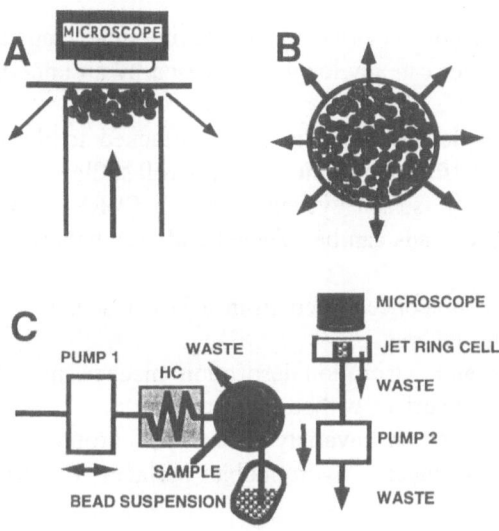

Figure 3. The jet ring cell (A) and sequential injection flow scheme (B)designed for FI-RST. The side view of the JR cell at left shows the beads trapped by forward flow in a well defined geometry against an optical flat, so that change of their optical properties can be continuously monitored by microscopy (or by optical fiber). On the right is a top view of the JR cell as seen by the detector. The sequential injection system is operated in stop/flow-reversed flow mode, under computer control. Initially pump 2 (a peristaltic pump) is stopped allowing the sequentially aspirated zones to be propelled by pump 2 into the JR cell where the beads are trapped, perfused and monitored. By stopping pump 1, and by operating pump 2 shortly, beads are instantaneously removed to waste. HC is the holding coil which separates the principal pumping system pump (1) from the reagent and sample solutions [11,12, 13]

defined volume of bead suspension, transport of the beads into the JR cell, perfusion of the bead layer under continuous monitoring, and removal as well as discharge of the beads into the waste Thus in contrast to all (bio)sensor systems, the sensing surface does not need to be regenerated, nor can it deteriorate during use, since it is automatically replaced by a new layer of beads whenever necessary Presently 10,000 to 20,000 beads (about 0 1mg) are used per measurement, which considering the low cost of these materials, is acceptable

Further miniaturization of the cell will, however be worthwhile since FI-RST have a number of advantages

* the reactive surface needs not to be regenerated, saving time, reagents and solvents
* no hysteresis, nor carryover can any longer be caused by an imperfect regeneration of the sensing surfaces,
* reactive groups need no longer be permanently attached to the reactive or sensing surfaces - often physical adsorption of the reagents will suffice
* the instrument or chemical sensor system based on FI-RST will be versatile, since different kinds of reactive beads can be selected and automatically introduced into the monitored area at will
* the analyte will become preconcentrated from a larger sample volume onto a small surface within the monitored area
* the chemical reactions and adsorption/desorption mechanism will not need to be reversible since the reacted surface will be disposed of
* a wider range of suitable beads with variety of functional groups are commercially available, including ion exchangers, hydrophobic surfaces (C-18), dextran and its derivatized forms
* by chance rather than by design, many chromatographic materials have suitable optical properties, being either transparent or highly reflective

6. Conclusion.

Overall miniaturization, however conceptually appealing ("Small is beautiful " E F Schumacher), is still facing technical obstacles in the design of functional valving and propulsion systems, which must be able to operate on real life samples, that often contain suspended particles And now, when it is actually *desirable* to manipulate suspensions - of polymeric beads, or of cells - the challenge is to be met The past history of μFIA (Fig 1,2), indicates that a partial miniaturization of the system is a way to go The prime candidate is a sensor integrated into a flow cell and system of service channels Valveless injection [3] is then easy to include The pumping system, (most likely stepper motor driven microsyringes, or electroosmotic pumping) will remain external, and since it can remotely operate the system trough columns of liquid held in the holding coil [1,7] the pumping element does not need to come in contact with aggressive chemicals Thus the instrument becomes logistically divided into a disposable (mass produced, inexpensive) and a non-disposable section

There is much to be gained if μFIA and μTAS will have capability to operate on real life samples containing suspended matter Disposable micro reactors with solid reagents where only few beads would serve as carriers for reagents and catalysts could be constructed Or let us consider ultramicrosystems, where perhaps only a single bead could be monitored by UV-VIS, fluorescence of FT-IR microscopy - or by fiber optics In such a ultramicrosystem even adherent cells grown on a single 150 micron Cytodex bead might be perfused, optically probed, and even their metabolism could be explored, since their metabolic products could be confined in a miniaturized flow cell and thus become sufficiently concentrated for an assay Perhaps even single non adherent cells may be identified , selected and sorted out in a micro fabricated fluidic system There are indeed ample opportunities in biotechnology and life sciences for μFIA and μTAS to become indispensable

Acknowledgement. The author expresses his gratitude to C H Pollema for reviewing this manuscript

References.

1   J Ruzicka, *Anal Chim Acta 261* (1992) 3
2   J Ruzicka, *Anal Chem 55* (1983) 1040 A
3   J Ruzicka and E H Hansen Flow Injection Analysis, 2nd Ed , Wiley, New York 1988 Ch 4 12 "Integrated Microconduits"
4   W E van der Linden , *Trends in Anal Chem , 6* (1987) 37
5   R Tijssen, *Anal Chim Acta 114* (1980) 71
6   J Hungerford, *Thesis* Univ of Washington , 1986
7   P K Dasgupta and S Liu, *Anal Chem 66* (1994) 1792
8   U Jonsson, L Fagerstam, B Ivarsson B Johnsson, R Karlsson, K Lundth, S Lofas, B Persson. H Roos. I Ronnberg S Sjolander, E Stenberg R Stahlberg, C Urbaniczky. H Ostlin and M Malmquist. *BioTechniques 11*(1991) 620
9   R Karlsson A Michaelsson and L Mattsson *J Immunol Methods 145* (1991) 229
10 *BIA Journal ,*vol 1 (1994) Pharmacia Biosensor Uppsala, Sweden,
11 J Ruzicka . C H Pollema. and K M Scudder, *Anal Chem 64 ,*(1993) 3566
12 C H Pollema and J Ruzicka *Anal Chem 66* (1994) 1825
13 J Ruzicka, *Analyst* (in press)

MICROMACHINED FLOW-THROUGH MEASUREMENT CHAMBERS USING LAPS CHEMICAL SENSORS

Luc Bousse and Richard McReynolds
Molecular Devices Corporation
1311 Orleans Drive
Sunnyvale, CA 94089, U.S.A.

0. Abstract

Micro Total Analysis Systems (µTAS) are envisioned as small analytical systems with incorporated sample handling. We discuss the concept of a system consisting of a microflow chamber in which a biological component is immobilized, and a chemical sensor which is part of the chamber. A particularly suitable sensor is the Light-Addressable Potentiometric Sensor (LAPS), due to its compatibility with micromachined surfaces and structures. Flow chambers are made by anisotropically etching channels in silicon, at the bottom of which one or more LAPS devices are defined. Eight separate flow channels are present on a 23 mm square chip. Data is presented on two types of biological components: enzymes and living cells.

1. Introduction

Work on chemical sensors made with planar microfabrication techniques has been ongoing for almost 25 years [1, 2]. In that period of time many sensor concepts have been proposed and developed, and some of them have reached the level of commercially available products [3-5]. Yet many difficulties have been encountered in the process of turning microfabricated chemical sensors into useful products such as: unreliable packaging; low lifetimes; lack of wafer-level definition of ion-selective membranes, etc... Due to the efforts of many groups of researchers, solutions to many of these problems are now available to a large extent [6-9].

However, another basic issue in making novel microfabricated sensors has gradually emerged. In many applications, it is not only the sensor that must be miniaturized, but the entire system. An individual microsensor, once packaged, has largely lost its size advantage over conventional devices. For instance, a packaged ISFET is not smaller than a micro glass electrode. As a result, pH-sensitive ISFETs are being sold more for applications where robustness is required than for their size. The user's perspective is to evaluate the entire system for answering his analytical questions. In addition to chemical sensors or transducers, that includes operations such as: sampling; moving samples; mixing reagents; incubation; calibration of sensors; separation steps; equilibrating liquids with gases (degassing). For a system to qualify as "micro" all the

127

A van den Berg and P Bergveld (eds), Micro Total Analysis Systems, 127–138
© 1995 *Kluwer Academic Publishers*

components needed to carry out the required additional operations must be equally miniaturized, and that often presents a technical challenge greater than the fabrication of a microsensor.

An important benefit of this systems approach is that it opens up many more applications. Only a portion of analytical chemistry involves measurements with electrodes, for instance, and a much greater part uses separation methods such as chromatography and electrophoresis. One of the well-known pioneering efforts in sensor microfabrication is the gas chromatograph on a wafer by Terry *et al.* in 1975 [10, 11]; this represents one of the earliest attempts to integrate and miniaturize many system components. So if separation methods are included, the number of potential applications of microfabricated systems increases dramatically. In this context, the recent work on capillary electrophoresis in microfabricated quartz channels is especially significant [12]. This paper will examine what is meant by a micro total analysis system, and describe our efforts in the area of bioanalysis.

2. Requirements for miniaturized analysis systems

2.1 DESIRED SYSTEM CHARACTERISTICS

The concept of a Micro Total Analysis System (μTAS) was first defined by Manz *et al.* [13] as an analysis system with incorporated sample handling "extremely close to the place of measurement." In a general fashion, it is logical to ask what characteristics we are aiming for in a μTAS, and which benefits are derived from them. We view the following characteristics as being desirable:

1. The analytical performance (detection limit) is equal or better than with conventional technology, or the measurement principle has no conventional equivalent. The improvement of analytical performance is the main focus in Ref. [13].
2. The entire system must be small enough so that it can be transported to the point where the measurement is needed. Examples of this capability are: hand-held clinical measurement devices that can be used at the bedside, or portable environmental monitoring systems.
3. The system must be complete, in that all required elements such as reagents are included.
4. It must be sufficiently fast. A portable system that takes a day to give an answer would not be more useful than transporting a sample to a central laboratory, for instance. How fast is enough depends on the particular application.
5. Internal operations such as sample handling, mixing, dilution, calibration, etc... should be transparent to the user.
6. It includes multiple capabilities, so that the user can extract all useful information from a single sample in one operation.

Not all μTAS can have all of these characteristics, but for a system to be called "micro" it probably should have many of them. This list can be summarized as saying that the value of a μTAS is that it provides precise answers where and when they are most needed.

One of the means to make systems with such characteristics is to use microfabrication methods that make many components in parallel. The prime example is of course all methods involving lithography on planar substrates such as glass, silicon, or ceramics. Another example of such a technique is screen printing. Some techniques are used on entire wafers, but not in parallel: laser drilling, for instance.

2.2 BASIC OPERATIONS NEEDED

Operations potentially needed in a μTAS include the following:
1. Taking of samples.
2. Adding and mixing reagents.
3. Waiting or incubating.
4. Gas/liquid equilibration, such as degassing, or removing bubbles.
5. Separation steps (including chromatography or electrophoresis).
6. Detection (typically using electrochemical or optical sensors).

A given method uses a subset of these. For instance, a simple chromatographic or electrophoretic method involves operations 1, 5, and 6. The sandwich immunoassay with filtration capture described in [3] involves taking a sample (1), mixing with reagents including an antibody/enzyme complex (2), incubating (3), filtration capture followed by a wash step to remove the unbound reagents (5), adding the enzyme substrate (2), and detection (6). ELISAs with optical detection often are more complex, and use multiple wash steps.

2.3 METHODS OF MINIATURIZATION

A μTAS must miniaturize the steps listed above (and any we may have omitted). There are two types of methods to do this. First, try to miniaturize the existing components used to accomplish these steps. Second, find new methods that accomplish the same result in a novel way appropriate to a μTAS. In the first family would be efforts to make microvalves and pumps, that move solutions around, mix them, and thereby miniaturize conventional chemistry. Homogeneous immunoassays involve efforts to find non-solution equivalents, such as controlled release of encapsulated reagents [14], or virtual separation using an evanescent wave optical signal [15], and are thus in the second category.

3. Microvolume chambers

3.1 PRINCIPLE

After these general comments on the principle of a μTAS, it is time to examine how the work done at Molecular Devices fits into this theme. Our efforts have focused on measuring the activity of a biological component by immobilizing it in a microvolume chamber [16]. The component can be either an enzyme [17] or living cells [18]; in both cases a chemical change in the chamber is measurable. These techniques have wide application in biochemistry and biology, and have been described in several reviews and publications [4, 19-24].

The principle we will focus on is to measure the activity of a biological component with a microflow system. This can be represented by the following Figure:

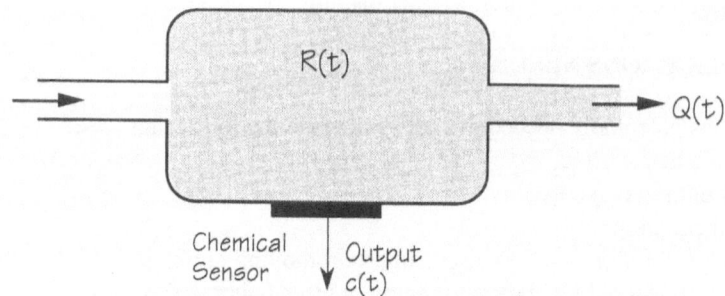

Figure 1. Concept of a micro flow chamber/ chemical sensor combination.

R(t) represents the rate at which a chemical is being generated by the biological component. Examples would be H^+/OH^- ions, K^+ ions, or redox compounds. In general, the aim is to measure R(t) with a certain accuracy and time resolution, by measuring the concentration c(t) in the chamber, and controlling the flow rate Q(t) of solution through the chamber. Note that this diagram above represents a system, not a discrete sensor.

A simple case is to assume that a species X is generated, and that no X is present in the incoming flow. We further neglect the time constant with which the generated X mixes in the chamber and reaches the sensor. If we call $c_x(t)$ the measured concentration, then the rate of outflow is $c_x(t)Q(t)$. This must balance the generation rate, and thus:

$$R(t) = Q(t) \cdot c_x(t) \tag{1}$$

In this simple case, the measured concentration directly measures the generation rate, provided the flow rate is either precisely held constant, or is also measured with high time resolution and accuracy.

A more practically useful problem is that the generated species is H^+ or OH^- ions, and that the incoming flow has pH = pH_0, and a buffer capacity β. Then the change of pH in the system is given by [16]:

$$\frac{d\text{pH}}{dt} = \frac{R_{net}}{V\beta} = \frac{R(t) + \left(10^{-\text{pH}} - 10^{-\text{pH}_0}\right) \cdot Q(t)}{V\beta} \tag{2}$$

where V is the chamber volume, and R_{net} the net rate of OH^-/H^+ generation due to both the biological activity and the inflow/outflow of ions, taking OH^- generation as positive and proton generation as negative. (This equation neglects surface buffering; to include it, see Ref. [16]). It follows that R(t) is:

$$R(t) = \frac{dpH}{dt} \cdot V\beta - \left(10^{-pH} - 10^{-pH_0}\right) \cdot Q(t) \tag{3}$$

This equation still neglects the diffusion time constant in the chamber between generation and sensing; Miller *et al.* have derived the more general case [25]. To obtain high sensitivity, the volume V and the flow rate Q must be kept low. It is also necessary to accurately measure the difference of the chamber pH and the incoming pH. There are some practical difficulties in using eq. (3): the operation of differentiating the measured pH(t) will amplify noise; and the requirements of accurately known absolute pH, and of stable and known Q are difficult to meet. One way to simplify the problem is to measure the rate of change of pH while the flow is zero. Then eq. (3) reduces to:

$$R(t) = \frac{dpH}{dt} V\beta \tag{4}$$

It is now only necessary to measure the rate of change of pH, not a small pH difference. The absolute pH value is involved indirectly, since β depends on pH. Since it is not permissible to let the pH change indefinitely, after enough data points have been read to determine dpH/dt, we turn the flow on long enough to return the chamber to pH_0. So, Q(t) becomes a square wave function. This is the method used for measuring the rate of cellular metabolism [18].

3.2 FABRICATION WITH SILICON MICROMACHINING

One way to make microvolume chambers is with micromachining in silicon. We have chosen this method for several reasons: it is the material for which the greatest variety of micromachining methods are available; and it is also a good sensor material, so that it is possible to place a chemical sensor in the chamber.

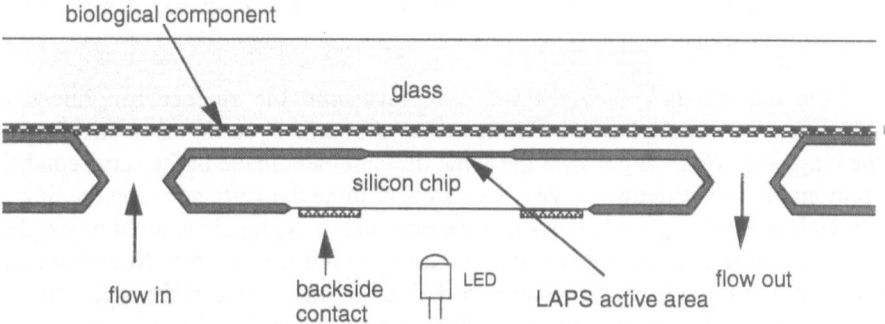

Figure 2. Schematic cross-section of a silicon micromachined microvolume flow chamber, with incorporated LAPS.

Our micromachining method of choice is anisotropic etching, using an alkaline etchant designed to provide very smooth surfaces. The chamber consists of a channel etched into the front surface of a <100> silicon wafer, which is covered by a glass cover slip [26, 27] To provide access to the chamber, holes are etched all the way through the silicon chip by double-sided anisotropic etching. Figure 2 shows a diagram of such a structure. Although the purpose is different, a similar method of making chambers is used by Wilding *et al.* [28]. The SEM photograph shown in Figure 3 shows a top view of channels etched in silicon, with the access holes at the end.

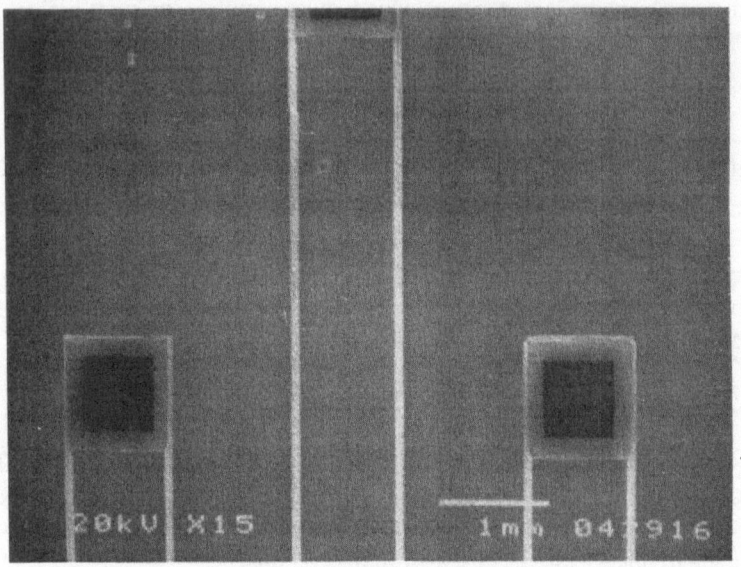

Figure 3. SEM top view of a chip with microvolume flow chambers, showing the flow openings etched through the chip. The chip has 8 separate flow channels.

On one of these devices we have measured the surface roughness with a profilometer (Tencor Alphastep). By measuring the surface profile on an unetched portion of the chip, and comparing it with the same measurement made in the same conditions at the bottom of a 100 μm deep cavity, we can determine the amount of roughening caused by the etch process. Figure 4 shows that the etch adds a negligible amount of roughness to the surface profile; the average deviation (Ra) in both traces is 5 nm. Note that this result is only valid at the horizontal resolution defined by the radius of the stylus used in the profilometer (12 μm). At horizontal length scales much smaller than this, roughness could be smaller or larger.

Another possible technique for making chambers is plasma etching, which we have used to make many small cavities in which cells can be sedimented [29]. It allows straighter sidewalls, but the chamber bottoms are considerably rougher, and therefore the sensor properties aren't as good.

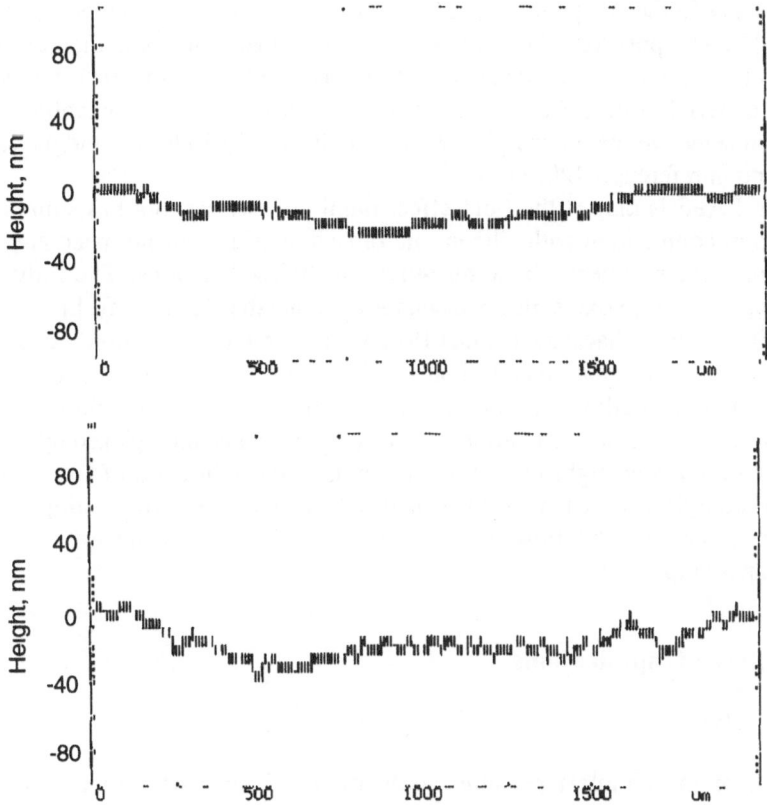

Figure 4. Top trace: Alphastep scan on unetched portion of an oxidized silicon chip.
Bottom trace: similar scan in 103.5 μm deep etched cavity on same chip

4. LAPS as part of a microvolume chamber

The sensor we use as part of the microvolume chamber is a light-addressable potentiometric sensor (LAPS) [30]. It consists of an electrolyte solution/insulator/silicon capacitor illuminated by an intensity-modulated light source, at a wavelength short enough to generate carriers in silicon and long enough to penetrate sufficiently. A convenient wavelength for which efficient LEDs are available is 940 nm. The illumination often occurs at the backside of the device, so that the frontside is free to form one surface of the sensing chamber When the capacitor is biased in inversion, the depletion region at the frontside surface collects light-generated carriers, generating an AC photocurrent When the capacitor is biased in accumulation, there is no photocurrent The transition between these two states in the AC photocurrent/bias curve turns out to be quite steep (a width of 0.2 to 0.3 V). The position of the inflection point in these curves is then used to track changes in the potential at the insulator surface. Depending on the material used as the outer gate material of the LAPS, it can be used to detect pH, redox potential, or cations using an ion-selective membrane [30]. More details on LAPS physics and data reduction can be found in references [20, 31].

The LAPS is part of the field-effect family of devices, but has some important advantages in conjunction with silicon micromachining. It does not need any frontside metalizations, interconnects, bonding wires, or diffused regions. The only frontside patterning used is a thin oxide/nitride insulator to define the active areas. In that sense, the LAPS is similar to the backside contact ISFET, although its fabrication process is much simpler. A LAPS can be placed at the bottom of a cavity, provided the silicon surface there is smooth enough to allow a good quality field-effect device. In addition, a LAPS is intrinsically a multisensor, since different sensing spots can be multiplexed by sequentially shining AC-modulated light on them. This allows the fabrication of multiple LAPS sensors in a single flow channel. One of the devices we are now testing for cellular applications is a chip with 8 flow channels, each containing 4 sensing areas, for a total of 32 sensors per chip.

5. Examples of applications

5.1 ENZYMES

One of the simplest possible applications of the LAPS/microflow chamber combination is the measurement of enzyme activity. One way to demonstrate this is to immobilize an enzyme in a chamber, and provide it with its substrate in the flow medium. As an example, let's consider acetyl cholinesterase, which catalyzes the hydrolysis of acetylcholine to acetate and choline, liberating protons in the process. Acetyl cholinesterase-coated agarose beads (Sigma) were immobilized between two thin polycarbonate membranes in a Cytosensor [4] chamber, and the pH response measured when an acetylcholine-containing medium is flowed through. Figure 5 shows the data from this experiment. Rates of about 120 μV/s are obtained. Note that the presence of the membranes slows down the time constant of the return to baseline of the pH during the flow-on periods. This demonstrates the measurement of enzyme activity, with possible applications to immunoassays.

5.2 CELLS

Most of the work at Molecular Devices Corporation on the microflow chamber/LAPS combination has used living cells as the biological component. The rate R(t) of pH change in the chamber is a measure of the metabolic rate of cells, which is coupled directly or indirectly to almost all events that affect the cells [19]. The result is a functional bioassay with applications in pharmacology [32], characterizing receptor/ligand interaction [33],

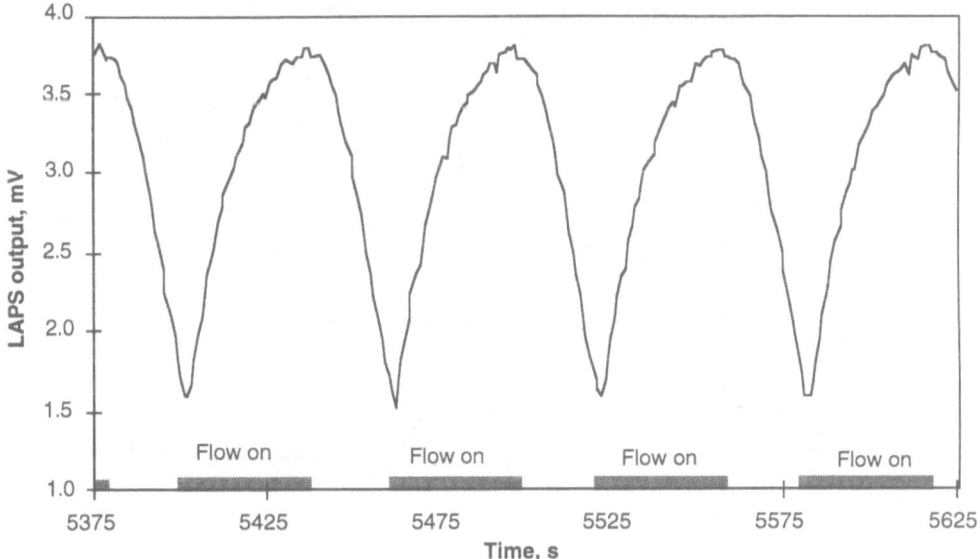

Figure 5. Sensor response to flow on/flow off cycles with a Cytosensor chamber loaded with acetyl cholinesterase coated beads. The flow medium contains 8 mM acetylcholine.

elucidating signal transduction pathways in cells [34], *in vitro* toxicology [35, 36], testing antiviral agents [18], and evaluating susceptibility to chemotherapeutic agents [18].

As an example of the ability to detect receptor/ligand interactions in real time, Figure 6 shows the response of m1WT3 cells to carbachol. These cells are a mammalian cell line (CHO-K1), which have been transfected with the m1–muscarinic acetylcholine receptor, a receptor which is not naturally present in CHO cells [32]. Each cell contains approximately 2×10^5 muscarinic receptors. Carbachol (carbamylcholine chloride), the agonist used to activate the cells, is a stable and soluble analog of acetylcholine which acts as an agonist for acetylcholine receptors. Note that carbachol is not selective for the subtypes of acetylcholine receptors: it acts as an agonist for both muscarinic and nicotinic subtypes. However, the transfection procedure guarantees that only the m1 muscarinic receptor is present in the cells we used. Figure 6 shows that the system can measure metabolic responses with a time resolution of about 100 seconds. The response to carbachol is extremely rapid, and the maximum occurs at the first data point after carbachol introduction, which confirms our earlier data [32].

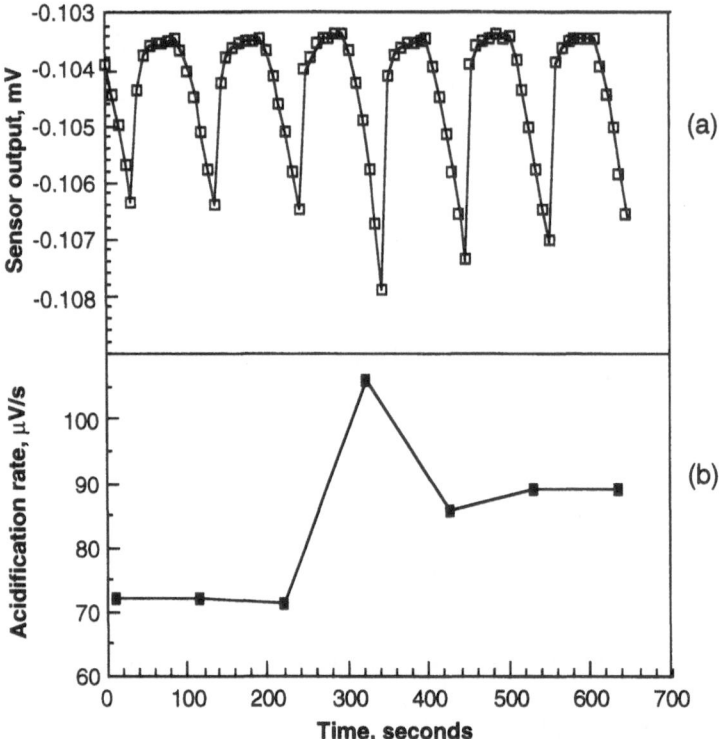

Figure 6. (a) Sensor output from a single channel of an eight-channel flow-through chip as a function of time, for CHO cells transfected with a muscarinic acetylcholine receptor. Carbachol , an acetylcholine analog, is added at about 250s. (b) Slopes of the sensor output/time data during the flow off periods, obtained by least-squares fitting. The increase in metabolic rate caused by the addition of carbachol can be seen.

6. Further elements needed

The system used to generate the data of Figure 6 uses a silicon sensor, micromachining, and a multichannel microflow path. Yet the other components are conventional, and the entire system is therefore not miniaturized. To approach the ideal of a μTAS as described above, all system components must be miniaturized. This will require a considerable amount of work on integrated and miniaturized valves, thermal regulation, degassing, and fluidic drive. Most of the existing work on microfabricated fluidics [37] is intended for gases, and does not achieve the component density we require. We are currently working on new methods for fabricating valves, degassers, and fluid drive, that will move us closer to the μTAS concept. However, those efforts are not yet as mature as the fabrication of sensors in micromachined flow chambers. Developing all these components and integrating them in a system is a considerable undertaking

Acknowledgment

We thank Calvin Chow and Timothy Dawes for assistance in providing experimental data. This work was supported by ARPA contract number MDA972-92-C-0005

References

1   P Bergveld, Development of an ion-sensitive solid-state device for neurophysiological measurements, *IEEE Trans Biomed Eng*, BME-17 (1970) 70–71
2   P Bergveld, Development, operation, and application of the ion-sensitive field effect transistor, a tool for electrophysiology, *IEEE Trans Biomed Eng*, BME-19 (1972) 342
3   J Briggs, *et al*, Total DNAassay system, *Amer Biotech Lab*, 7 (1989) 34–38
4   H M McConnell, J C Owicki, J W Parce, D L Miller, G T Baxter, H G Wada, and S Pitchford, The cytosensor microphysiometer biological applications of silicon technology, *Science, 257* (1992) 1906–1912
5   G Davis, Development of a commercial multichannel clinical sensor chip, *Third World Congress on Biosensors New Orleans, 1994*
6   K Potjekamloth, J Janata, and M Josowicz, Electrochemical encapsulation for sensors, *Sensors and Actuators, 18* (1989) 415–425
7   H H van den Vlekkert, N F de Rooij, A van den Berg, and A Grisel, Multi-ion sensing system based on glass-encapsulated pH-ISFETs and a pseudo-REFET, *Sensors and Actuators B, 1* (1990) 395–400
8   J R Haak, P D van der Wal, and D N Reinhoudt, Molecular materials for the transduction of chemical information by CHEMFETs, *Sensors and Actuators B, 8* (1992) 211–219
9   P D van der Wal, A van den Berg, and N F de Rooij, Universal approach for the fabrication of Ca^{2+}, K^+ and NO_3^- sensitive membrane ISFETs, *Sensors and Actuators B, 18-19* (1994) 200–207
10   S C Terry, A gas chromatography system fabricated on a silicon wafer using integrated circuit technology, *Ph D Dissertation*, Stanford University, 1975
11   S C Terry, J H Jerman, and J B Angell, A gas chromatograph air analyzer fabricated on a silicon wafer, *IEEE Trans Electron Devices*, ED-26 (1979) 1880
12   D J Harrison, K Fluri, K Seiler, Z Fan, C S Effenhauser, and A Manz, Micromachining a miniaturized capillary electrophoresis-based chemical analysis system on a chip, *Science, 261* (1993) 895–897
13   A Manz, N Graber, and H M Widmer, Miniaturized total chemical analysis systems a novel concept for chemical sensing, *Sensors and Actuators B, 1* (1990) 244–248
14   S M Barnard and D R Walt, Chemical sensors based on controlled-release polymer systems, *Science, 251* (1991) 927–929
15   M T Flanagan, A M Sloper, and R H Ashworth, From electronic to opto-electronic biosensors an engineering view, *Anal Chim Acta, 213* (1988) 23–33
16   L Bousse, J C Owicki, and J W Parce, Biosensors with microvolume reaction chambers, in S Yamauchi (ed), *Chemical Sensor Technology,* Vol 4, Kodansha/Elsevier, Tokyo and Amsterdam, 1992, pp 145–166
17   L J Bousse, G Kirk, and G Sigal, Biosensors for detection of enzymes immobilized in microvolume reaction chambers, *Sensors and Actuators B, 1* (1990) 361–367
18   J W Parce, J C Owicki, K M Kercso, G B Sigal, H G Wada, V C Muir, L J Bousse, K L Ross, B I Sikic, and H M McConnell, Detection of cell-affecting agents with a silicon biosensor, *Science, 246* (1989) 243–247

138

19 J C Owicki and J W Parce, Biosensors based on the energy metabolism of living cells the physical chemistry and cell biology of extracellular acidification, *Biosensors and Bioelectronics*, 7 (1992) 255–272

20 J C Owicki, L Bousse, D G Hafeman, G L Kirk, J D Olson, H G Wada, and J W Parce, The light-addressable potentiometric sensor principles and biological applications, *Ann Rev Biophys Biomol Struct*, 23 (1994) 87–113

21 H M McConnell, P Rice, H G Wada, J C Owicki, and J W Parce, The microphysiometer biosensor, *Current Opinion in Structural Biology*, 1 (1991) 647–652

22 J D Olson, P R Panfili, R Armenta, M Femmel, H Merrick, J Gumperz, M Goltz, and R F Zuk, A silicon sensor-based filtration immunoassay using biotin-mediated capture, *J Immunological Methods*, 134 (1990) 71–79

23 J M Libby and G W Wada, Detection of Neisseria Meningitidis and Yersinia Pestis with a Novel Silicon Based Sensor, *J Clin Micro*, 27 (1989) 1456–1459

24 K Dill, M Lin, C Poteras, C Fraser, J C O Owicki, D G Hafeman, and J Olson, Determination of solution phase antibody-antigen binding constants with the threshold system equilibrium binding constants for anti-fluorescein, anti-saxitoxin, and anti-ricin Antibodies, *Anal Biochem*, 217 (1994) 128–138

25 D L Miller and J C Owicki, PMA induces a change in phenotype of TE671 cells from a smooth muscle-type towards a skeletal or cardiac muscle-type, *33d Annual Meeting of the ASCB New Orleans, 1993*

26 L Bousse, R J McReynolds, G Kirk, T Dawes, P Lam, W R Bemiss, and J W Parce, Micromachined multichannel systems for the measurement of cellular metabolism, *International Conference on Solid State Sensors and Actuators, Yokohama, 1993*, pp 916–920

27 L J Bousse, R J McReynolds, G Kirk, P Lam, and J W Parce, Integrated Fluidics for Biosensors Used to Measure Cellular Metabolism, *Proceedings of the Symposium on Chemical Sensors II, Proceedings of the Electrochemical Society Hawaii, 1993*, pp 742–745

28 P Wilding, J Pfahler, H H Bau, J N Zemel, and L J Kricka, Manipulation and flow of biological fluids in straight channels micromachined in silicon, *Clinical Chemistry*, 40 (1994) 43–47

29 L J Bousse, J W Parce, J C Owicki, and K M Kercso, Silicon micromachining in the fabrication of biosensors using living cells, *Technical Digest IEEE Solid State Sensor and Actuator Workshop, Hilton Head S C ,1990*, pp 173–176

30 D G Hafeman, J W Parce, and H M McConnell, Light-addressable potentiometric sensor for biochemical systems, *Science*, 240 (1988) 1182–1185

31 L Bousse, S Mostarshed, D Hafeman, M Sartore, M Adami, and C Nicolini, Investigation of carrier transport through silicon wafers by photocurrent measurements, *J App Phys*, 75 (1994) 4000–4008

32 J C Owicki, J W Parce, K M Kercso, G B Sigal, V C Muir, J C Venter, C M Fraser, and H M McConnell, Continuous monitoring of receptor-mediated changes in the metabolic rates of living cells, *Proc Natl Acad Sci USA*, 87 (1990) 4007–4011

33 C Bouvier, J A Salon, R A Johnson, and O Civelli, Dopaminergic activity measured in D_1 and D_2-transfected fibroblasts by silicon-microphysiometry, *J Receptor Res*, 13 (1993) 559–571

34 G T Baxter, D L Miller, R C Kuo, H G Wada, and J C Owicki, PKCε is involved in GM-CSF Signal transduction Evidence from microphysiometry and antisense oligonucleotide experiments, *Biochemistry*, 31 (1992) 10950–10954

35 P Catroux, A Rougier, K G Dossou, and M Cottin, The silicon microphysiometer for testing ocular toxicity in vitro, *Toxicology in Vitro*, 7 (1994) 465–469

36 L H Bruner, K R Miller, J C Owicki, J W Parce, and V C Muir, Testing ocular irritancy in vitro with the silicon microphysiometer, *Toxicology In Vitro*, 5 (1991) 277–284

37 P Gravesen, J Branebjerg, and O S Jensen, Microfluidics - a review, *J Micromech Microeng*, 3 (1993) 168–182

DEVELOPMENT OF A PCR-MICROREACTOR

M. Allen Northrup, Carlos Gonzalez, Stacy Lehew and Rob Hills
Engineering Research Division, Microtechnology Center, L-222
Lawrence Livermore National Laboratory,
P O Box 808, Livermore, California 94551, USA

Abstract

The application of microfabrication technology to the development of miniaturized analytical and clinical instrumentation is an area of active interest and research As a part of this effort, we are developing miniaturized devices and instrument components including reaction chambers, microfluidic devices (i e pumps, valves, and flow systems) and detection systems for biomedical applications Specifically, a microfabricated thermal cycling instrument for application to the polymerase chain reaction (PCR) is being developed The miniaturization of a PCR thermal cycler and associated analytical system will allow for a portable, low-power, rapid, and highly efficient bioanalytical instrument We have successfully amplified several DNA targets from different biological systems in silicon-based microfabricated reaction chambers These include human immunodeficiency virus (HIV) and β-globin DNA targets Verification of the amplified target has been provided by standard agarose gel electrophoresis Thermal modeling and infrared imaging have helped delineate optimal designs leading to efficient multiple-heater reaction chambers Instrument efficiencies and PCR amplification results from the micro devices compare favorably to the commercial benchtop PCR systems In the present report, we will discuss the most recent micro-PCR DNA amplification results, reaction chamber and thermal cycler optimization, fluidic manipulation and detection strategies, and the overall direction and advantages of the application of microfabrication to DNA-based microinstruments

Work performed under the auspices of the U.S. Department of Energy by the Lawrence Livermore National Laboratory under contract number W-7405-ENG-48

APPLICATION OF MINIATURE ANALYZERS: FROM MICROFLUIDIC COMPONENTS TO μTAS

Jens Branebjerg[1,2], Birgit Fabius[1] and Peter Gravesen[1]
[1] Danfoss A/S, DK-6430 Nordborg, Denmark
[2] Mikroelektronik Centret, The Technical University of Denmark,
Build. 345 East, DK-2800 Kopenhagen

0. Abstract

The process from the fabrication of individual micromechanical components to the realisation of a micro-TAS is discussed. Possibilities and limitations in the choice of the dimensions of the channels in the flow system are investigated experimentally with the emphasis on mixing aspects. The experimental evaluation of valves especially designed for use in chemical analysis systems is presented. Finally, the importance of focusing on the chemical aspects of micro-TAS is discussed.

1. Introduction

Total analysis systems in microtechnology, Micro-TAS, have long been singled out as a strong application for microtechnology and micromechanics in particular. Many new processes have been developed and many new components for fluidic applications have been designed and manufactured [1]. The microfluidic components have usually been characterised as individual components. However, some micro-TAS have been manufactured and their system characteristics and chemical performance have been evaluated [2] [3]. In micro-TAS, different concepts have been investigated such as gas chromatographs [4], systems involving separation by electrophoresis [2], and systems based on chemical reactions between a sample and one or more reagents [3] [5].

There is a need for more knowledge of the system characteristics in micro-TAS, and we find it important to pay more attention to system aspects in micro-TAS such as dosing, mixing, separation and detection. These aspects and their interaction strongly influence the performance of the chemical analysis in the micro-TAS, and therefore we propose a change in the direction of the research from microfluidic components to system aspects in micro-TAS.

We expect that micro-TAS will be used in industrial sensors and are investigating the possibilities and limitations in systems with a liquid consumption of a few micro-litres per minute.

A. van den Berg and P. Bergveld (eds.), Micro Total Analysis Systems, 141–151.

Of all the different micro-TAS concepts, we are mainly interested in systems which include chemical reactions in liquids. The measurement is based on the mixing of the liquid sample with one or more reagents and the detection of characteristic parameters in the reaction product.

Some of the important system parameters are mixing, under different conditions, and dead volumes and leakage in switching valves. To perform an experimental investigation of these system parameters, micro-TAS valves and pumps, and miniaturised mixing channels were fabricated. The results of the experiments are presented in this paper.

2. Mixing in micro-systems

The understanding of mixing is essential for the optimization of a micro-TAS. Mixing procedures, well known from the macroscopic world, such as stirring or creation of turbulent flow cannot be scaled down due to the strong tendency of microfluidic flows to be laminar.

Miyake et al. [6] describe mixing in macroscopic systems as a two-step process - segregation of two or more liquids by stirring or turbulent flow followed by inter-diffusion between small domains of varied composition. In order to obtain a similar two-step mixing process in a micro-fluid component, they proposed [6] injecting one liquid through 400 20 μm x 20 μm holes into a small volume containing a second stationary liquid. In this way segregation was obtained under laminar flow conditions, and a significant reduction in mixing time was demonstrated.

This paper deals with the mixing of two liquids flowing in parallel in miniaturized channels, at continuous flow rates, and with Reynolds numbers in the range from 0.01 to 150 - well below the limit for fully developed turbulent flow. Flow and mixing characteristics have been investigated in channels of different cross- sectional dimensions and different flow lay-outs. Straight channels were used to study mixing under undisturbed laminar flow conditions and meander- or zigzag-shaped channels were studied in order to identify laminar flow disturbances and their influence on mixing.

It is well known [7] that a secondary flow is created at bends with a small radius of curvature, due to inertial forces acting on the central part on the flow with the largest velocity. This results in a redistribution of the two liquid components and possibly a larger contact area. At sharp corners one could even expect local turbulence, since the sharp corner may act as a slit-type orifice if the Reynolds number exceeds the corresponding transition number 15.

Either of the two disturbances described above will not initiate turbulent flow down-stream at small Reynolds numbers. Instead a laminar flow will develop through re-laminization. However, if a sufficiently high number of bends or sharp corners are spaced at a mutual distance shorter than the entrance length for development of laminar flow, one should expect mixing properties significantly different for those of a straight channel of similar length.

3. Mixing experiments in mini-channels

Two miniaturized test channels were fabricated to study the flow and mixing conditions. The first allowed mixing controlled by the introduction of a secondary flow or local turbulence and included approximately 80 zigzag-shaped curves and the other channel was straight, as sketched in figure 1. The channels were manufactured in an acrylic plastic block using traditional machining, and both had a total length of 100 mm and a cross-section of 300 μm by 600 μm.

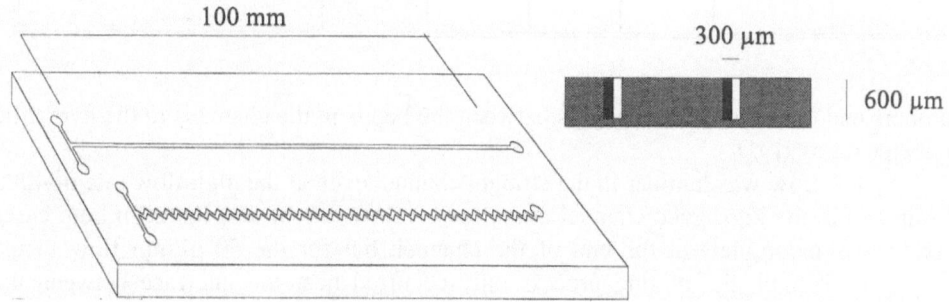

Figure 1. The plastic block with two mini-channel mixers are sketched to the left. To the right is shown the cross-section of the block with two different liquids in each mini-channel.

The evaluation of the mixing process was carried out in a test set-up based on monitoring the flow of coloured liquids in the channels. The transparent acrylic plastic block with the channels was placed on an X, Y, Z stage underneath a video camera with microscope optics. The inlets of the channels were connected to liquid reservoirs and the flow rates were controlled by hydrostatic pressure. The video signals were recorded and later analyzed in slow-motion play-back.

A solution of the pH-indicator bromothymol-blue (BTB), slightly yellow, and a transparent solution of sodium hydroxide were used in the experiments to visualise the mixing process. The two liquids react instantaneously and the observed time of the colour change is interpreted as the mixing time. The reaction product has a dark blue colour and is easily seen in the video recordings.

Mixing experiments at flow rates of 50 μl/min, 900 μl/min and 4000 μl/min were performed in the channel with 80 zigzags. The 4000 μl/min flow was also investigated in the flow system with the straight channel. The data of the flow experiment are summarized in table 1, together with calculated values of the Reynolds

Table I. Data on mixing experiments.

Mini-channels	Channel type	Total length (mm)	Length between bends L (mm)	Max liquid speed (mm/sec)	Reynolds number	L/D_h	Comments
Flow rate 50 μl/min	zigzag	100	1	5	1 85	2 5	Partial mixing in progress at the end of the channel
Flow rate 900 μl/min	zigzag	100	1	83	33	2 5	Mixing completed at the end of channel
Flow rate 4000 μl/min	zigzag	100	1	370	148	2 5	Mixing completed after one-third of the channel
	straight	100	100	370	148	250	No observable mixing in the channel

numbers and the ratios of the lengths between the bends in the channels to the hydraulic diameter ratios (L/D_h).

The flow was laminar in the straight channel even at the high flow rate of 4000 μl/min and in the zigzagged channel at the lowest flow rate of 50 μl/min. In both cases mixing was incomplete at the end of the channel, but for the 50 μl/min flow in the zigzagged channel, the mixing process only occurred near the interface between the liquids in the centre of the channel. In the zigzag-shaped channel the flow pattern changed when passing the zigzags at larger flow rates. At 900 μl/min the introduction of secondary flow or local turbulence by the zigzag structure was sufficient to obtain a homogeneous mixing at the end of the channel. At 4000 μl/min the mixing was complete after passing one-third of the channel.

4. The micro-TAS concept

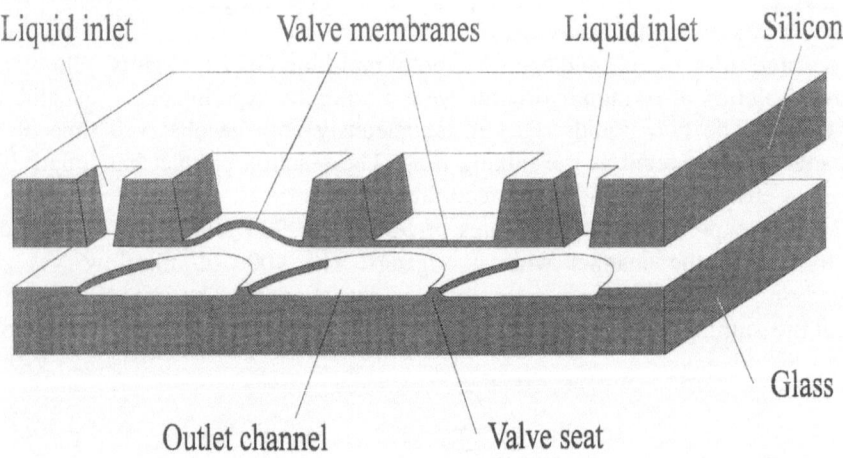

Figure 2. A micro-TAS switching valve with two inlets and one common outlet channel.

A micro-TAS concept was developed based on a silicon wafer bonded to a glass wafer. Channels were etched in the glass wafer and holes for liquid inlets as well as membranes for valves and pumps were made in the silicon wafer. The actuation of the membranes was based on pneumatics and not integrated in the structure.

Valves are one of the basic elements in a micro-TAS concept and should be designed with the operation of a chemical system in mind. Therefore, system aspects such as minimal dead volume, leakage and simplicity play a major role in the design of the valves. The valves can work individually or two valves can be coupled together with a membrane to form a pump. Two valves can also be placed next to each other with one common outlet channel and in this way form a switching valve as shown in figure 2.

Micro-valves are formed in areas where two disconnected channels in the glass wafer are placed underneath a silicon membrane. When the membrane is pressed down on the glass surface the liquid flow between the two channels is disconnected, but when the membrane is deflected and pulled away from the glass surface the channels are interconnected by the space underneath the membrane and the liquid will flow.

5. Flow experiments in micro-TAS structures

The experiments with the micro-TAS elements were carried out to evaluate liquid flow and mixing in micro-channels, and the characteristics of the switching valves. The evaluation relies on the possibility of studying the liquid flow in the micro-TAS elements through the transparent glass wafer, and the experimental set-up used for investigating liquid mixing in mini-channels was also used for the characterisation of the micro-TAS elements.

5.1 MIXING IN MICROCHANNELS

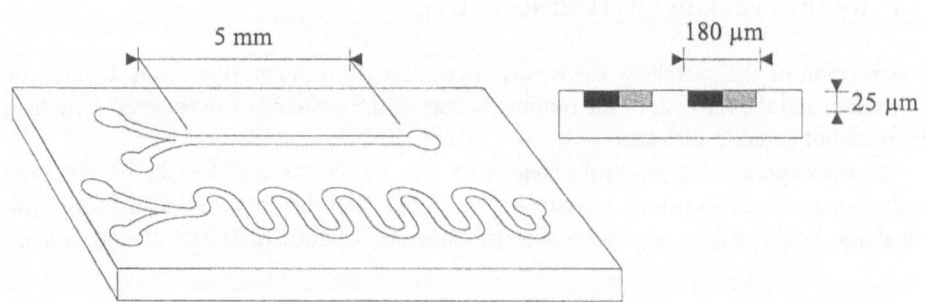

Figure 3. Micro-channels etched in the glass wafer are sketched to the left. In the cross-sections, to the right, is shown how the two liquids flow in each channel.

In the flow and mixing experiments two different channel structures were used In both structures the width and the height of the two test channels were 280 mm by 25 mm One channel was straight and had a length of 5 mm, and the other was meander-shaped with ten curves and a total length of 25 mm The flow rate was varied between 0 06 ml/min and 6 ml/min

The observed flow was clearly laminar with no tendency towards disturbance even when passing very sharp corners Mixing was observed at the border between the two liquids in the channels The mixing band was homogeneously broadened down the channel The lower the flow rate the broader the mixing band Observations from the experiment are summarised in table II

Table II. Data on mixing experiments in micro-channels

Micro-channels	Channel type	Total length (mm)	Length between bends L (mm)	Max liquid speed (mm/sec)	Reynolds number	L/D_h	Comments
Flow rate 6 ml/min	meander	25	2 5	22	1 0	57	Undisturbed laminar flow No observable mixing
	straight	5	5	22	1 0	114	Undisturbed laminar flow No observable mixing
Flow rate 0 06 ml/min	meander	25	2 5	0 22	0 01	57	Undisturbed laminar flow Mixing by diffusion
	straight	5	5	0 22	0 01	114	Undisturbed laminar flow Mixing by diffusion

5 2 OPERATION OF THE SWITCHING VALVE

The evaluation of the switching valve was concentrated on monitoring the leak rate, the dead volume and the flushing of the output channel of the switching valves when switching from one input liquid to the other

In the experimental set-up an area of 25 mm by 25 mm could easily be observed with the resolution of the video camera and its optics, and this feature was used to look for leakage in the valve We were able to observe volumes of 0 015 nl and still no leakage was detected, indicating that any possible leaks were dissolved by diffusion. On the basis of this observation the leak rate was calculated to be less than 5 nl/min for a 1 mm wide valve seat.

An experiment to detect dead volumes in the switching valve was also performed. The replacement of the liquid in the outlet channel of the switching valve was monitored, while actuating one of the two inlet valves. When closing one of the inlet valves the liquid entering through the other inlet valve smoothly flushed the common outlet channel. It was evident that the previous liquid volume was displaced from the switching valve, leaving no detectable carry-over.

With the same test set-up but with a monochromatic light source the absolute movements of the valve membrane were studied. Interference fringes were observed as a result of reflection from the surface of the glass valve seat and from the valve membrane. The absolute height of the valve orifice was measured by counting the number of fringes when opening the valve. The measured distance between the valve seat and the membrane was 9 μm for an actuation pressure of -1 bar. With this opening the cross-sectional area of the valve orifice was approximately 0.01 mm^2.

Additionally, the shape of the membrane was studied while actuating the valve. The pattern of the fringes showed how the liquid between the membrane and the valve seat was pressed out when closing the valve. When the valve was fully closed no interference fringes could be observed and the gap between the membrane and the valve seat was less than 0.3 μm. This measurement was used to calculate a leak rate in the valve of less than 1.4 nl/min, at a pressure difference of 3000 Pa and a valve seat with a length of 30 μm and a width of 1 mm.

6. Discussion

6.1 MIXING IN MINIATURISED CHANNELS

In microfluidics the flow is normally laminar but non-laminar flow can occur in components with short channels or orifices, at Reynolds numbers as low as 15 [1]. In our experiments the Reynolds numbers were all below 150 and we observed non-laminar flow conditions where secondary flow and presumable local turbulence was generated at sharp corners in the zigzags of the mini-channel. The maximum Reynolds numbers of the mixing experiments are plotted as a function of the ratio of the length between bends to the hydraulic diameter (L/D_h) in figure 4.

It can be seen that the results from the two experiments in which secondary flow or local turbulence resulted in mixing are plotted near to the line marking the transitional Reynolds number. The rest of the experiments are clearly in the laminar flow region. For the experiment plotted just below the line in figure 4 the mixing was completed after 80 zigzags at the end of the channel, while for the experiment plotted above the line the mixing was already completed one-third of the way down the channel after approximately 25 zigzags.

In the micro-channels mixing was observed at a very low flow rate and at the end of the zigzagged mini-channel at a flow rate of 50 µl/min. The mixing was caused

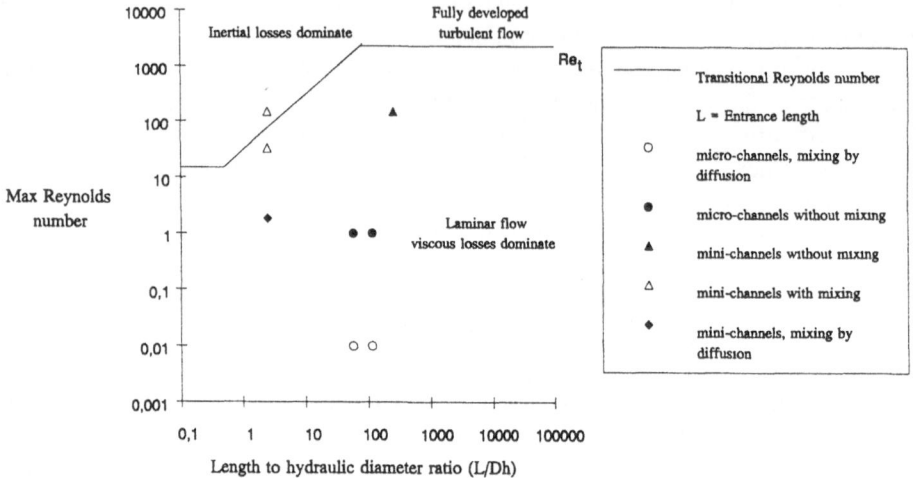

Figure 4. Plot of the maximum Reynolds numbers af the mixing experiments as a function of the ratio of the length between bends to the hydraulic diameter.

by diffusion because of the relatively long transit time in the channels at these small flow rates. The transit time in the micro-channels was 22 seconds and 114 seconds at 60 nl/min and in the mini-channels it was 20 seconds at 50 µl/min.

For the experiments in the laminar flow regime with transit times of less than 1.2 seconds the liquids flowed alongside each other through the channels without detectable mixing. The mixing experiments showed that both mixing processes are possible in the mini-channels. At high flow rates the zigzags initiate mixing by secondary flow and at low flow rates the transit time is long enough to allow mixing by diffusion. In the micro-channels the flow is always laminar and mixing is dominated by diffusion.

6.2 INFLUENCE OF SWITCHING VALVES ON THE PERFORMANCE OF THE MICRO-TAS

Control of the dosing in a micro-TAS is especially important in the region between the point at which the sample enters the mixing channel to the point at which the reaction product is detected. The switching valve based on the developed micro-TAS concept is intended to be placed between the sample inlet and the detector. Hence, the leak rate and dead volumes are critical parameters, which have been evaluated in detail.

From the results of two independent experiments, the leak rate was calculated to be less than 1.4 nl/min in a perfectly functioning valve. This results in a leak ratio of 1:1000 for a system flow of 1 µl/min in a micro-TAS, and such a leak ratio is not expected to cause measurement problems. But if one particle of 1 µm in diameter is trapped on the valve seat and prevents closing of the valve, a slit with an expected width of 100µm, will occur and result in a leak rate of about 5 nl/min. In a system with a flow rate of 1 µl/min, a leak rate of 5 nl/min may add an error to the chemical signal. To prevent leak problems in future micro-TAS, particle filtration and well-designed valve seats should be included in the design of the system.

Flow-rate depended dead volumes and carry-over problems usually encountered in macroscopic systems originating from back-flow regions with no liquid exchange, do not occur in the micro-TAS. As all exchange of liquid in the micro TAS takes place by either laminar flow or diffusion, every segment of the flow volume must either be flushed or be within diffusion range from liquid flowing with a velocity near the maximum flow velocity. Hence, care must be taken to avoid sharp corners and dead ends. In the manufacturing of the valves, isotropic etching was used to make the curved channels in the glass wafer, and during the evaluation of the switching valve no detectable liquid was trapped in dead volumes when switched between two liquids of different colour.

7. Conclusions and further research

The mixing process was experimentally studied in mini-channels manufactured by traditional mechanical machining as well as micromachined channels. In the characterisation it was shown that mixing is possible by the introduction of secondary flow or local turbulence in miniaturised channels at high flow rates. In both mini-channels and micro-channels, mixing by diffusion was observed at low flow rates when the transit time through the channel exceeded 20 seconds.

The leak rate in the fabricated valves was approximately 1.4 nl/min or less. Leak rates of this low level are necessary for reproducible measurements in micro-TAS, and to achieve this, particle filters and valves designed to prevent particle trapping must be integrated in future systems.

Dead volumes in structures which are not flushed generate carry-over and reduce the accuracy of the measurement. Hence, a micro-TAS concept was developed and different system elements fabricated. In the evaluation of a switching valve the outlet channel of the valve was completely flushed and no dead volume was detected.

In the near future chemical reactions and their detection in micro-TAS, micro-chemistry for short, need more attention. Micro-chemistry activities must address the fundamental problems of whether standard physio-chemical parameters do apply to the micro-TAS, or if parameters such as solubility and viscosity change when the channel dimensions approach the size of the molecules. These are essential steps towards the implementation of known methods for chemical analysis in micro-TAS.

It is important to focus attention on chemical detectors for microsystems. In microsystems the detection volume is reduced drastically compared to almost all existing detectors. CHEMFETs and micro-electrodes are two of the very few examples of detectors in which the general benefit of an ultra-small sample volume may be utilized also in the detection [8] [9].

Chemical durability of materials is another research topic of importance. The tiny structures which are exposed to the harsh chemical environment inside the micro-TAS must be extremely stable chemically. Etching rates must be well below 1 Å/h if 1 μm thick structures or a thin film coating are to be used for more than one year. The stability of the reagents used in the system must also be investigated if maintenance-free periods of a half year or more are to be achieved by the system.

To improve the research in micro-chemistry a micro-TAS concept must be developed in which different measurement conditions are easily obtained. A micro-TAS concept based on hybridisation, as proposed by the MESA Institute [10], could be a way to improve flexibility and it also features the re-use of micro-TAS components.

Acknowledgements

This work was partly performed in the Materials Centre for Microelectronics (MCM) program, supported by the Danish Agency for Trade and Industry, The Danish Natural Science Research Foundation, and the Danish Technical Science Research Foundation under the Materials Development Program.

References

1. Peter Gravesen, Jens Branebjerg and Ole Søndergård Jensen. Microfluidics - a review. *J. Micromech. Microeng.* 3 (1993) 168-182.
2. Andreas Manz, D. Jed Harrison, Elisabeth Verpoorte, H.Michael Widmer. Planar chips technology for miniaturization of separation systems: A developing perspective in chemical monitoring. *Advances in chromatography.* 33 (1993) 1-65.
3. Elisabeth Verpoorte, Andreas Manz, H.M.Widmer, Bart van der Schoot, Nico F de Rooij. A three-

dimensional micro flow system for multi-step chemical analysis *The 7th international conference on solid-state sensors and actuators, Yokohama, Japan,1993*

4   S C Terry, J H Jerman and J B Angell A gas chromatographic air analyzer fabricated in silicon *IEEE Tran Electron Devises ED-26* (1979) 1880-1886

5   Bart van der Schoot and Piet Bergveld An ISFET-based microlitre titrator integration of a chemical sensor-actuator system *Sensors and Actuators* 8 (1985) 11-22

6   Ryo Miyake, Theo S J Lammerink, Miko Elwenspoek, Jan H J Fluitman Micro mixer with fast diffusion *The 1993 Workshop in Micro Electro Mechanical Systems* Fort Lauderdale, florida, USA, 1993

7   Henrik Soeberg Viscous flow in curved Tubes-I Velocity profiles *Chemical Engineering Science* Vol 43, No 4 (1988) 855-862

8   P Bergveld Future applications of ISFETs *Sensors and Actuators B*,4 (1991) 125-133

9   H Ping Wu Fabrication and characterization of a new class of microelectrode arrays exhibiting steady-state current behaviour *Anal Chem* 65 (1993) 1643-1646

10   Jan H J Fluitman, Albert van den Berg and Theo S J Lammerink, Proceedings Workshop on Micro Total Analysis Systems, Enschede, The Netherlands, (1994),73

MICROANALYSIS SYSTEMS
FOR GASES

I. Lundström, A. Lloyd Spetz, H. Sundgren and F. Winquist
Laboratory of Applied Physics, Linköping University,
S-581 83 Linköping, Sweden

0. Abstract

This contribution deals with the use of chemical sensor arrays to develop microanalysis systems for gases which may serve as "electronic noses" One particular sensor technology based on field effect devices is described in some detail The principles of electronic noses are reveiwed Some results obtained with noses incorporating field effect devices are presented Different forms of integrated sensor arrays capable of generating detailed information about gas mixtures or odours are discussed Micromachining can e g be used to create a temperature gradient along a sensor array Furthermore, continuous sensing surfaces with a spatially resolved detection principle provide new possibilities regarding the microanalysis of gases

1. Introduction

Microanalysis systems for gases cover a wide range of different developments including microfabricated valves, columns and distribution systems as well as chemical sensors and chemical sensor arrays In this contribution we concentrate on gas sensors based on field effect devices Such devices have the advantage that they can easily be miniaturized and integrated into arrays containing a large number of them They are, therefore, interesting candidates in microanalytical systems, not only for ions as decribed in other parts of this book, but also for gases

Field-effect devices sensitive to (gas) molecules have been around for about 20 years Their properties have been thoroughly studied and discrete sensors for a large number of species have been developed [1-4] A field effect structure is in principle a "metal"-insulator semiconductor structure where the gate, the "metal", can be any conducting layer or medium (e g metal, semiconductor, electrolyte, or polymer) The conducting layer can furthermore be allowed to have a rather small conductivity and still work well as a gate material for a field effect device The possibility to find suitable gate materials for a given sensing situation is therefore quite large Our development has, however,

153

A van den Berg and P Bergveld (eds), Micro Total Analysis Systems, 153–163
© 1995 *Kluwer Academic Publishers*

mainly been made on field effect devices with gates of catalytic metals like Pd, Ir and Pt.

The principles of field effect devices with special regard to chemical sensing will be briefly described in the next section. It will be demonstrated how such devices can be used in sensor arrays eventually together with other types of gas sensors to make electronic noses for the identification, classification and analysis of gas mixtures (or odours) [5-7]. A truly microanalytical system, an "olfactory camera" based on a field effect transistor array is also suggested.

Finally, we mention the potential use of continuous sensing surfaces to analyze gas mixtures, through a spatially resolved detection of the response along the surface. This idea has been implemented using a scanning light pulse technique to analyze spatially resolved gas induced electrical polarization phenomena in "large" (~ 10 mm^2) catalytic metal gates on field effect structures [8-10].

2. Gas sensing with field effect devices

2.1. DEVICE PHYSICS

In metal (conductor)-insulator-semiconductor (MIS) field effect structures, the electrical properties of the semiconductor surface are controlled by its surface potential, φ_s, which determines the number of charge carriers at the semiconductor surface, the width of the depletion region, etc. The surface potential in turn is a function of the externally applied voltage on the gate (the metal), V_G, and the difference in the work functions of the gate material, W_m, and the semiconductor, W_s, that is

$$\varphi_s = f_1 (V_G - W_{ms}/q) \tag{1}$$
and
$$I_D \text{ or } C = f_2 (\varphi_s) = f_3 (V_G - W_{ms}/q) \tag{2}$$

where $W_{ms} \equiv W_m - W_s$, I_D is the drain current of a field effect transistor and C the capacitance of an MIS capacitor (see Fig. 1).

Equations (1) and (2) suggest that all phenomena which influence the surface potential of the semiconductor can be used for chemical sensing. In field effect devices, this change is caused by work function changes of the gate material which occur as a bulk effect (a change in Fermi level) in the gate material and/or as a polarization at the gate material-insulator interface. To keep a given surface potential, φ_s at the semiconductor surface we have to apply a gate voltage, $V_G - \Delta V$, where

$$\Delta V = -\Delta W_{ms}/q \tag{3}$$

if a change, ΔW_{ms}, occurs in the difference in the work functions between the gate material and the semiconductor as illustrated in Fig. 1. More detailed descriptions of the device physics are found for example in refs. [1,4].

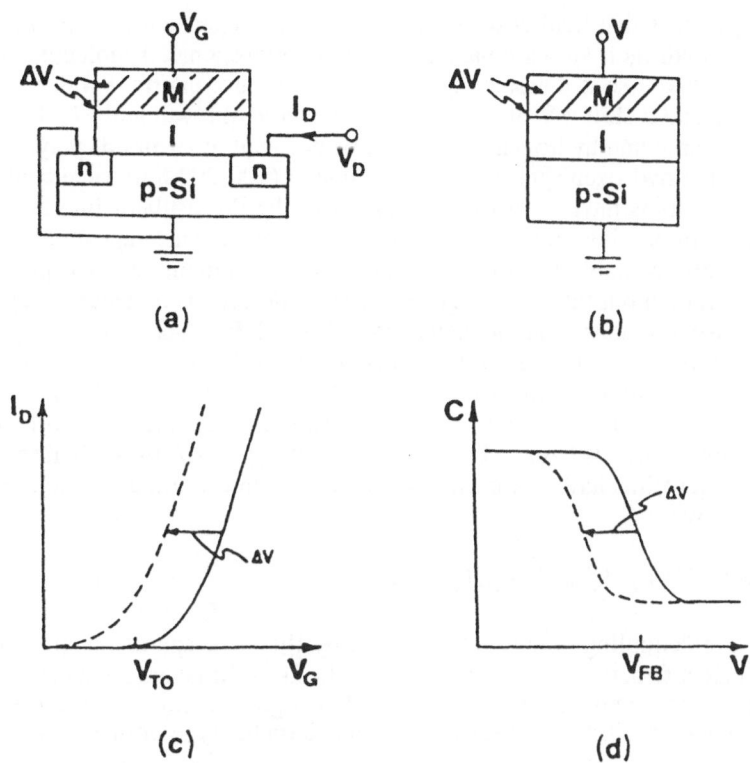

Figure 1. (a) Schematic drawing of a metal-insulator (SiO_2)-semiconductor field effect transistor (MISFET). The insulator and gate metal are typically 100 nm thick. The channel length, the distance between the n-regions, is typically of the order of 1-10 μm and the channel width (the dimension perpendicular to the paper) 10-100 μm.

(b) A metal-insulator-semiconductor capacitance (MISCAP) which is a simpler test structure yielding similar information as the MISFET. The sinsulator and metal thickness as in (a). The area of the metal contact may vary from about 100 μm² to several mm².

(c) and (d) schematic illustration of the influence of work function changes on the electrical properties of a MISFET (c) and MISCAP (d).

The gate "metal" may be any material with large enough conductivity to serve as a dc-equipotential surface. ΔV in equation (3) displaces the electrical characteristics along the voltage axis. It changes (decreases) the threshold voltage, V_{TO} of the FET or the flatband voltage, V_{FB}, of the capacitor. V_{TO} is the gate voltage necessary to give a conducting channel along the semiconductor surface. V_{FB} is the voltage necessary to make the energy band of the semiconductor flat. Drawing taken from [6].

The prototype gas sensitive field effect device was the hydrogen sensitive Pd-gate FET [1]. In this case hydrogen atoms emanating from dehydrogenated molecules on the palladium surface give rise to a dipole layer at the Pd-insulator interface which causes a decrease in W_{ms} and a shift ΔV of the $I_D V_G$-curve along the voltage axis. Molecules like hydrogen, ethanol and hydrogen sulfide can be detected in air in this way. The Pd-layer in the original hydrogen sensor was thick enough (100-200 nm) to be continuous and the hydrogen atoms have to diffuse through it to the Pd-insulator interface. With such a layer it is not possible, for example, to detect ammonia since apparently no free hydrogen atoms are available from the reaction between ammonia and oxygen on the metal surface. It turned out that a more general way to detect gas molecules was to use thin (3-20 nm) discontinuous catalytic metal gates [1-4,12]. For such layers it is possible to detect not only ΔV's due to polarization phenomena at the metal insulator interface but also due to contributions from surface potential changes of the metal, ΔV_s, (due to adsorption and chemical reactions) and from adsorbates on the insulator surface between the metal islands, ΔV_i. The details of the polarization phenomena in discontinuous metal films have not been worked out yet, but in principle the measured ΔV will be given by

$$\Delta V = g_i(\gamma)\,\Delta V_i + g_a(\gamma)\,\Delta V_a + g_s(\gamma)\,\Delta V_s \qquad (4)$$

where the g´s are "coupling" factors determined by the coverage of metal, γ, on the surface. Field effect structures with gates of (thin) Pt, Ir or Pd layers have been used to detect e.g. ammonia and amines, alcohols, hydrogen sulfide, and unsaturated hydrocarbons. Their sensitivity and selectivity depend on the type of metal used and on the operation temperature of the device which normally lies between 50 and 200°C. The smallest detectable amounts (in air) are of the order of .1-1 ppm. It has been found that in pulsed operation, where the gas to be tested is supplied as (short) pulses in a carrier gas (air), the pulse response is rather reproducible over long periods of time.

2.2. EXAMPLES OF APPLICATIONS

There are several (commercial) applications for discrete gas sensitive devices. Two classical examples are the tinoxide based sensors for combustible gases ("Taguchi or Figaro sensors") used in e.g. gas alarms and the so called λ-sond (based on zirconia) used to control the combustion in cars.

In case of the field effect devices the hydrogen sensitive Pd-gate devices are used in hydrogen leak detectors and monitors [13]. Furthermore, several other applications for these devices have been developed and tested, a recent review is found in [4]. We like to mention one "microanalytical" application here where an ammonia sensitive device is used behind a gaspermeable membrane to detect gas production during enzymatic reactions either in small (flow through) liquid cells, in a probe or in solid sample and reagent carriers [14]. The idea is that during several enzymatic reactions ammonium ions are produced which are in equilibrium with ammonia molecules. These diffuse through the gaspermeable membrane. (The equilibrium is shifted towards NH_3 with increasing pH.) The idea has e.g. been used to detect small amounts of Hg^{2+} through its inhibition of the enzymatic cleavage of urea by urease, in which reaction ammonia is produced. The detection limit is of the order of 0.1 - 1 nM of Hg^{2+} depending on the type of detection system used. Sample volumes down to 2-4 μl are enough e.g. with the solid carrier system [14].

Another area where small (and integrated) sensors will be of large interest is in the field of "artificial olfactory senses" or "electronic noses" which is described below.

3. Electronic Noses

3.1. PRINCIPLES

Gas sensors are in general not completely selective, i.e. sensitive only to one species, but can be made with different selectivity patterns as mentioned above for the field effect devices. Such devices can be used in arrays together with pattern recognition routines to analyze complex gas mixtures (odours). The principle of the "electronic nose" is schematically illustrated in Fig. 2. This concept has been around for some time (see e.g. [15]). It is believed that electronic noses will be able to provide real time analysis of odours for quality and process control. Electronic noses can of course be constructed by the use of several types of chemical sensors. The gas sensitive field effect devices are interesting in this connection since they are fabricated by silicon technology, which makes it easy to process miniaturized sensor arrays. Furthermore, by the use of different chemically sensitive gate materials, different catalytic metals, conducting polymers, metal oxides, etc., and by the use of different operation temperatures, several selectivity patterns and sensitivity regions are obtained.

The pattern recognition is performed using statistical methods and/or so called artificial neural networks [15]. These methods are implemented using (commercially) available software, but also dedicated signal treatment chips are developed. Electronic noses based on different technologies are now being commercialized.

The complexity of the nose depends to a large extent on its application, which may be to identify a gas (e.g. "ethanol or methanol"), to classify a product ("good, average, bad") or to quantify the components of a gas mixture (e.g. "10 ppm NH_3; 4 ppm H_2S; 37 ppm ethanol"). The requirements on the nose become more and more demanding in the applications above.

Chemical sensors (three in the example) with different
selectivity patterns towards different classes of molecules

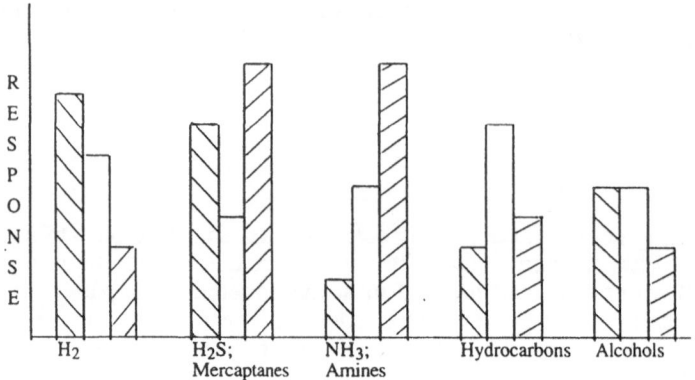

give response patterns of gas mixtures (odours)

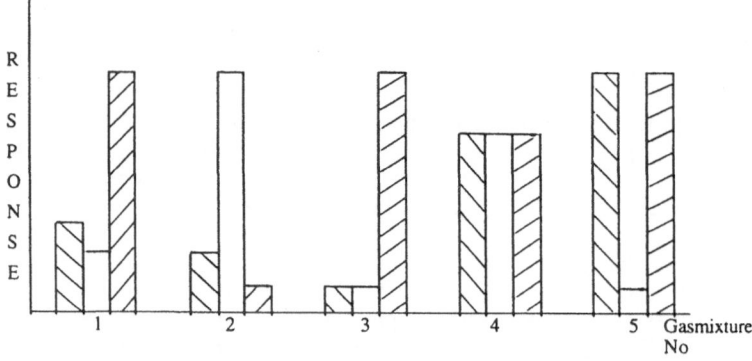

which together with pattern recognition are used for

- classification

- identification

- quantification

of the gas mixtures (odours)

Figure 2. Schematic illustration of the principles of an electronic nose. (Drawing adapted from Sensors and Actuators [17].)

3.2. ELECTRONIC NOSES WITH FIELD EFFECT DEVICES

Several sensor arrays consisting of field effect devices have been tested for the identification, classification and quantification of odours from different materials and gas mixtures [3-7,16-17]. These arrays have contained 6-10 field effect devices sometimes combined with other chemical sensors like those based on SnO_2 ("Taguchi", "Figaro" sensors). The selectivity patterns of the field effect devices were controlled by the choice of the catalytic metal, its thickness and the operation temperature of the device. As an example, we show the results obtained on the identification of cheeses from their odours. In this case the array consisted of 8 FET's and 4 Taguchi devices. Samples of ten different kinds of cheese, each sample approximately 25 g, were stored at room temperature for 40 hours in 400 ml glass beakers covered with parafilm. For each cheese two beakers were identically prepared. After the storage time, the atmosphere in each beaker was analyzed by passing it over the sensor array for 60 seconds. For a given sample the response with clean air as a reference, from each of the sensors in the array was stored. Each beaker was analyzed five times and one measurement from each of the beakers was used for the training of a neural network. Table 1 shows the result for a network with twelve input nodes (the sensors), ten hidden nodes and ten output nodes (the different cheeses). It is observed that the predictions were quite good except for one of the cheeses. It was found that a smaller network (array) consisting of only 6 FET devices made a correct prediction of 7 of the 10 cheeses and the 4 Taguchi sensors alone made a correct prediction of 3 cheeses [17].

Table 1. Identification of cheeses (from [17])

Cheese type in test set	Relative output values from the different output nodes. (The sum of the outputs was normalized to one in each test. Outputs smaller than 0.01 are not shown.)									
	Node No.									
	1	2	3	4	5	6	7	8	9	10
1 Svarta Sara, Denmark	**.80**							.20		
2 Gouda, Germany		.30				.02			**.67**	.01
3 Havarti, Denmark			**1.0**							
4 Gorgonzola, Italy			.07	**.72**		.13		.05	.03	
5 Danablue, Denmark					**.94**	.05				
6 Brie Mareillat, France				.12		**.78**		.02	.07	
7 Brie de Meaux, France							**1.0**			
8 Vachero, France			.01	.04				**.94**		
9 Torparost, Sweden		.02				.09			**.88**	
10 Sheep cheese, Bulgaria					.06				.04	**.89**

4. Integrated sensor arrays

4.1. PRINCIPLES

So far we have mainly used discrete devices in our sensor arrays. It is, however, easy to integrate several field effect devices on one single chip as illustrated in Fig. 3. The only difference from the normally used fabrication methods is that we like to create different selectivity patterns at different parts of the array which means that different gate materials should be used at different parts and/or temperature gradients created along the chip. This is facilitated by making the integrated array "large" on the scale of normal integrated electronics. Still the chip in Fig. 3, which contains 20 gas sensitive FET's, represents a miniaturized system for chemical analysis, which, in principle, will be able to replace e.g. gas chromatography in many applications. The temperature gradient may be obtained by removing the silicon from (part of) the back of the chip and applying a heat source at the thin end of the chip. A temperature change from about 100-200°C along the chip is desired.

Sensor arrays similar to that in Fig. 3 are also developed in collaboration with research groups in Latvia [18].

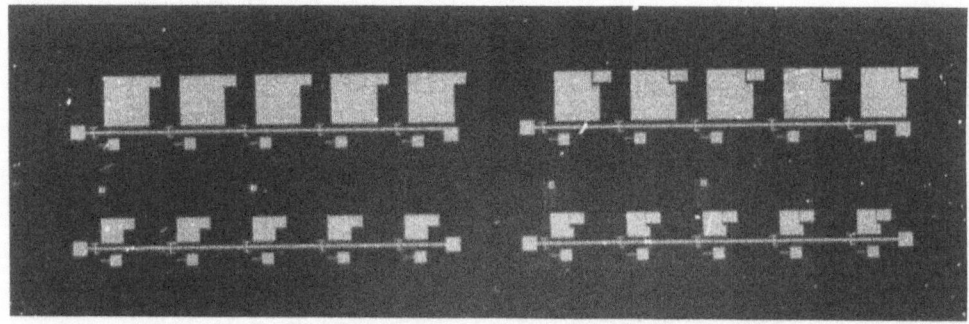

Figure 3. Picture of an integrated sensor array consisting of 20 FET's (of two different sizes). The transistors cover an area of 4x15 mm². The layout allows e.g. a temperature gradient and different gate materials to be applied along the array. The chip is mounted on a header which may be provided with a lid with gas inlet and outlet to obtain a miniaturized analytical system for small volumes and flows.

4.2. OLFACTORY CAMERA

We suggested a more spectacular, two dimensional, development of integrated sensor arrays a few years ago to create what could be called an "olfactory camera" [6]. The olfactory camera is simply a chip with an array of field effect transistors, heaters and temperature sensors. The sensing transistors are coupled as diodes into which a constant current is injected and the necessary gate voltage V_G measured. The proposed olfactory camera chip resembles the CCD-chip used e.g. in solid state vidicons.The chemical selectivity over the sensor array can be varied in several ways, for example by temperature gradients and a compositional variation of the gates along the chip. It should be pointed out that also other materials than catalytic metals can be used as gates, for example conducting polymers or a combination of materials. The camera chip or a separate chip should also contain (standard) driving and reading circuits. An olfactory camera should be able to produce an image identifying the composition of a gas mixture in real time. Further signal processing should also yield the concentration of the species in the gas mixture. The fact that the result is obtained as a two dimensional response map makes classical image processing possible for the treatment of the data obtained.

4.3. CONTINUOUS SENSING SURFACES

Another way to produce "ofactory images" is to use a large (continuous) sensing surface and a laterally resolved method to "read" the surface. Field effect structures also provide some possibilities here. If the metal gate is thin enough to let light penetrate to the semiconductor, the local variations of the surface potential of the semiconductor can be probed with a pulsed light beam scanning over the surface [8-10]. In a metal-insulator semiconductor structure the photocapacitive current induced by the light pulses is a function of the extension of the depletion region at the semiconductor surface, and hence the surface potential of the semiconductor. The photocapacitive current is thus large for a positive bias on a MIS-structure with a p-type semiconductor and small at negative bias. Since the width of the depletion region is a function of the surface potential of the semiconductor, we can use the same arguments for the photocapacitive current i_{ph} as for the capacitance or current in Equation (2), which means that the i_{ph} curve shifts along the voltage axis according to Equation (3) upon a change in W_{ms}. By using a focussed light beam it is now possible to obtain ΔV locally for a large sensing surface and thus to produce maps of $\Delta V(x,y)$ over the sensor surface.

The details of such maps depend both on the chemical sensitivity of the catalytic metals used and the fact that for sufficiently large catalytic surfaces (of the order of 1-10 mm^2), the combustion of molecules on the surface can not be neglected. These two parameters, the sensitivity and combustion rate, will therefore determine the observed response patterns [10]. We have thus with the laterally resolved detection the possibility to influence the selectivity also by changes in combustion rate. Temperature gradients along the sensing surface as well as areas of different catalytic metals e.g. make it possible to obtain response images which are unique for a given molecule or an odour [8-10].

An example of such an olfactory image is shown in Fig. 4. This response map was obtained from a sensing surface with in principle two active catalytic metal bands, thin Pd and Pt, and a temperature of 100-110°C at the cool end and 170-180°C at the warm end.

162

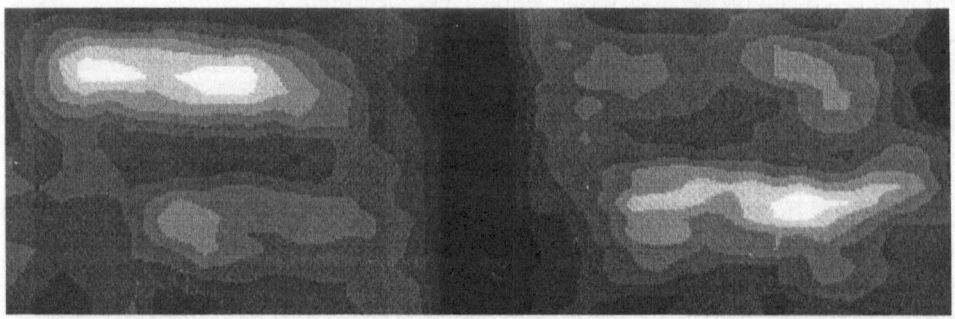

Figure 4. Illustration of "olfactory" images obtained with a test surface consisting of 6 mm long bands of Pd and Pt (1 mm wide and about 6 nm thick). The images show areas of isophotocapacitive current changes when the surface was exposed to the odours from gorgonzola (left) and camembert (right). The brighter the area the larger the current change. See the text for further description.

5. Conclusions

It can be concluded that gas sensors can be made very small and with small consumption of the molecules to be detected. They can, therefore, be used to analyze gas evalution from small samples and with small gas flows with both industrial and medical applications. Furthermore, the use of sensor arrays gives the possibility to construct electronic noses for more complicated situations related to identification, classification and quantification of gas mixtures or odours. The final microanalysis system for gases may perhaps consist of microfabricated gas handling, an integrated sensor array and a dedicated chip for signal treatment and pattern recognition.

Acknowledgements

Our research on chemical sensors and electronic noses is supported by grants from the Swedish National Board for Technical and Industrial Development, the Swedish Engineering Sciences Research Council and the Center for Industrial Information Technology at Linköping University.

163

References

1 I Lundstrom, M Armgarth and L -G Petersson, Physics with catalytic metal gate chemical sensors, *CRC Crit Rev Solid State Mater Sci*, *15* (1989) 201-278
2 I Lundstrom, A Spetz, F Winquist, U Ackelid and H Sundgren, Catalytic metals and field effect devices - a useful combination, *Sensors and Actuators, B1* (1990) 15-20
3 I Lundstrom, C Svensson, A Spetz, H Sundgren and F Winquist, From hydrogen sensors to olfactory images - twenty years with catalytic field effect devices, *Sensors and Actuators B, 13-14* (1993) 16-23
4 A Spetz, F Winquist, H Sundgren and I Lundstrom, Field effect gas sensors in *Gas Sensors* (G Sbeweglieri, ed), Kluwer Academic Publishers, Dordrecht, The Netherlands, 1992, 219-279
5 H Sundgren, F Winquist, I Lukkari and I Lundstrom, Artificial neural networks and gas sensor arrays quantification of individual components in a gas mixture, *Meas Sci Technol*, *2* (1991) 464-469
6 I Lundstrom, E Hedborg, H Sundgren and F Winquist, Electronic noses based on field effect structures, ref 15, pp 303-319
7 F Winquist, E G Hornsten, H Sundgren and I Lundstrom, Performance of an electronic nose for quality estimation of ground meat, *Meas Sci Technol 4* (1993) 1493-1500
8 I Lundstrom, R Erlandsson, U Frykman, E Hedborg, A Spetz, H Sundgren, S Welin and F Winquist, Artificial ' olfactory" images from a chemical sensor using a light pulse technique, *Nature, 352* (1991) 47-50
9 F Winquist and I Lundstrom, Geruchsbilder, *Technische Rundschau Heft 19,* (1993) 54-56
10 I Lundstrom, H Sundgren and F Winquist, Generation of response images of gas mixtures, *J Appl Phys 74* (1993) 6953-6961
11 I Lundstrom, M S Shivaraman, C M Svensson and L Lundkvist, Hydrogen sensitive MOS field effect transistor, *Appl Phys Lett*, *26* (1975) 55-57
12 F Winquist, A Spetz, M Armgarth, C Nylander and I Lundstrom, Modified palladium metal-oxide-semiconductor structures with increased ammonia gas sensitivity, *Appl Phys Lett*, *43* (1983) 839-841
13 Sensistor AB, Linkoping, Sweden
14 F Winquist, I Lundstrom and B Danielsson, Three level analysis of mercury using urease in combination with an ammonia gas sensitive semiconductor structure, *Analytical Letters 21* (1988) 1801-1816
15 J W Gardner and P N Bartlett (eds), *Sensors and Sensory Systems for an electronic Nose, NATO ASI Series E Applied Sciences,* Vol 212, Kluwer, Dordrecht, 1992
16 F Winquist, H Sundgren, E Hedborg, A Spetz, M Holmberg and I Lundstrom, Identification of molecules with field effect structures, *Technical Digest 11th Sensor Symposium, Japan* (1992) 257-262
17 I Lundstrom, T Ederth, H Karlis, H Sundgren, A Spetz and F Winquist, Recent developments in field effect gas sensors, *Sensors and Actuators B* (in press)
18 A Lusis, J Kleperis, V Eglitis , A Lloyd Spetz, I Lundstrom, H Sundgren, F Winquist, O Strautmanis, I Slaidins, P Rozukalns and S Sjulzics, First steps of µTAS in Latvia, *this issue*

BONDING AND ASSEMBLING METHODS
FOR REALISING A μTAS

Shuichi Shoji and Masayoshi Esashi[1]

Department of Electronic and Communication Engineering, Waseda University,
3-4-1 Ohkubo, Shinjuku-ku, Tokyo 169, Japan
[1]Department of Mechatronics and Precision Engineering, Tohoku University,
Aza Aoba, Aramaki, Aoba-ku, Sendai 980, Japan

0. Abstract

Bonding and assembly methods for realising a μTAS are summarised. A multi-level stack of wafers on which functional devices are formed is a useful structure to construct a sophisticated μTAS. Glass-silicon stacks fabricated by an anodic bonding have been widely used for micro flow control devices and systems. Bending of the glass-silicon structure sometimes observed prevents subsequent bonding. The reasons of this problem were studied to realise multi-level glass-silicon stacks. The anodic bonding using intermediate glass thin layer was also very useful for constructing silicon stacks. The required condition and the mechanism of this bonding were studied. A microconnector and a separable channel type microvalve useful in a μTAS are also described.

1. Introduction

Most microvalves and micropumps developed so far consist of stacked wafers in which diaphragms, cavities, channels, etc. are formed. The structures, bonding methods, sizes, and flow ranges of the microvalves and the micropumps are listed in Table 1 and Table 2 [1-17]. In many cases, glass-silicon stacks have been used because of the ease of bonding. In fact, it is easy to align the wafers having structures due to the glass transparency to optical wavelengths. The glass-to-silicon anodic bonding method has been used for this purpose [1-8]. Silicon-silicon stacks have been also used in some microvalves and micropumps [13,14]. In this case, fine micro structures can be formed on each wafer by an anisotropic etching. A simple gluing, a high temperature fusion bonding, an anodic bonding using a thin intermediate glass layer have been used [9-14]. Some microvalves and micropumps were also fabricated without any bonding processes [16,17].

165

A. van den Berg and P. Bergveld (eds.), Micro Total Analysis Systems, 165–179.
© 1995 *Kluwer Academic Publishers.*

Table 1 Structures and features of the microvalves

Type	Actuator	Structure	Bonding Method	Size	Flow Range
Bulk Silicon [1] (Normally Closed)	Stack Type Piezoelectric	Si-PG	AB	10mmx10mm x10mm	40ml/min (N_2 gas 0 5kgf/cm^2)
Three-way [2]	Stack Type Piezoelectric	PG-Si-PG 1 2	1 AB 2 AB	10mmx10mm x10mm	15 µl/min (H_2O 0 17kgf/cm^2)
Microheater [3]	Thermo-pneumatic	PG-Si-PG-Si 1 2 3	1,2,3 AB	6 3mmx6 6mm 2 0mm	15l/min (N_2 7 1kgf/cm^2)
Silicone [4] Rubber Seat	Cantilever Type Piezoelectric	PM+Si-PG 1	MC 1 AB	24mmx12mm x5mm	9µ l/min (H_2O 0 2kgf/cm^2)
Conductive [9] Film	Electrostatic	Si-IM-Si	Silicon Resin (IM)	25mmx25mm x1 3mm	
Pressure [11] Balance	Electrostatic	Si-Si-PG 1 2	1 FB 2 AB	1mmx1 8mm x1 4mm	160ml/min (Air 1 34kgf/cm^2)
Diaphragm [12]	Bimetallic	Si-Si	FB	2 5mmφ	300ml/min (Air 7kgf/cm^2)
Active [13] Check	Electromagnetic /electrostatic	Si-IM-Si	AB (IM,SPG)	3mmx8mm	3ml/min (Air 0 16kgf/cm^2)
Separable [16]	Stack Type Piezoelectric	Si+PM-Si 1	MC 1 AB	12mmx10mm x10mm	90µl/mm (H_2O 0 5kgf/cm^2)

Structure.PG,Pyrex Glass, M,Metal, IM;Intermediate Layer, SPG; Sputtered PG, PM;Polymer Membrane
Bonding Method. AB; Anodic Bonding, FB, Fusion Bonding, MC,Mechanical Contact

Table 2 Structures and features of the micropumps

Type	Actuator	Structure	Bonding Method	Flow Range	Maximum Output Pressure
Reciprocating	Disk Type [5] Piezoelectric	PG-Si-PG 1 2	1 AB 2 AB	8µl/min (100V,1Hz)	1 0mH$_2$O
	Stack Type [6] Piezoelectric	PG-Si-PG 1 2	1 AB 2 AB	40µl/min (100V,40Hz)	1 5mH$_2$O
	Thermo-pneumatic [7]	PG-Si-PG 1 2	1 AB 2 AB	58µl/min (2 5W,5Hz)	0 3mH$_2$O
	Electrostatic [14]	Si-Si-Si-Si 1 2 3	1,2,3.SGAB or EB	350µl/min (170V,400Hz)	2.5mH$_2$O
	Disk Type [15] Piezoelectric	BR-BR		3 3ml/min H$_2$O (141V,307Hz) 35ml/min Air (20V,6kHz)	2.4mH$_2$O 0.23mH$_2$O
	Pneumatic [17]	Gl-Au-Ti	(LIGA Process)	80µl/min (5Hz)	0 47mH$_2$O
Peristaltic	Disk Type [8] Piezoelectric	PG-Si-PG 1 2	1 AB 2 AB	100µl/min (80V,15Hz)	0 6mH$_2$O
	Thermopneumatic[10] (Laser Light Driven)	Si-IM-Gl	IM:Photoresist	90µl/min (3Hz)	0 03mH$_2$O

Structure: PG;Pyrex Glass, BR,Brass, Gl,Glass Plate,IM;Intermediate Layer
Bonding Method: AB,Anodic Bonding, SGAB,Sputtered Glass Anodic Bonding, EB; Eutectic Bonding

Sophisticated micro flow systems have been fabricated by combining micro flow control devices, microsensors, etc. The structures and bonding methods of these systems are listed in Table 3 [1,2,7,18-20] A micro flow injection analysis system as shown in Fig 1 consists of silicon-glass-silicon-glass-stacks [2] Bonding methods are the key technology for fabricating such complicated and high performance micro systems Considering the yield, hybrid systems consisting of some blocks of the flow systems whose inlets and outlets are connected hermetically sealed will be also realistic The reduction of the dead volume of the connecting port is most important in these systems

Table 3 Structures and features of the microsystems

System	Object	Structure	Bonding Method
Mass Flow Controller [1]	Gas	PG-Si	AB
Gas Chromatograph [18]	Gas	PG-Si	AB
Flow Regurated [19] Micropump	Liquid	PG-Si-PG 1 2	1,2 AB
Micro Dosing [7] System	Liquid	PG-Si-PG-Si-PG 1 2 3 4	1,2,3,4 AB
Flow Injection [2] Analysis	Liquid	Si-PG-Si-PG 1 2 3	1,2,3 AB
Stacked Chemical [20] Analysis System	Liquid	Gl-IM-Gl---IM-Gl	(IM Polymer)

Structure PG, Pyrex Glass, Gl, Glass Plate, IM, Intermediate Layer
Bonding Method AB, Anodic Bonding

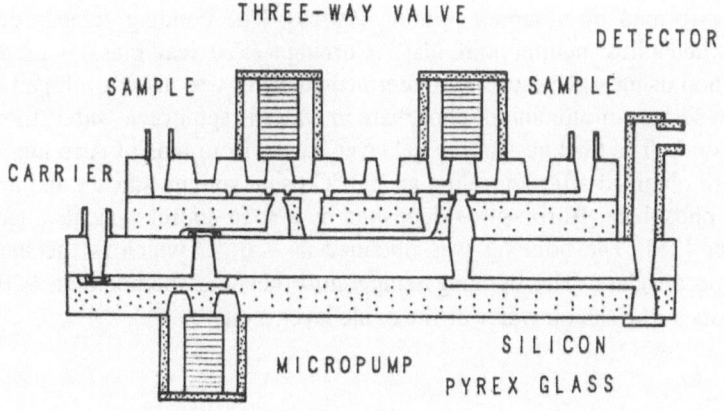

Fig.1 Structure of the integrated FIA system

The bonding methods between wafers are described and discussed in the following section A microconnector and a separable microvalve useful for a μTAS are described in the following sections

2. Bonding methods

There are many kinds of bonding methods as shown in Fig 2 The features of these methods are listed in Table 4

2 1 GLUING

The most simple bonding methods are gluing the wafers with an adhesive A thin adhesive can be formed by spin coating when the viscosity of it is controlled In most cases, no high temperature annealing process is necessary There are many available combinations of wafers by choosing an adequate adhesive This method has, however, a problem that channels or holes fabricated in the wafers are plugged by the adhesive A photosensitive dry film (Hitachi Kasei Co) is applicable to avoid this problem [21] After a dry film is laminated on a wafer and patterned by photolithography, the other wafer is aligned and contacted applying a pressure of about 0 2 kgf/cm^2 The bonding completed after annealing at a temperature of around 150 °C for 30 min The thickness of the film can be chosen from 50 μm to 200 μm These methods have a loss of dimensional tolerance at curing process The chemical durability of the adhesive and the film must be tested carefully

2 2 LOW TEMPERATURE GLASS BONDING

Some silicon-to-silicon low temperature thermal bonding methods using thin glass intermediate layers were reported One of these is a bonding, so called 'frit seal', using a patterned glass layer formed by a screen printing [22,23] The bonding temperature is higher than 415 °C when a low melting point glass (Corning #7570 lead glass) is used

Another method using a glass gel as an intermediate layer was also developed [24] A solution of sodium silicate or aluminium phosphate in water is spun on a wafer, then the other wafer is faced on it The thickness of the gel layer ranges from tens of Å to hundreds of Å Good bonds are obtained after annealing at 200 °C using sodium silicate and at 350 °C using aluminium phosphate Boron-glass deposited by a solid-source was also used as an intermediate layer [25] The bonding was obtained at 450 °C which is the melting temperature of the boron-glass The bonding temperature becomes higher than 900 °C, when phosphorus exits in the silicon wafer or the oxide layer on it

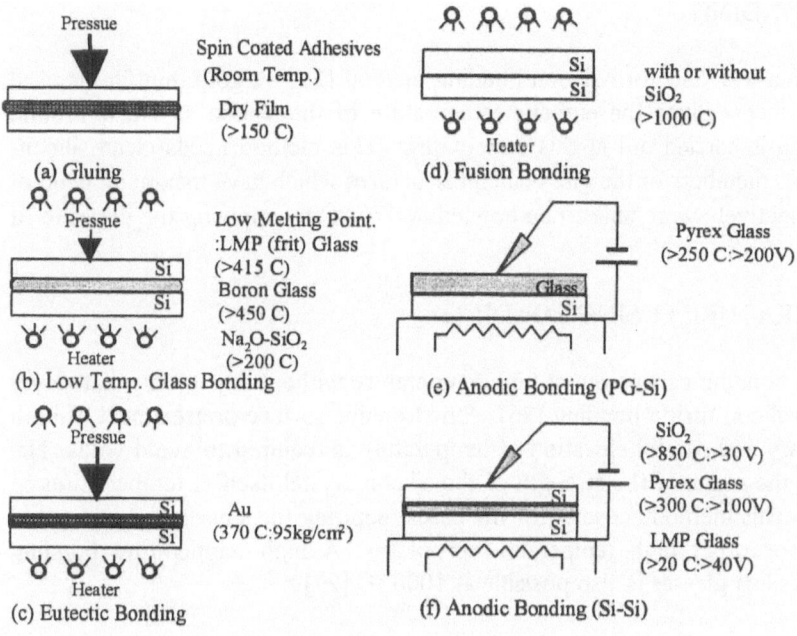

Fig.2 Set-ups of the bonding methods

Table 4 Features of the bonding methods

Bonding Method	Substrates	Intermediate Layer	Temperature (°C)	Applied Voltage (V)	Applied Pressure (kg/cm³)	Selective Bonding (Patternning)	Reference
Gluing	Si-Si Glass-Si Glass-Glass	Spin Coated Adhesives	R.T.	—	>0.2	X	
		Photosensitive Dry Film	>150	—	>0.5	Photolithography	[21]
Low Temperature Glass Bonding	Si-Si Glass-Si Glass-Glass	Frit Glass	>415	—	—	Screen Printing	[22,23]
		Liquid Glass (Na₂O-SiO₂)	>200	—	—	X	[24]
	Si-Si	Boron Glass	>450	—	—	X	[25]
Eutectic Bonding	Si-Si	Au	370	—	95	Photolithography	[22,23]
Fusion Bonding	Si-Si	—	>1000	—	—	X	[26]
Anodic Bonding	Glass-Si	—	>250	>200	—	[Sio (>1μm)] [TiW/Au]	[22,28,29]
	Si-Si	SiO₂	>850	>30	—	Photolithography	[30]
	Si-Si	Sputtered Pyerx Glass	>300	>100	—	Lift off	[31]
	Si-Si Si-Al /Glass Si-ITO/Glass	Sputtered Low Melting Point Glass	R.T.	>40	>0.8	Lift off	[32]

2 3 EUTECTIC BONDING

This method is a classical silicon-to-silicon bonding method [23] A gold thin film is used as an intermediate layer Since the eutectic temperature of the silicon an Au is around 370 °C, the bonding is carried out at this temperature This method needs clean silicon-gold interfaces Two members of the glass-silicon structures which have maximum bend of 2 μm an 5 μm respectively were able to be bonded at 370 °C by applying the pressure of about 95 kgf/cm^2

2 4 HIGH TEMPERATURE FUSING BONDING

A silicon-to-silicon bonding carried out at high temperature without any intermediate layer is called silicon-to-silicon fusion bonding [26] A hydrophilic surface pretreatment of both members is necessary and careful elevation of temperature is required to avoid voids The bond strength is in the range of the strength of the silicon crystal itself at temperatures of 1000 °C or higher This method is useful for the cases requiring the subsequent integrated circuit fabrication or other high temperature processes A high temperature bonding between two fused silica glasses is also possible at 1000 °C [27]

2 5 ANODIC BONDING

2 5 1 *Glass-to-silicon anodic bonding*

This method have been frequently used for fabricating microvalves and micropumps as described in the section 2 The perfect flatness of a glass wafer as well as a silicon wafer is required A pre-treatment of the glass, a removal of the damaged surface layer caused by the mechanical fishing is useful This is done by etching the glass surface using diluted HF for a few minutes Since the bonding takes place at more than 250°C that is necessary for alkaline ion migration, the thermal-expansion matches between glass and silicon are required to avoid bending after cooling The bend causes the unexpected deformation of a diaphragm or the sticking between structures fabricated in the wafers

Movable parts like a diaphragm or a valve seat can be prevented from bonding if it is covered with a thick oxide (1μm) layer or a TiW/Au metal layer Many glasses like a Pyrex glass (Corning #7740) are available for use The HOYA SD-2 glass shows almost perfect thermal-expansion matching to silicon The thermal expansions of silicon and these glassed are shown in Fig 3 Since the SD-2 glass has low alkaline atom content, a higher bonding temperature (>350 °C) and a higher applied voltage is necessary

171

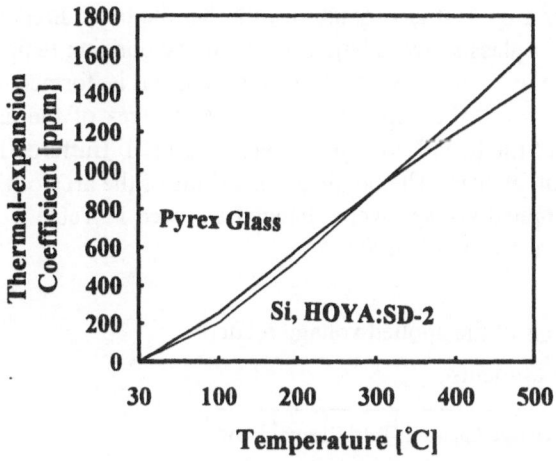

Figure 3. Thermal expansions of Si, Corning #7740 and HOYA #SD-2

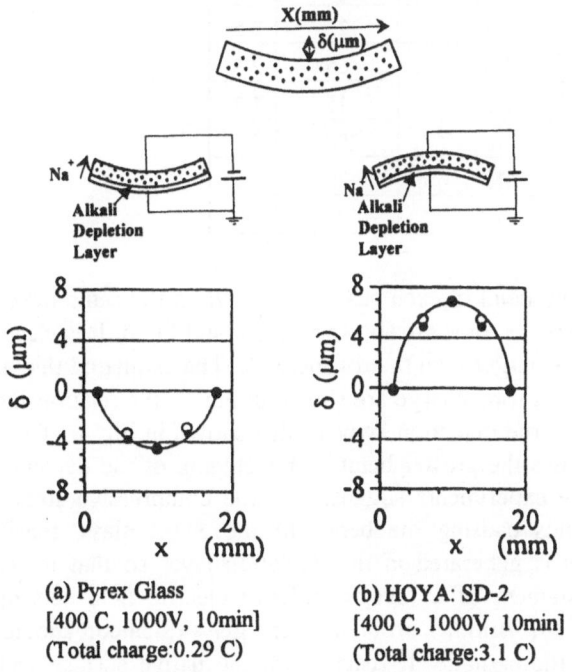

(a) Pyrex Glass
[400 C, 1000V, 10min]
(Total charge:0.29 C)

(b) HOYA: SD-2
[400 C, 1000V, 10min]
(Total charge:3.1 C)

Figure 4. Bend profiles of the glasses after applying a voltage of 1000 V at 400 °C.

Bending was observed just applying a voltage to the glass at the bonding temperature. The bend profiles of a Pyrex glass and a SD-2 glass of 2 mm square and 300

172

m thick are shown in Fig 4 It is very interesting that the bend direction are opposite It is obvious that a thinner glass shows a larger bend At the bonding temperature, alkaline ions like sodium move the cathode so that depletion layer is formed near the anode To estimate the thickness of the depletion layer, the changes of the etching rate from the surface to the bulk of the bend Pyrex glass were measured Buffered HF (50% HF NH$_4$F = 1 5) was used as an etchant The bonding conditions of the area of the cathode electrode and the period of applied voltage were changed as listed in Table 5 The applied voltage and temperature were 100 V and 400 °C

Table 5 Conditions of the applied voltage related
bend experiments

	Area of Cathode Electrode (mm^2)	Voltage Applying Period (min)	Total Charge (mC)	Bend (µm)
A	10x10	30	400	6 3
B	10x10	10	220	3 8
C	10x10	2 5	125	2 5
D	10x10	1 0	65	2 3
E	10x10	0 5	15	1 3
F	18x18	10	290	4 5
G	Probe	10	25	0 5
H	-	-	-	0

The total charge and bend of each cases are shown in the same table The etch depth vs time and the etch rate vs time are shown in Fig 5 and Fig 6 It is clear that the etch rate of the depletion layer is larger than that of the bulk The estimated thickness of the depletion layer and the bend vs total charge are shown in Fig 7 the relationships between the bend and the thickness of the depletion layer is also shown in Fig 8 The glass having greater depletion layer shows the greater bend After etching of the depletion layer, the bending disappeared These experiments suggest that the compressive stress is generated in the depletion layer, thus causing the bend In the SD-2 glass, tensile stress instead of compressive stress is generated in the depletion layer so that it bends to the opposite direction The testament of a reverse polarity electric field subsequent to the anodic bonding is useful [28] Another problem of the glass-to-silicon anodic is the deposition of alkaline atoms to the cathode It reacts with the native surface water to form alkaline hydroxide which in turn attacks the glass When another stack of silicon is required, this layer has to be removed by a diluted HF solution A time parallel bonding is a useful method to obtain the glass-silicon-glass stack [29]

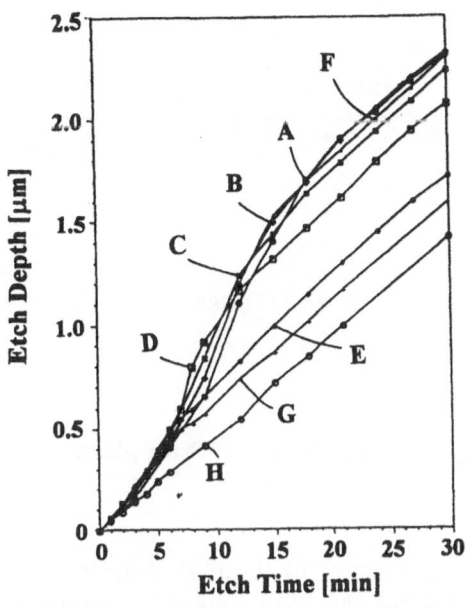

Fig.5 Etch depth vs. time of the Pyrex glasses after applying voltage

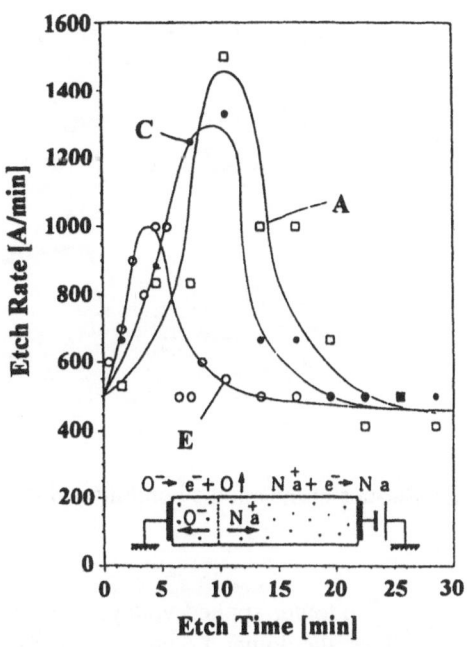

Fig.6 Etch rate vs. time of the Pyrex glasses after applying voltage

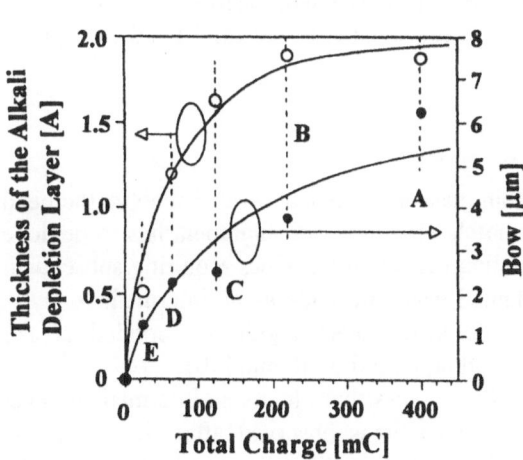

Fig.7 Relationships between the bends and the total charges

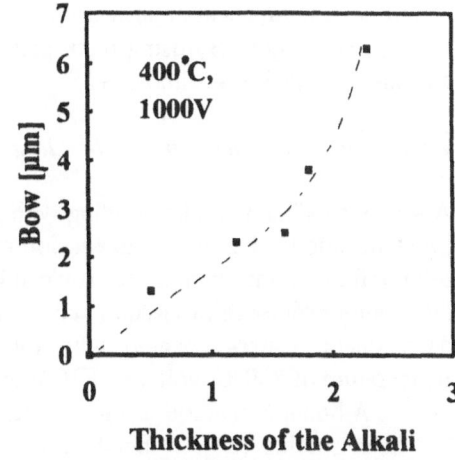

Fig.8 Relationships between the bends and the thickness of the depletion layer

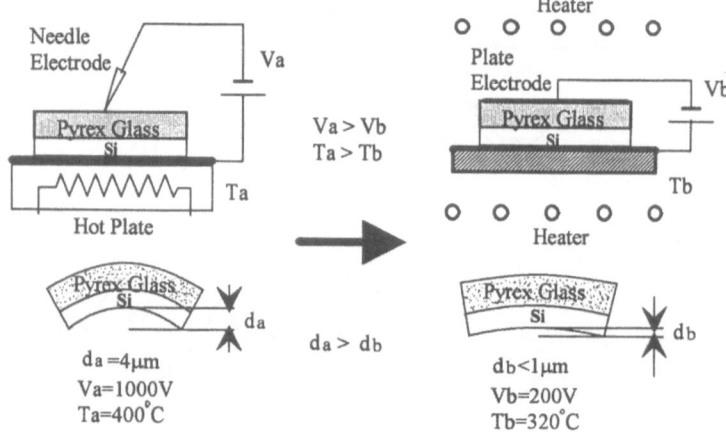

Fig.9 Set-ups of the anodic bonding and the bend profiles of the silicon-Pyrex glasses

The lower applied voltage reduces the bend caused by the alkaline depletion layer as well as the lower bonding temperature reduces the bend caused by the thermal-expansion mismatch. Since the temperature distribution in the glass could be a reason for the stress, the whole system should be heated up to the same temperature, by taking a wide cathode area and using a heating chamber as shown in Fig. 9, both the bonding temperature and the applying voltage were reduced to 320°C and 500V. The wide cathode is also effective to make the necessary bonding period shorter. A segmented annular electrode structure will be also useful [28]. The improved set-up enables smaller bend as mentioned in Fig.9. By using thick glass of 1.2 μm thickness, the total bend of a glass-silicon-glass stack was almost zero.

2.5.2. Silicon-to-silicon anodic bonding

A silicon-to-silicon anodic bonding with a thin glass intermediate layer can solve the bend problems due to the thermal-expansion mismatch. However, the alignment has to be done using infrared light microscopy A clean bonding suitable for devices requiring subsequent fabrication processes of circuits can be obtained using an oxide as an intermediate layer. Two silicon wafers covered with thin oxide layers were anodically bonded at the temperature of 850°C, voltage of 30 V and bonding period of 45 min [30].

A bonding method using a sputtered Pyrex glass thin film as an intermediate layer has been studied [31]. A bend of the wafer sometimes observed after the sputtering deposition. It is originated in the residual stress in the Pyrex film. The residual stress in the film is related to the gas [Ar (90%) + O2 (10%)] pressure during the sputtering. A compressive stress is generated at low gas pressure (<20 mTorr) and tensile stress at high pressure (>100 mTorr). No bend was observed under gas pressure of 30 mTorr even if the film was thicker than 1 um. The subsequent steam annealing under 565°C is effective to make the silicon rich Pyrex glass to be stoichiometric. To avoid discharge during bonding, a silicon oxide under layer is necessary. A selective bonding versus time relationships during the anodic bonding (at 375°C and the applied voltage of 100 V) are shown in Fig. 10. The current I is divided into a transient current I_C and a constant current $I_{R'}$. The transient current is considered to be displacement current of alkali ions (Na+) while the constant current is originated in the electrons flow through the gap.

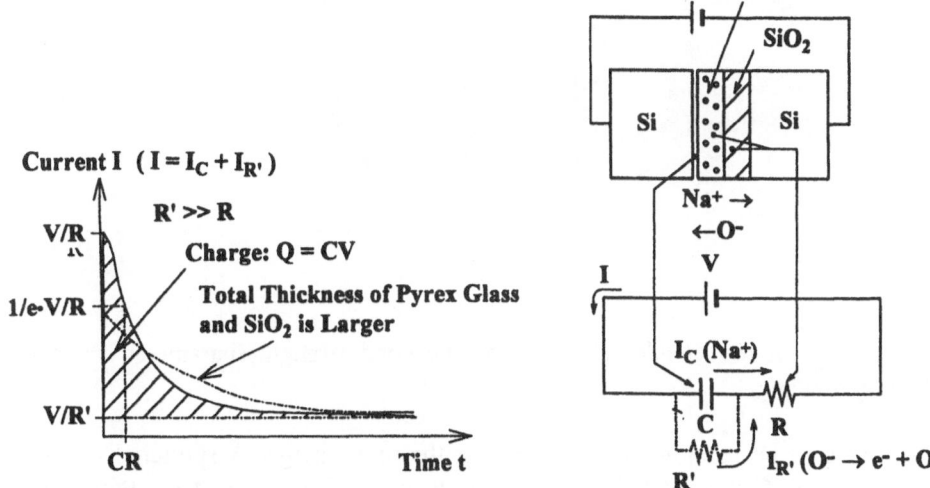

Fig.10 Typical current vs. time relationships during the anodic bonding

Fig.11 Equivalent circuit model of the anodic bonding

The migration of oxygen ion and oxygen generation at the interface is considered to be concerned. The total charge, integration of the transient current, was almost constant in cases of the different thickness of the oxide and Pyrex glass. From this point of view, both the Pyrex glass and the oxide act like a resistance at high temperature and the observed capacitance is existed only the gap between the Pyrex glass and the silicon. The equivalent circuit of this system is shown in Fig. 11. Numbers of the bonded samples classified by the bond strength with the parameter of the total charge during bonding are shown in Fig. 12. The bond strength depends on the total charge but do not depend on the thickness of the Pyrex glass. A reasonable bond strength is obtained at the charge par area greater than 3.0 mC/cm^2. The minimum thickness of the Pyrex layer which can be bonded was about 2000Å. This experiment indicated that the required temperature and applied

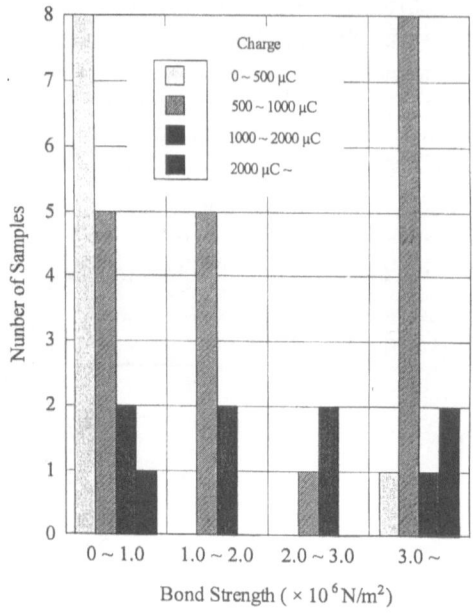

Fig 12 Numbers of the samples classified by the bond strength (parameter the total charge during bonding)

voltage are able to be recognised by monitoring the total charge A sputtered thin low melting point glass (Corning #7570) film is able to be used in stead of the Pyrex glass [32,33] The low viscosity of the #7570 at low temperature promotes the deformation of the glass so that expands the intimate contact The large relative applying voltage (40 V), the bonding was obtained The treatment of the wafer after sputtering is, however, critical due to the poor chemical stability of the film Perfect flatness and a very clean surface is required at room temperature bonding

3. Microconnector

Hybrid systems consist of some functional blocks mad on different wafers connecting their inlets and outlets by hermetically sealed ports A modified conventional connector using an O-ring can be applicable for this purpose Fig 13 shows a structure of a microconnector using an O-ring made of a photosensitive dry film This can be made by simple

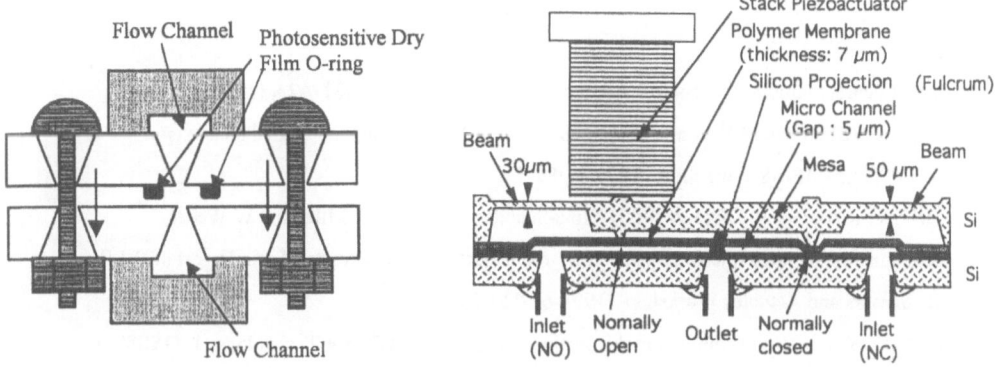

Fig. 13 Structure of the microconnector

Fig. 14 Structure of the separable channel type 3-way microvalve

photolithography Here, mechanical fixing with screws is necessary This method will be useful in larger scale systems, because the relative area of the screws becomes smaller

4. Separable channel type microvalves

Low cost and disposable systems are required in medical use In this case, microsystems consists of actuator parts and separable channel parts will be useful The channel part can be disposable while the actuator parts should be reused many times It is very economical because the actuating parts are generally more expensive than the channel part The structure of the separable channel type three-way microvalve is shown in Fig 14 [16] This structure is applicable to fabricate a sample injector or a chemical analysing cell [34] Large scale analysing systems can be realised on the wafer by applying this structure

5. Conclusion

Multi-level wafer stacks are one of the most useful structures for realising a μTAS Bonding methods reviewed here are the key technologies for the fabrication The anodic bonding of the glass-to-silicon and the silicon-to-silicon will still be the most promising methods A microconnector and a separable channel microvalve provide another possibility on the integration of flow devices Further studies of assembly methods are still necessary for realising a sophisticated μTAS

Acknowledgements
The authors wish to thank Mr Y Shoji, Mr K Suzuki and Mr T Shin in Tohoku University for their help

References

1) M Esashı, Integrated micro flow control systems, Sensors & Actuators, A21-A23, (1990), 161-167

2) S Nakagawa, S Shojı and M Esashı, A micro chemical analyzing system integrated on a silicon wafer, Proc of IEEE-MEMS Workshop, (1990-2), 89-94

3) M J Zdeblıck, R Anderson, J Jankowskı, B Klıne-Schoder, L Chrıstel, R Mıles and W Weber, Thermpneumatically actuated microvalves and integrated electro-fluidic circuits, Tech Digest of IEEE Solid-State Sensors and Actuator Workshop, (1994-6), 251-255

4) S Shojı, M Esashı,, Micromachınıng for chemical sensors, Chemical Sensor Technology, 1, (1988), 179-193

5) H T G van Lıntel,, F C M van de Pol and S Bouwstra, A piezoelectric micropump based on micromachıng of silicon, Sensors & Actuators, 15, (1988), 153-167

6) S Shojı, S Nakagawa and M Esashı, Micropump and sample-ınjector for integrated chemical analyzing systems, Sensors & Actuators, A21-A23, (1990), 189-192

7) T S J Lammerınk, M Elwenspoek and J H J Fluıtman, Integrated micro-lıquıd dosing system, Proc of IEEE-MEMS Workshop, (1993-2), 254-259

8) J G Smıts, Piezoelectric micropump with three valve workıng perıstaltıcally, Sensors & Actuators, A21-A23, (1990), 203-206

9) M Shıkıda and K Sato, Characterıstıcs of an electrostatically-drıven gas valve under high-pressure condıtıons, Proc of IEEE-MEMS Workshop, (1994-1), 235-240

10) H Mızoguchı, M Ando, T Mızuno, T Takagı and N Nakajıma, Design and fabrication of light drıven micropump, Proc of IEEE-MEMS Workshop, (1992-2), 31-36

11) M A.Huff, J R Gılbert and M A.Schmıdt, Flow characterıstıcs of a pressure-balanced microvalve, Tech Dıg of Transducers'93, (1993-6), 98-101

12) H Jerman, Electrıcally-actıvated, micromachıned diaphragm valves, Tech Dıg IEEE, Sensors & Actuators Workshop, (1990-6), 65-69

13) D Bosch, B Heımhofer, G Muck, H Seıdel, U Thumser and W Welser, A silicon microvalve with combined electromagnetıc/electrostatıc actuatıon, Sensors & Actuators A, 37-38, (1993), 684-692

14) R Zengerle, W Geıgel, A.Rıchter, Y Ulrıch, S Klııge and A.Rıchter, Application of micro diaphragm pumps ın microfluıd systems, Aktuator 94, (1994)

15) E Stemme and G Stemme, A novel piezoelectric valve-less fluıd pump, Tech Dıg of Transducers'93, (1993-6), 110-113

16) S Shojı, van der B H van der Schoot, N F de Rooıj and M Esashı, Smallest dead volume microvalves for integrated chemical analyzing systems, Tech Dıg of Transducers'91, (1991-6), 1052-1055

17) R.Rapp, W.K.Schomburg, D.Maas, J.Schulz, W.Stark, LIGA Micropump for gases and liquids, Sensors & Actuators A, 40, (1994), 57-61

18) S.C.Terry, J.H.Jerman and J.B.Angell, A gas chromatographic air analyzer fabricated on a silicon wafer, IEEE, Trans. Electron Device, ED-26, (1979), 1880-1886

19) V.Gass, B.H.van der Schoot and N.F.de Rooij, Integrated flow-regurated silicon micropump, Sensors & Actuators A, 43, (1994), 335-338

20) E.Verpoorte, A.Manz, H.M.Widmer, B.H.van der Schoot and N.F.de Rooij, A three-dimensional micro flow system for a multi-step chemical analysis, Tech. Dig.of Transducers'93, (1993-6), 939-942

21) S.Shoji and M.Esashi, Micro flow cell for blood gas analysis realizing very small sample volume, Sensors & Actuators B, 8, (1992), 205-208

22) C.D.Fung, P.W.Cheung, W.H.Ko and D.G.Fleming, Micromachining and micropackageing of transducers, Elsevier Science Publishers B. V., Amsterdam, (1985), 41-61

23) T. A. Knecht, Bonding techniques for solid state pressure sensors, the 4th Int. Conf. on Solid-state Sensors and Actuators, (1987-6), 95-98

24) H.J.Quezer, W.Benecke and C.Dell, Low temperature wafer bonding for micromechanical applications, Proc. IEEE-MEMS Workshop, (1992-2), 49-55

25) L.A.Field and R.S.Muller, Fusing silicon wafers with low melting temperature glass, Sensors & actuators, A21-23, (1990), 935-938

26) M.A.Schmidt, Silicon wafer bonding for micromechanical devices, Tech. Digest of IEEE Solid-State Sensors and Actuator Workshop, (1994-6), 127-131

27) D.Sobek, S.D.Senturia and M.L.Gray, Microfabricated fused silica flow chambers for flow cytometry, Tech. Digest of IEEE Solid-State Sensors and Actuator Workshop, (1994-6), 260-263

28) M.Harz, Anodic bonding for the third dimension, J. Micromech. Microeng. 2, (1992), 161-163

29) T.Rogers, Considerations of anodic bonding for capacitive type silicon/glass sensor fabrication, J. Micromech. Microeng. 2, (1992), 164-166

30) T.A.Anthony, Dielectric isolation of silicon by anodic bonding, J. Appl. Phys. 58, 3, (1985), 1240-1247

31) A.Hanneborg, M.Nese and P.Ohlckers, Silicon-to-silicon anodic bonding, Pro. of MME'90, Tech. Dig., (1990), 100-107

32) M.Esashi, A.Nakano, S.Shoji and H.Hebiguchi, Low-temperature silicon-to-silicon anodic bonding with intermediate low melting point glass, Sensors & actuators, A21-23, (1990), 931-934

33) M.Esashi, S.Shoji and A.Nakano, Normally closed microvalve and micropump fabricated on a silicon wafer, Sensors & Actuators, 20, (1989), 163-167.

34) S.Shoji, M.Esashi, N.F.de Rooij and B.H.van der Schoot, Micro flow control devices for integrated medical and chemical analyzing systems, Proc. of IMACS/SICE Int. Symp. on Robotics, Mechatronics and Manufacturing Systems'92, (1992-9), 617-622

MICROSYSTEMS FOR ANALYSIS
IN FLOWING SOLUTIONS

Bart H. van der Schoot[1], Elisabeth M.J. Verpoorte[2], Sylvain Jeanneret[1], Andreas Manz[2], Nico F. de Rooij[1]
[1]Institute of Microtechnology, University of Neuchâtel,
Rue A -L Breguet 2, CH-2000, Neuchâtel, Switzerland
[2]Ciba-Geigy, Corporate Analytical Research, CH-4002 Basel, Switzerland

Abstract

A three-dimensional modular setup for a miniaturized analysis system for flowing streams is presented The system uses silicon micromachined pumps and flow manifolds in combination with electrochemical sensors or optical detection Applications range from simple ion concentration measurements with ISFETs to a multi-step chemical analysis of phosphate Miniaturization of the flow systems leads to a substantial reduction in reagent consumption

Introduction

Miniaturization of systems for chemical analysis can lead to an improved efficiency with respect to sample size, response time and reagent consumption The extent of miniaturization that can be achieved with conventional machining techniques has its limits and therefore a growing interest exists in the use of microfabricated elements for fluid handling The technologies to make micromachined silicon structures such as pumps and valves and those to produce chemical sensors are in principle compatible [1] and thus it would be viable to integrate complete analytical instruments on a single silicon substrate However, a hybrid assembly is preferable over full integration as it provides a much greater flexibility and is more cost effective

For the automation of chemical analysis, incorporation of sample handling, treatment and detection into a flow system is a popular approach [2] These flow systems can be built up from a number of standard components such as reaction coils, T-junctions, pumps, valves, detectors, etc The use of standard modules allows for a flexible setup of the systems and can easily be adapted to a specific analysis Silicon micromachining lends itself very well for the fabrication of both passive and active components of small dimensions required for the miniaturization of flow systems As the base material is usually a silicon wafer, it will be clear that the thus fabricated components are in principle planar structures

A van den Berg and P Bergveld (eds), Micro Total Analysis Systems, 181–190
© 1995 *Kluwer Academic Publishers*

In order to minimize the volume associated with interconnections and to come to a robust setup, a vertical arrangement of elements is chosen [3] The construction of three-dimensional flow manifolds out of planar structures allows for a maximum flexibility in the interconnection of the elements Instead of using valves to introduce samples into the system, multiple pumps are used to control the flow in the system Although this increases the number of pumps required compared to a normal Flow Injection Analysis (FIA) system, it eliminates the need for a micromachined sample injection valve that is more difficult to realize

The basic setup of the analysis systems presented here is depicted in figure 1 A detector is placed at the inlet of the assembly A sample can be introduced by operating the sample pump and subsequently be washed out by using a second pump working in the opposite direction For more complex analysis schemes, the sample can be drawn into a flow manifold and mixed with one or more reagents After the reaction is completed, the reaction product can be transported back to be measured into the detector by the wash pump before being flushed out of the system completely

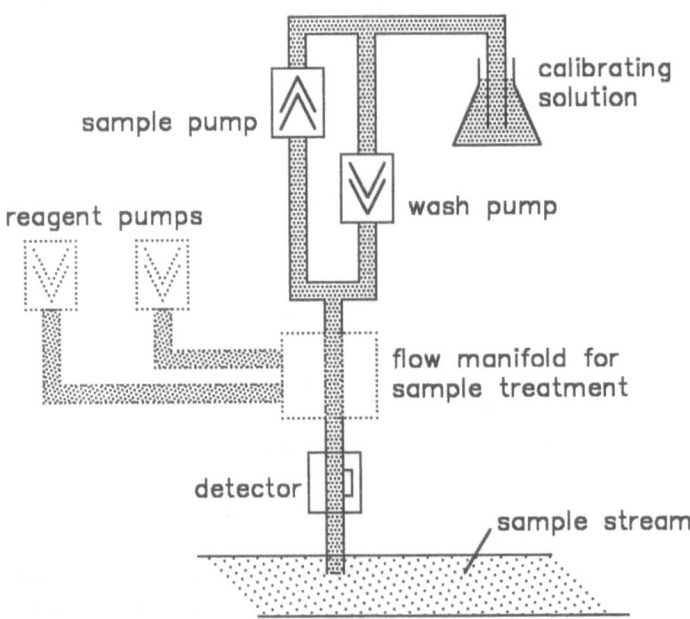

Figure 1   Basic setup for a miniaturized analysis system

The advantage of the setup as depicted above is that the sample does not enter the system beyond the flow manifold and does not reach the pumps where unfiltered solutions might interfere with the proper operation of the microfabricated pump valves

Hardware

The module size for the stacked analysis system is 22×22mm². This size is above all determined by the size required for the micro pumps described below. Also, this size allows the arrangement of 4 chips on a 3" wafer or 9 pieces on a 4" wafer with minimal loss of material.

Pumps

The micropumps used in construction of the analysis systems described here are piezo-electrically driven membrane pumps. The design is similar to that of Van Lintel *et al.* [4]. It consists of a micromachined silicon part anodically bonded between two Pyrex glass plates. The thicker of the glass plates (1.5mm) serves as the base plate on which the valves close. The thinner glass plate (0.3mm) forms the pump membrane, driven by a ceramic piezo disc (Philips PXE5, 10mm diameter, 0.2mm thickness) which is glued onto the glass using a conductive epoxy. The silicon part forms two passive valves on the glass base plate. A 1μm layer of silicon dioxide on the valve seats prevents bonding to the glass when the pumps are assembled. The operating principle of the pump is presented in figure 1. The pumps are operated with a driving voltages up to 300 Volt at frequencies up to several hundred Hz. Although pump rates of 1mL/min are achievable, the applications described typically use flow rates in the order of 20-200 μL/min. The operation of the pumps at elevated frequencies (>20 Hz) ensures that the influence of remaining flow pulsation on the detector signal can easily be removed by appropriate filtering.

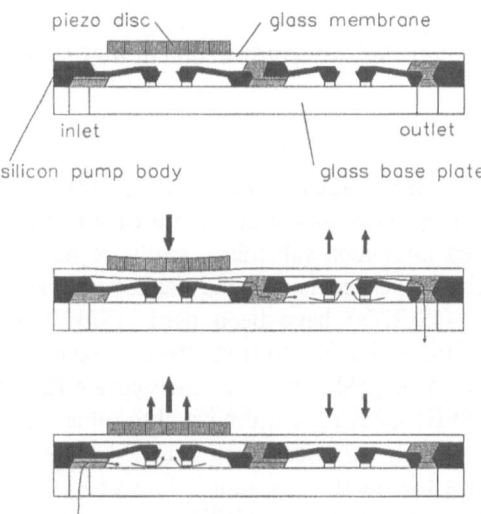

Figure 2.   *Operation of the piezo-electric micropump*

Flow Manifolds

As discussed, flow manifolds for the miniaturized analysis systems are constructed by stacking of planar elements Next to the pumps, a number of different elements with liquid channels and through holes have been realized by anisotropic etching of silicon All elements measure $22 \times 22 mm^2$, channels are 600μm wide across the top and 200μm deep, the through holes are $1 \times 1 mm^2$ at their widest points Fluid connections to the pumps are made at the corners The channel structures have eight holes around the circumference and a ninth hole in the center, thus allowing a maximum flexibility in the setup Figure 3 shows some examples of the elements that are used to construct the flow manifolds

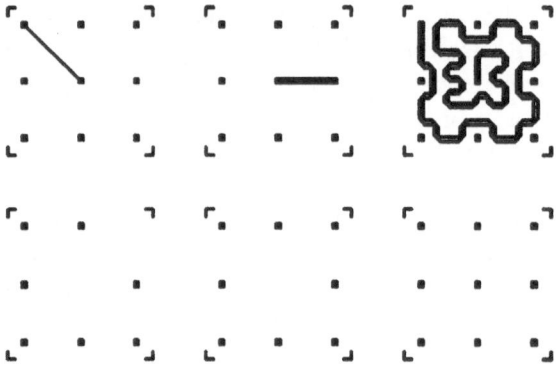

Figure 3.   Layout of some planar silicon chips for construction of flow manifolds

Detectors

Most important for the application in miniaturized systems is that the detector is capable of measuring in small volumes Electrochemical sensors are easy to miniaturize and can, especially when microfabricated on silicon substrates, easily be built into the modular setup presented here For the first experiments with the stacked analysis system, Ion Sensitive Field Effect Transistors (ISFETs) have been used ISFETs have an intrinsically rapid response and are therefore well-suited to study the time response of the system as a whole, as this is not limited by a sluggish sensor To incorporate ISFETs into the system, silicon chips with multiple ISFETs, having a width equal to other elements in the stacked system, are equipped with a polysiloxane sealing ring [5] Thus, a flow-through channel with a cross section of a few tenths of a mm is formed by placing the detector cell against the next element in the stack The ISFETs can be used for the measurement of pH or can be adapted with ion sensitive membranes for the measurement of other species [6]

Optical detection is another method that can well be adapted to the measurement in small volumes [7]. With the example that will be described here, the analysis of phosphate, the detector has not yet been incorporated in the stack of elements. Instead, a capillary detector cell with a 1 mm optical path length is placed at the entrance of the system.

Measurements

Ion concentration measurements with ISFETs

In its most elementary version as depicted in Figure 1, a miniaturized analysis system has been used for the measurement of ion concentrations with ion sensitive field effect transistors (ISFET). The system comprises two pumps and two multi-ISFET detector cells, one of which acts as a reference. Preliminary experiments have been performed with a functional model of the system where the individual elements were connected with lengths of silicone rubber tubing. Measurement of potassium concentrations using an ISFET with a valinomycin / PVC membrane in a glass flow through cell showed the feasibility of the setup [8].

A truly stacked version of the analysis system has been used for the measurement of pH. The setup for these measurements is shown in figure 4.

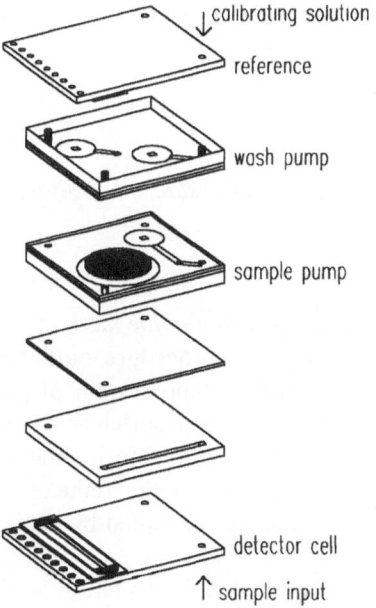

Figure 4.   *Arrangement of the modules in the stack for pH-measurements*

Figure 5 gives the response of the system for samples of pH 3, 4, 5 and 6 with a baseline solution of pH 7 Sample time is 8 seconds, wash time is 14 seconds and the flow rate for both pumps is approximately 200 μl/min The peak height is equal to about 50 mV/pH and the sample time is long enough to reach equilibrium Sample size is thus about 27μL and the consumption of baseline solution equals 3 3 mL/h This is up to two orders of magnitude less than for a more conventional flow injection system However, sample size and flow rate are considerably larger than with earlier reported measurements in a glass detector cell [8] Proper design of the detector cell is thus of the greatest importance

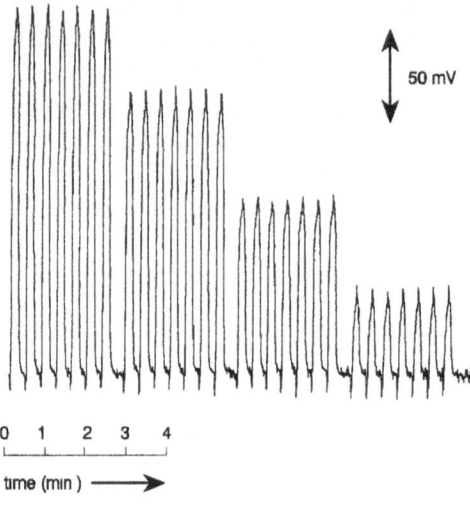

Figure 5.   *Output signal for the measurement of pH in the stacked analysis system depicted in Fig.4.*

Measurement of phosphate concentration using the molybdenum blue method
        Phosphate in water may be determined according to a procedure outlined in [9], known as the "molybdenum blue method" It involves the complexation of phosphate with molybdate, with subsequent reduction of the complex with ascorbic acid The result is a complex having an intense blue color The overall reaction rate is limited by the complexation step, with maximum conversion of phosphate to the reduced complex requiring about 10 minutes This analytical procedure has been adapted by many groups for phosphate analysis in flow systems (see, for instance, [2, 10]) In one instance, a system was developed to monitor phosphate concentrations in fermentation broths [11] The flow manifold employed in that application is the model for the phosphate analysis using a stacked system described in this paper

The micro flow system used for this measurement is basically the same as given in figure 1. Next to the sample and wash pumps, two additional pumps are used to supply solutions of molybdate and ascorbic acid. Between the pumps and the detector, a mixing coil of approximately 30 microliters is located. Here the reaction mixture is held for 40 seconds to let the reaction proceed before it is flushed back through the detector. The detector is a capillary detector cell of 1 mm pathlength, interfaced to a UV-visible spectrometer by means of optical fibers. The setup of the system is given in Figure 6.

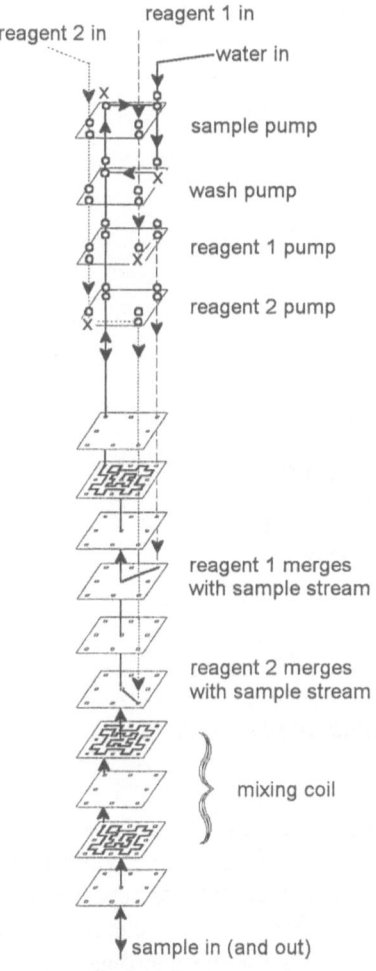

Figure 6   *Stacked modular system for the measurement of phosphate. The optical detector is located at the bottom of the stack.*

Since the analysis is being carried out under non-equilibrium conditions, measured peak intensity should show a strong dependence on the amount of time the sample spends in the system Analysis is carried out as follows 1) During 60 seconds the system is flushed with water from the wash pump 2) A sample of approximately 90 μL is drawn into the system by the sample pump 3) Wash and reagent pumps are operated simultaneously for 7 seconds during which period 25 μL of the sample is mixed with the reagents 4) The flow is stopped for 40 seconds to let the reaction proceed 5) The reacted sample plug is washed out of the system and is measured while passing the detector The total measurement cycle thus takes about 4 minutes Response to phosphate was found to be linear over the range of concentrations given in Figure 7, with a correlation coefficient, r, of 0 9991

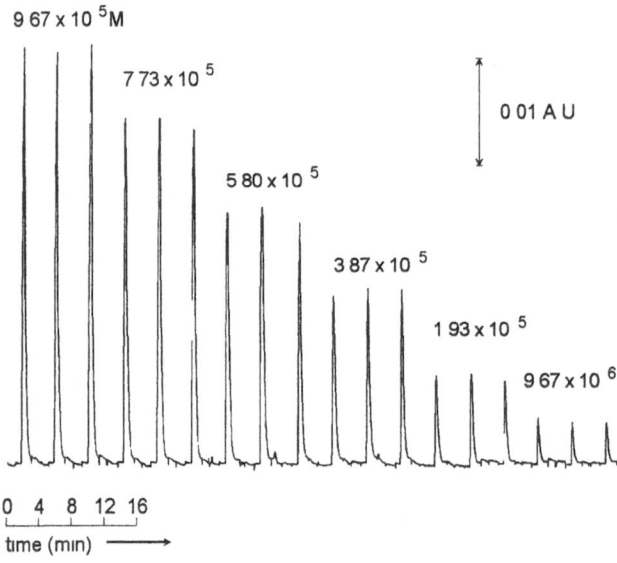

Figure 7.   Analysis results for the determination of phosphate

The phosphate analysis described here is characterized by an analysis time of four minutes, when system flushing is included This is in contrast to the conventional flow injection analysis of phosphate using the molybdenum blue method reported in [2] and [10] These two studies indicate analysis times of 30 and 80 seconds, respectively Several factors play a role in the longer analysis times for the micromachined system First is the incorporation of a time-based injection technique in the system, as opposed to the volume-defined methods using valves described in [2] and [10] While elimination of the need for an injection valve simplifies the system, time-based injection schemes by definition contribute to longer analysis times A second factor is the design of the stacked manifold itself, which requires that sample both enter and exit it at the same location This differs from conventional systems, where a sample is introduced at one end and detected at the other Thorough flushing of the stack is therefore required between samples to avoid carryover from one sample to the next This precludes injection of a second sample while the first is still being detected, a time-saving measure implemented in [2] Optimization of the stacked system's design, taking these factors into consideration, as well as a further reduction in overall system volume, will lead to improved analysis times

A comparison of the flow manifold reported in [11] and the μ-TAS reveal that dramatic decreases in reagent and solution consumption can be achieved upon implementation of the μ-TAS Whereas the system in [11] has a volume of between 1 and 2 mL, that of the μ-TAS reported here is about 90 μL However, the amount of reagent required for an analysis drops by a factor of 270 This is due in part to the flow rates in the μ-TAS being one-tenth of the corresponding rates in the conventional system, as mentioned above The additional decrease in reagent consumption is a result of the stop flow procedure applied, which saw addition of reagents over a period of only 7 seconds during an analysis In contrast, the nature of the system in [11] dictated that reagent pumps ran continuously throughout an analysis Thus, where 1 7 mL of ascorbic acid was needed in the conventional system (0 56 mL/min over 3 min), only 6 2 μL was required in the μ-TAS analysis Similarly, only 3 0 μL of molybdate reagent were added in the μ-TAS, as opposed to 0 8 mL in the larger system Use of piezoelectrically driven micro pumps allows reproducible addition of microlitre amounts of solution in small, sub-microlitre steps This is difficult to impossible to achieve with most commercially available pumps, whose lowest volume resolutions lie in the microlitre range

Conclusion

Initial experiments have demonstrated the feasibility of flow manifolds having a stacked configuration for chemical analysis systems A substantial reduction in system size may be accomplished when micromachined silicon elements are employed Implementation of the concept of merging zones of sample and reagents and the valveless injection scheme result in more efficient consumption of reagents In order to benefit fully from the possible advantages of miniaturization, a careful design of the components as well as the analysis methods is required

Acknowledgments

The authors thank M Garn, A Spielmann, and S Haemmerli for several very helpful discussions and P Thiébaud for performing the pH measurements Part of this work has been financially supported by the Commission for the Promotion of Scientific Research (CERS) and the Swiss Foundation for Microtechnology Research (FSRM)

References

1 H H van den Vlekkert, N F de Rooij, A van den Berg, A Grisel, Multi-ion Sensing System Based on Glass-encapsulated pH-ISFETs and a Pseudo-REFET, *Sensors and Actuators, B1* (1990) 395-400

2 J Ruzicka, E H Hansen, Flow Injection Analysis (2nd Ed), New York John Wiley and Sons, 1988

3 J C Fettinger, A Manz, H Ludi, H M Widmer, Stacked modules for micro flow systems in chemical analysis concept and studies using an enlarged model, *Sensors and Actuators B*, 17 (1993) 19-25

4 H T G van Lintel, F C M van de Pol and S Bouwstra, A piezoelectric micro pump based on micromachining of silicon, *Sensors and Actuators* 15 (1988) 153-167

5 B H van der Schoot, S Jeanneret, A van den Berg, N F de Rooij, Modular setup for a miniaturized chemical analysis system, *Sensors and Actuators B*, 15-16 (1993) 211-213

6 P D van der Wal, A van den Berg, N F de Rooij, Universal approach for the fabrication of Ca^{2+}-, K^{+}- and NO^{3-}- sensitive membrane ISFETs, *Sensors and Actuators B*, 18-19 (1994) 200-207

7 E M J Verpoorte, A Manz, H Ludi, A E Bruno, F Maystre, B Krattiger, H M Widmer, B H van der Schoot and N F de Rooij, A silicon flow cell for optical detection in miniaturized total chemical analysis systems, *Sensors and Actuators B*, 6 (1992) 66-70

8 B H van der Schoot, S Jeanneret, A van den Berg, N F de Rooij, A silicon integrated miniature chemical analysis system, *Sensors and Actuators B*, 6 (1992) 57-60

9 J Murphy, J P Riley, A modified single solution method for the determination of phosphate in natural waters, *Anal Chim Acta*, 27 (1962) 31-36

10 P Linares, M D Luque de Castro, M Valcarcel, Sequential automatic on-line determination of aquiculture nutrients phosphate and nitrate, *Journal of Automatic Chemistry*, 14 (1992) 173-175

11 A Spielmann, M Garn, S Haemmerli, A Manz, H M Widmer, Connecting analytical systems to bioreactors the monitoring and control of biological processes, *Proceedings of the Seminar of the Swiss Committee of Analytical Chemistry*, Berne, Switzerland 16 October 1992, p An 49

COMBINED BLOOD GAS SENSOR FOR pO$_2$, pCO$_2$ AND pH

Ph. Arquint, B.H. van der Schoot and N.F. de Rooij
Institute of Microtechnology, University of Neuchâtel
Rue A.-L. Breguet 2
CH-2000 Neuchâtel, Switzerland

0. Abstract

A miniaturized chemical analysis system for extracorporeal monitoring of blood pO$_2$, pCO$_2$ and pH is presented, combining a silicon-based sensor chip and an integrated flow-through channel. Classical electrochemical sensing principles are used, realized in a planar form. The sensor is fabricated entirely on wafer level using IC-compatible processes. By integrating a flow-through channel directly on chip, the sample size and the reagent consumption are drastically reduced. The device characterization has been performed in aqueous solutions, blood intended for transfusion and in whole blood. The sensor exhibits an excellent linearity, low drift and a functional lifetime of more than 2 months.

1. Introduction

The rapid development of silicon technology has strongly stimulated the fabrication of miniaturized electrochemical sensors based on solid-state devices. The largest effort has been made in the biomedical area where the drive is to monitor at the bedside important parameters (blood gases, K$^+$, Na$^+$ and glucose) thus avoiding time consuming centralized laboratory analysis. For all these species miniaturized electrochemical sensors based on Ion-Sensitive Field Effect Transistors (ISFETs) [1] or thin-film planar amperometric cells [2] have already existed for several years, however, their practical applicability remains rather limited. This is mainly due to the lack of longer term sensor stability which requires frequent sensor recalibrations. This important drawback can be overcome by incorporation of the sensor into a total chemical analysis system [3]. Such a complex analysis system, consisting of silicon μpumps, μvalves, sensors and electronics, performs sampling, calibration and signal processing. Due to its extremely small size, it offers improved efficiency with respect to sample volume, reagent consumption and response time. Thus, it is a promising tool for a handheld or bedside analyzer suitable for clinical applications.

191

A. van den Berg and P. Bergveld (eds.), Micro Total Analysis Systems, 191–194.
© 1995 *Kluwer Academic Publishers.*

2. Sensor Chip Design and Fabrication

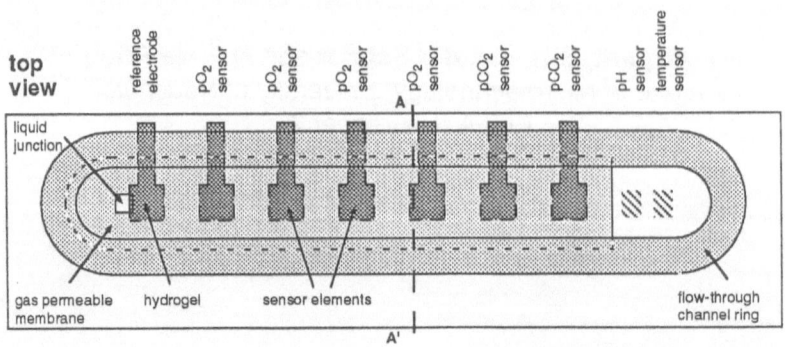

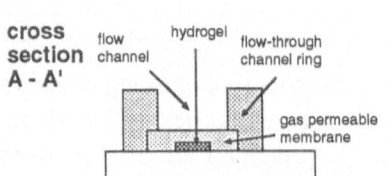

Figure 1: Top view and cross section of the pO_2, pCO_2 and pH sensor chip (22 mm x 6 mm), showing the nine sensing elements aligned in the flow-through channel

The sensor chip provides a flow-through channel directly on chip, which is defined by a 600 μm thick polymeric ring (Fig. 1). In the channel, having an internal volume of 15 μl, nine individual sensing elements are aligned in a row, featuring a reference electrode, four amperometric pO_2 sensors, two ISFET-based pCO_2 sensors, one pH-ISFET sensor and one temperature sensor. The sensor cells, apart from the pH sensor, which is not covered, consist of a 1 mm x 1 mm hydrogel layer (HG), which is centered on top of the solid-state transducing element.

The Clark-type pO_2 sensor provides a working electrode which is an array of 9 x 8 small Pt electrodes with 5 μm in diameter and with a spacing of 100 μm, connected in parallel. The working electrode, the Pt counter electrode and the Ag/AgCl reference electrode are covered with a HG layer, which is separated from the sample by a gas permeable membrane (GPM). The Severinghaus pCO_2 sensor consists of a pH ISFET and a Ag/AgCl reference electrode, both covered with the HG and the GPM. The reference electrode for the pH sensor has a similar layout as the one for the pCO_2 sensor, but has an opening in the GPM of 0.5 mm x 0.5 mm, which is used as the liquid junction. As pH transducing element, an uncovered pH-ISFET is used.

Figure 2: Photograph of the completed device

The fabrication of the sensor devices is performed entirely on wafer level in six main steps: First, the ISFETs are processed by standard IC technology Then, Pt is deposited by evaporation and structured The Pt is covered with a passivation layer (Si₃N₄). This layer is opened in the active zones of the Pt electrodes by H_3PO_4 etching. In the 3rd step, 10 µm of Ag is electrodeposited on certain of the Pt structures, which will later serve as Ag/AgCl reference electrodes The Ag layer is subsequently partially chloridized by electrochemical means In the 4th to the 6th step, the polyacrylamide HG (30 µm thick), the polysiloxane GPM (20 µm) and the polysiloxane ring (600 µm) are individually structured. Photolithographic patterning is used to deposit these three polymeric layers Prepolymeric solutions are cast on the wafer and polymerized by exposing to UV light using a photomask. In a subsequent development step, the unexposed parts are removed.

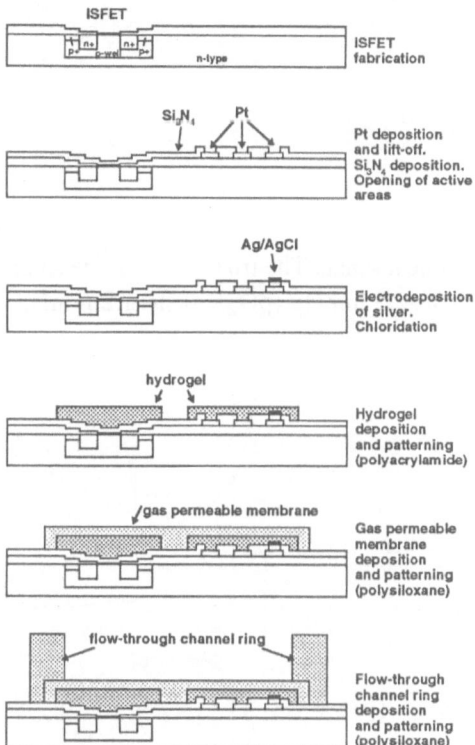

Figure 3. Main fabrication steps

3. Sensor Characterization

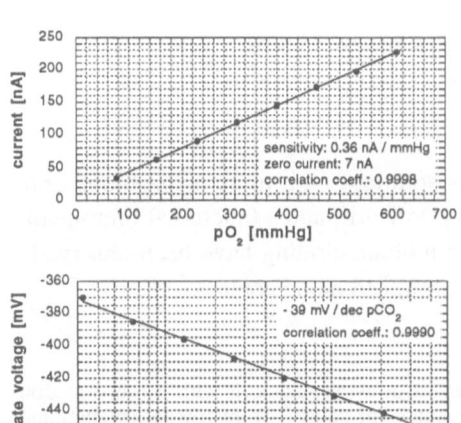

Figure 4. Sensor calibration curves

pO₂: *in 100 mM KCl solution equilibrated with O_2 / N_2 gas mixtures,*

pCO₂. *in phosphate buffer solution (pH 7 4) where NaHCO₃ is added to adjust the pCO₂,*

pH. *in phosphate buffer solution having different pH, but a constant chloride concentration of 50 mM KCl The signal is drift corrected, by means of an intermittent calibration at pH 7 8*

A high linearity is observed for all three sensors (Fig 4) in a broad range. The response time t_{95} (for a 95 % response) is

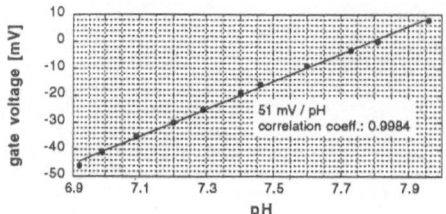

1 to 2 min. The device can be operated continuously for over 2 months. The stability is ±2 %/day (pO_2) and 1-2 mV/day (pCO_2, pH).

The simultaneous performance of the sensors is tested with commercial blood gas quality control solutions (Fig. 5) having defined pO_2, pCO_2 and pH values in the clinical range. The true values of the samples are reproduced (n=10) with an accuracy of -1.3/+2.5 mmHg (pO_2), -1.3/+1.2 mmHg (pCO_2) and -0.005/+0.006 (pH).

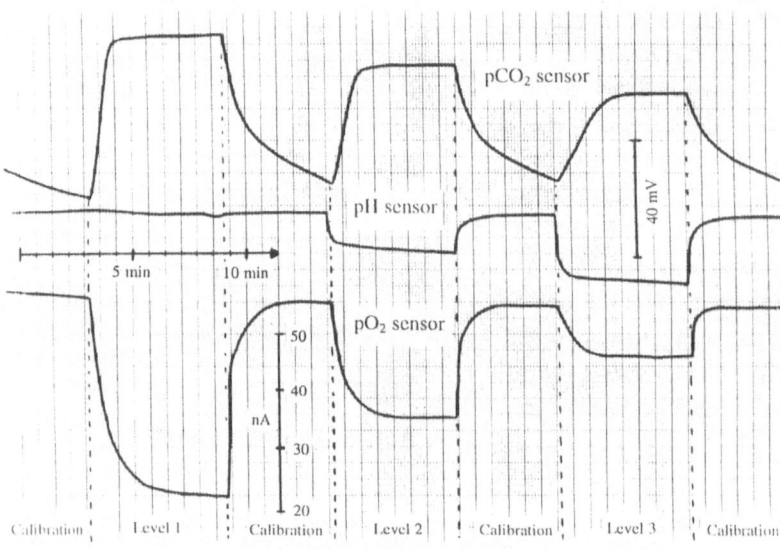

Figure 5: Simultaneous response using quality control solutions ("Level 1" to "Level 3"). Between the samples, the channel is rinsed with an air-saturated phosphate buffer of pH 7.2 ("Calibration")

Note: the pCO_2/pH sensor have a mV scale, the pO_2 sensor a nA scale

4. Conclusion

The sensor device has been tested with aqueous solution, with transfusion blood and with whole blood, exhibiting an excellent linearity, low drift and a functional lifetime of more than 2 months. Neither problems related with blood clotting have been observed, nor a degradation of the sensor performance with continued blood contact.

References
1 A. Sibbald, P.D. Whalley, A.K. Covington, A miniature flow-through cell with a four-function ChemFET integrated circuit for simultaneous measurements of potassium, hydrogen, calcium and sodium ions. Anal. Chim. Acta 1984; 159: 47-62.
2 W.J. Butner, G.J. Maclay, J.R. Stetter, Microfabricated amperometric gas sensors with an integrated design. Sensors & Materials 1990; 2: 99-106.
3 B.H. van der Schoot, S. Jeanneret, A. van den Berg and N.F. de Rooij, Microsystems for flow injection analysis. Analytical Methods & Instrumentation 1993; 1/1: 38-42.

A FLUID HANDLING AND INJECTION MICROSYSTEM FOR A µTAS

N. Croce, M. C. Carrozza, P. Dario
Scuola Superiore Sant'Anna
via Carducci, 40, 56127 Pisa, Italy

Abstract

A fluid handling and injection system for a micro total chemical analysis system (µTAS) for monitoring water conditions is presented in this paper. In general a µTAS is obtained by the integration of the following components: a) a fluid handling and injection system; b) a selective sensor to detect and measure the concentration of chemical species; c) an electronic unit to control pumps, valves and sensors. The fluid injection system described in this paper has been designed and realized in order to obtain optimal hydraulic behaviour, to simplify fabrication and to achieve good reliability. A basic configuration of the proposed µTAS comprises: micropumps, active microvalves, flow channels and conduits for reactors. The above components can be combined so as to obtain different configurations intended for sophisticated analysis processes.

Introduction

The realization of a micro total chemical analysis system (µTAS) permits to perform *in situ* chemical analysis with many advantages in terms of low reagent consumption, short response time and small sample volume. Many different approaches have been described so far to design and fabricate the single components and/or the whole system of a µTAS [1] [2]. In this paper we present a fluid handling microsystem realized by hybrid technology: stereolithography and silicon bulk micromachining. The main potential advantage of this approach is to provide high modularity and thus to allow to realize a more flexible and cheaper µsystem. A second (related) advantage is to allow more freedom in the optimization of the pure hydraulic performance of each component. The basic components of the hydraulic system we are designing are: micropumps, active microvalves, pressure tanks, flow channels and tubing for reactors, mixing

195

A. van den Berg and P. Bergveld (eds.), Micro Total Analysis Systems, 195–198.

chambers and detection chambers and tanks for sample, calibration solution and chemical reagents

The micropump and active microvalves are driven by a piezoelectric disk glued to a thin brass plate The micropump body is realized by stereolithography, whereas the active microvalve body and the channels are fabricated by silicon etching

The connection between the various components is obtained by stacking a number of silicon wafers in order to obtain a modular and compact three-dimensional structure In analogy with macro fluidic circuits, the microsystem is realized by assembling many fluidic microcomponents fabricated on silicon wafer in order to obtain a useful configuration

Description of Components

The micropumps and microvalves are driven by two piezoelectric unimorphs, each composed by a piezoelectric ceramic disk glued to a thin brass plate The shape of the actuator plate has been studied by FEM simulation in order to optimize the performance of the micropump and microvalves Results of this simulation have been presented elsewhere [3] The micropump actuator is driven by a sinusoidal voltage while the microvalve actuator is controlled by a dc voltage and a switch The power supply of the microsystem has been miniaturized and work is in progress to realize a dedicated control The pump body includes the chamber and two passive valves to direct the flow The pump chamber and the passive valves are made out of UV photocurable polymer material and manufactured by a single stereolithography process The design of the pump chamber has been based on optimum hydraulic criteria, and aimed to obtain long life and quick, cheap and easy manufacturing Stereolitography has been selected since it allows the microfabrication of 3D structures of any complex shapes even incorporating "integral" movable parts, that may require no assembly The pump has been tested experimentally [3] a flow rate of 45 μl/s at zero mm delivery head and a value of 2,460 mm delivery head with 4μl/s flow rate have been obtained (see Figure 1) These values are among the highest ever reported in literature for micropumps of similar size and working principle

Flow channels have been fabricated by wet etching of a silicon <100> wafer with a KOH aqueous solution The microvalve body is formed by a cylindrical chamber realized in a Plexiglas plate This plate is glued to a silicon wafer on which flow channels have been previously etched as depicted in Figure 2a The wet seal is obtained by a rubber plug located on the actuator plate

In order to obtain a regular and continuous flow rate a pressure tank has been connected to the hydraulic circuit A photograph of a part of the fabricated microsystem is shown in Figure 3

Conclusions

In this paper we have presented a fluid handling microsystem realized by hybrid technology: stereolithography and silicon bulk micromachining. The basic components of the microsystems have been fabricated and tested and some

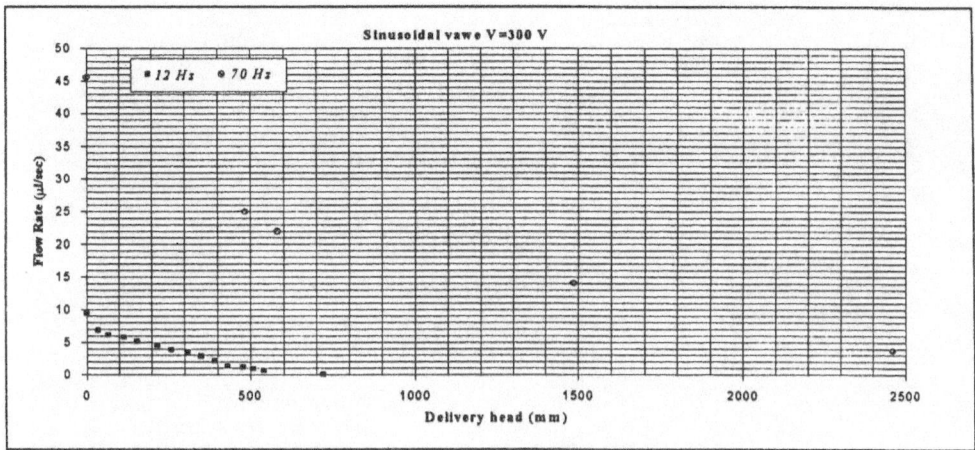

Figure 1. Flow rate vs frequency of the micropump

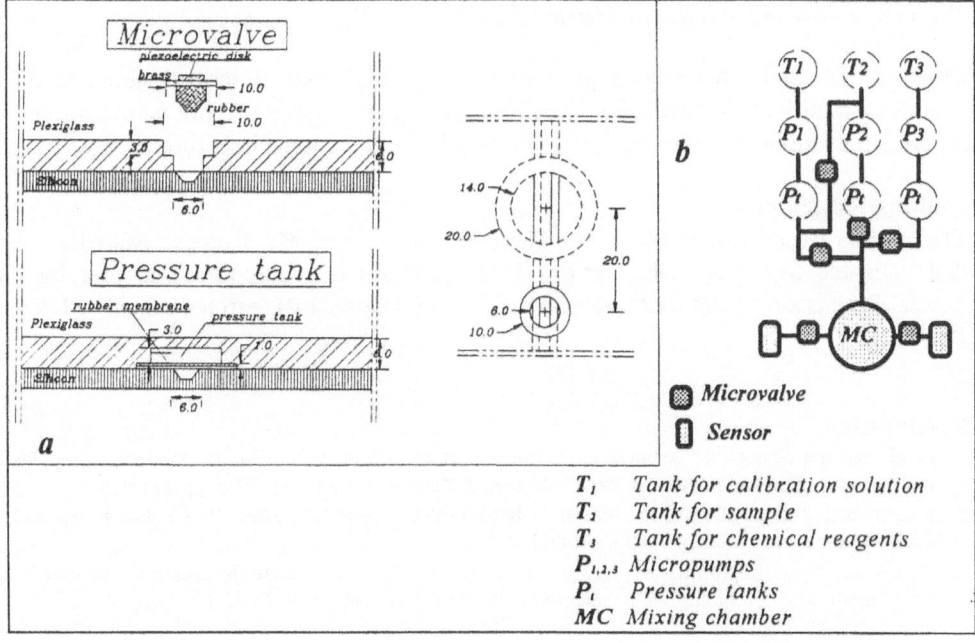

Figure 2. Microvalve and pressure tank. Hydraulic circuit of the integrated microsystem.

198

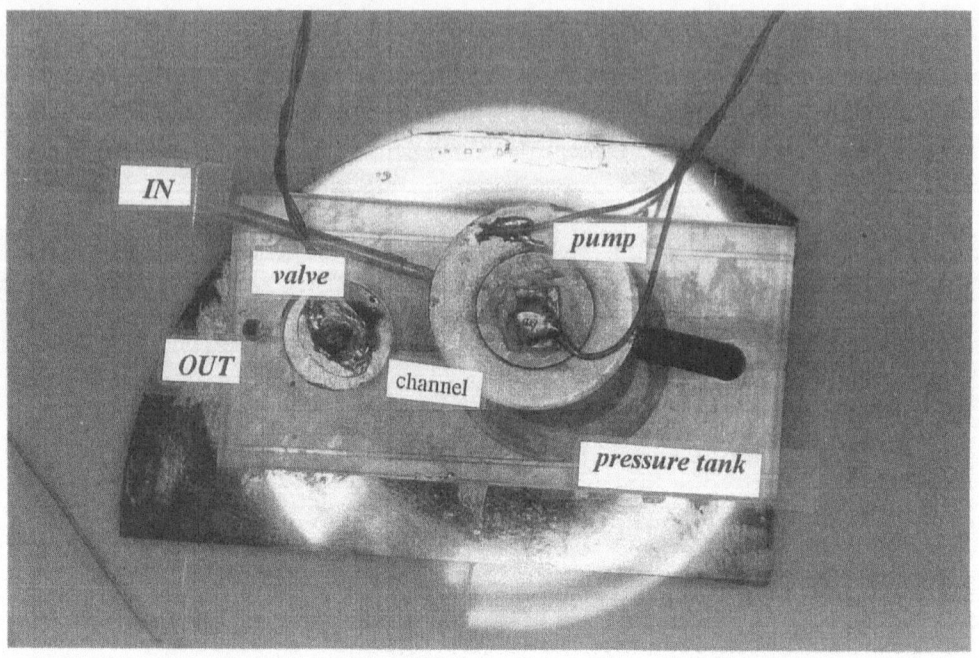

Figure 3. A view of microfluidic system

experimental results have been presented The components described above can be
assembled in order to realize a fluid handling and injection system Work is in progress to
realize a fully integrated microsystem, whose scheme is depicted in Figure 2b

Acknowledgements
The authors would like to thank Mr Carlo Filippeschi and Mr Roberto Martorana for
their valuable technical collaboration This work has been supported in part by the
Special Project on "Solid State Electronics" of the National Research Council (CNR) of
Italy

References
1 B H van der Schoot, S Jeanneret, A van den Berg and N F de Rooij, "Modular setup for a
   miniaturized chemical analysis system", *Sensors & Actuators B 15-16* (1993), pp 211-213
2 P Gravesen, J Branebjerg O S Jensen, "Microfluidics A Review", *MME'93 Workshop September
   7-8 1993, Neuchatel, Switzerland* pp 143-164
3 P Dario. M C Carrozza, N Croce, B Magnani,"A Piezoelectric Micropump Realized by
   Stereolitography *Actuator '94 Proceedings Bremen, Germany, June 15-17 1994,* pp 42-45

DESIGN OF AN ADAPTIVE UNSUPERVISED HYBRID MICROSYSTEM FOR ARTIFICIAL OLFACTION

Fabrizio A.M. Davide, Corrado Di Natale and Arnaldo D'Amico
Department of Electronic Engineering, University of Rome "Tor Vergata",
Via della Ricerca Scientifica, 00133 Roma, Italy

0. Abstract

The design of a microsystem for odor classification is introduced, which is composed of a self-organizing artificial neural network (SOM) and a QMB sensor array Fabrication requires micromachining, integrated optics and GaAs/Silicon technology High parallelism, novel pulse-stream circuitry and high degree of integration are the most remarkable qualities that are reported and commented

1. Introduction

This paper outlines a microsystem for odor identification and classification A self-organizing artificial neural network (SOM) and a sensor array for gas sensing [1] form the basis of the microsystem It is able to work as a stand alone system, suitable both for industrial and biomedical applications

Should the system be placed in an environment of interest, the learning algorithm of the SOM network processes the sensor outputs step by step, providing a statistical identification of the environment in an "unsupervised" fashion, i e without any supplementary information other than that supplied by the sensor array [2] Odor classification is accomplished exploiting a set of internal synthetic categories, which is recursively constructed by the network during the lifetime of the system This set contains a model of each kind of odor detected by the sensor array, whose accuracy is weighted on the basis of the statistical recurrence of the odor

The resulting performances [2] are qualitatively conform the biological standards and very encouraging if compared to those of non-adaptive systems

2. Microsystem design

Figure 1 shows the GaAs based microsystem, thought in principle as two separate sections which can eventually be integrated according to the perpendicular kind of applications Further, Si/GaAs technologies can be combined as well, depending on both the required cost/performance ratio and the specific environment where the microsystem has to be installed

A van den Berg and P Bergveld (eds), Micro Total Analysis Systems, 199–202
© 1995 *Kluwer Academic Publishers*

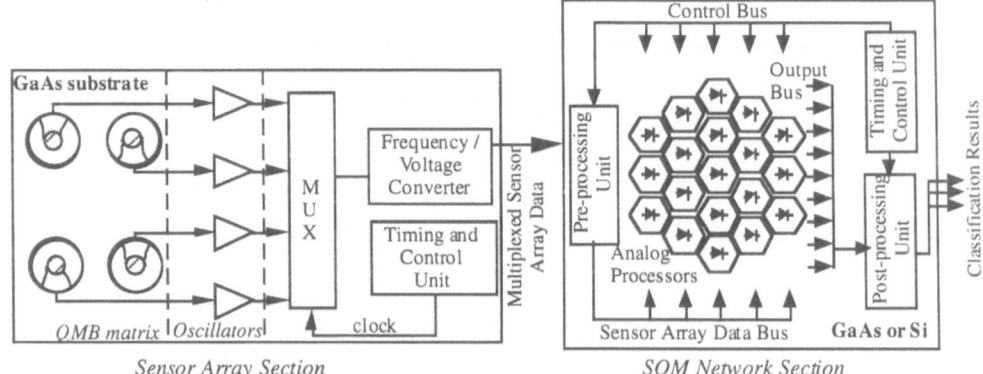

Fig.1. Schematic of the two substrates.

QMB sensors are placed in cylindrical cavities obtained by wet/dry etching procedures. Fig. 2 shows the quartz-substrate system together with the quartz metallization contacts (electrodes 1 and 2). Three peripheral points of the bottom surface (S_2) of each quartz rest on the substrate (among them only P_1 and P_2 are shown in the figure). Point P_2 electrically connects the bottom quartz electrode. In this arrangement a suitable quartz-cut orientation still allows the quartz structure to oscillate. Contacts C_1 and C_2 are then connected to the amplifiers which are integrated on the same substrate.

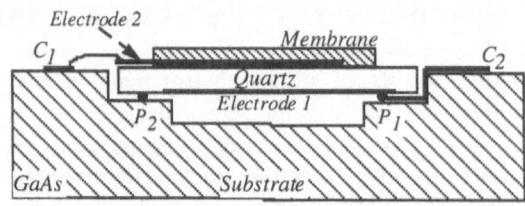

Fig. 2. Schematic of the quartz-substrate system
(each elements is explained in the text).

An analog opto-electronic micromachined implementation of the SOM network is considered, which employs a novel hybrid wireless mechanism for the activation of the neural groups.

The network is composed of several simple analog processors placed on a regular 2-D grid without any communication link between them. Furtherly each processor is provided with a photoresistor and a light emitting diode, upward oriented, which is surmounted by a silicon micromachined light reflector, downward oriented (Fig. 3). The "lateral" inter-neuron interaction, needed by the competitive learning algorithm, is performed through a hybrid electro-optical interaction: due to a specifically designed mirror shape, each cell can transmit an intensity modulated light signal to its neighbours, with a neighbourhood radius that can be controlled in real time tuning the intensity and the emission angle of the LED.

The light sensitive part is made by ad hoc hydrogenated amorphous silicon (α-Si:H) photoresistors deposited and lithographically located nearby each cell. The main advantages offered by the (α-Si:H) technology are: high intrinsic flexibility, relatively low deposition temperature and very high relative change of resistivity under optical illumination.

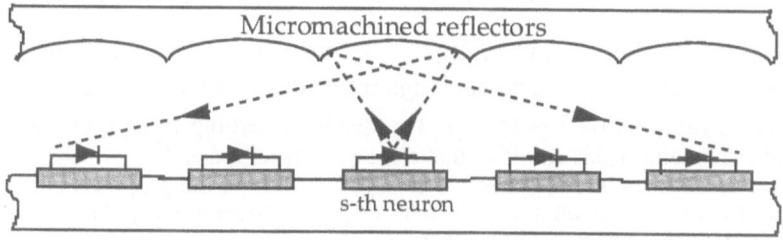

Fig 3 One-dimensional picture of the micromachined system of reflectors and of the GaAs chip

Fig. 4-a shows the basic circuitry of an elementary cell, composed of an array of analog storage elements for the synaptic weights, another array for temporary storage, a switching network and a monostable element, that is the core of the cell. Each analog storage elements is framed by a capacitor C_M carrying a voltage proportionally related to a synaptic weight value. The switching network is controlled by a number of command signals broadcast to all the cells. It is remarkable that specialized computational units are not employed, saving area and lowering circuital complexity.

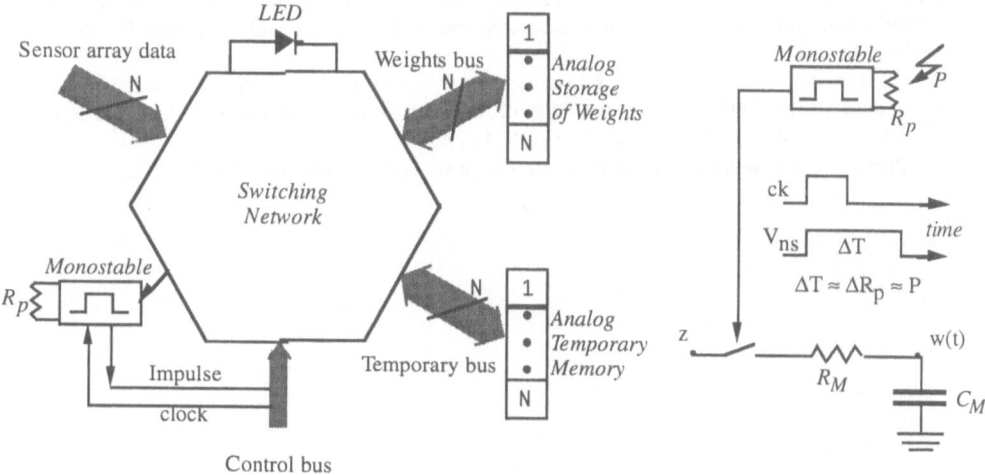

Fig. 4. a) Block diagram of an analog processor. b) Weight update controlled by a monostable unit

Changes in synaptic weights are done by connecting a suitable voltage to capacitors through a switch and a resistor (Fig. 4-b). The lasting of connection T is controlled by the monostable element and is proportional to the optical power that irradiates the cell's photoresistor R_M. Assuming z to be a sensor output and w the stored synaptic weigth, after a time T (short in comparison with $R_M C_M$) the new weight value becomes:

$$w(\Delta T) \approx w(0) + \frac{\Delta T}{R_M C_M}(z - w(0)) \tag{1}$$

which is just the form of the required "updating of the difference only" algorithm.

The competitive learning algorithm, involving the computation of the Euclidean distance measure $(z - w)^2$ for each cell, exploits the same mechanism. The monostable unit is used to generate a time period $T \propto (z - w)$. The result, according to eqn. (1), is the required squared difference (plus an offset that can be easily subtracted).
In comparison with most implementations (e.g. reported in [3]), the proposed scheme is characterized by a lower occupation area per cell, and allows the integration of 10 sensors and 100X100 network by a quite conservative technology.

3. Conclusions

The bio-inspired architecture and the hybrid technology here proposed ensure: the exploitation of the great potential of parallelism involved in the odour classification tasks; a great efficiency in the occupation area of the chip due to wireless communication; higher integration degree compared to that of pure optical implementations; continuous adaptation to environmental changes without requiring, at design stage, a strict preventive forecasting of the real operating conditions.

References

1 F A M Davide, C Di Natale, A D'Amico, Self-Organizing multisensor systems for odor classification internal categorization, adaption and drift, Sensors and Actuators B, 18, 1-3, (1994),244-258
2 F A M Davide, C Di Natale, A D'Amico, Sensor arrays and self-organizing maps for odor analysis in artificial olfactory systems, Proc of the International Conference on Artificial Neural Network '94 (ICANN '94), Sorrento, Italy, (1994)
3 M Glesner, W Pochmuller, Neurocomputers, Chapman and Hill, London, (1994)

INTEGRATION OF AN AMPEROMETRIC GLUCOSE SENSOR IN A μ-TAS

L. Forssén, H. Elderstig, L. Eng[1], M. Nordling[2]
Industrial Microelectronics Center, P O Box 1084, S-164 21 Kista, Sweden
[1]Department of Biochemistry and Biotechnology, Royal Institute of Technology, S-100 44 Stockholm, Sweden
[2]Department of Physical Chemistry, Box 532, S-751 21 Uppsala, Sweden

Abstract

A method for incorporation of an amperometric glucose sensor in a capillary system as a thin film on the inner surface of the channels has been demonstrated The capillary system chip was fabricated with VLSI technology on a quartz substrate The sensor was designed as an enzyme electrode consisting of glucose oxidase, immobilized in a redox polymer by a crosslinker and was bound to an Au-electrode The glucose solution was introduced in the capillary system with electroosmotic pumping Amperometry was used to measure the response Results from dipstick measurements with varying electrode areas, in order to study the enzyme electrode reaction, are presented

1. Introduction

Miniaturized Total Analysis Systems (μ-TAS) are multifunctional with functions such as chemical analysis, liquid handling, optical and electrical sensors, electronics for responses, actuators etc The μ-TAS used in this work is an integrated capillary system on a quartz substrate The system was fabricated with standard VLSI technology, which includes thin film deposition, photolithography, oxidation and etching [1]

An enzyme was immobilized in a redox polymer matrix on an Au-electrode in the capillary system Glucose oxidase (GO) was crosslinked into a redox polymer with polyethyleneglycol diglycidyleether, PEGDGE The redox polymer consists of polyvinyl pyridine units, with covalently bound osmium complexes and ammonium methyl groups The Os-complexes acts as redox mediators between the enzyme and the electrode The ammonium methyl groups works as handles for PEGDGE crosslinking [2] GO contains two FAD (flavine-adenine dinucleotide) groups, that are reduced by glucose to $FADH_2$ The glucose thereby oxidizes to gluconolactone, reaction (1)

$$GO(FAD) + glucose \rightarrow GO(FADH_2) + gloconolactone \qquad (1)$$

A van den Berg and P Bergveld (eds), Micro Total Analysis Systems, 203–207

The reduced enzyme is reoxidized by osmium redox couples on the polymer chain close to the active site on the enzyme The electrons from the reaction then propagates to the electrode surface by means of electron hopping between the redox sites in the polymer The polymer is crosslinked by PEGDGE to the electrode surface, where the osmium complex oxidizes and the electrons are transferred to the electrode The electron transfer procedure is shown in figure 1 The current produced is proportional to the rate of glucose conversion, and provides a measure of the substrate concentration

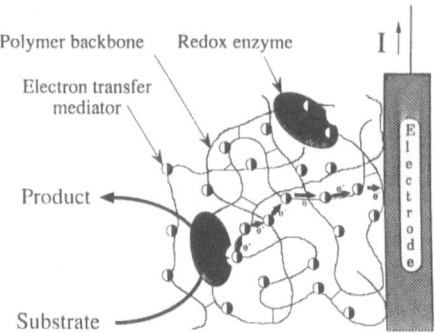

Figure 1. Polymer based enzyme electrode; electron propagation

2. Experimental

Dipsticks were fabricated in order to study the enzyme electrode In the first experiment we used a glassy carbon electrode as working electrode, Au-electrode as counter electrode, and an Ag/AgCl reference electrode The glassy carbon electrode was washed and polished before use The GO, 3 75 mg/ml, the osmium polymer, 3 75 mg/ml, and the crosslinker PEGDGE, 1 25 mg/ml, were mixed 1 1 1 15 µl of this mixture was placed on the polished area (2 x 6 mm) The solution was left overnight to dry and crosslink on the electrode The measurements were carried out in a phosphate buffer at pH 7 The glucose concentration was increased stepwise, and the responding current was measured amperometrically at 500 mV, which is high enough to oxidize the redox couples on the polymer

Gold dipstick electrodes with varying areas have also been studied On a 35 mm^2 electrode, 15 µl of the enzyme/polymer solution was applied, as a first test on gold surface Measurements were also carried out on 1 mm^2-, 0 01 mm^2- and 0 0001 mm^2 electrodes, where 1 µl of the solution was applied

Quartz wafers, 4 inch in diameter, were used in the experiments In the first step they were covered with polycrystalline silicon (poly-Si) using Low Pressure Chemical Deposition, LPCVD, technique In the photolithography step that followed the wafers were patterned in photoresist with a designed mask in a Canon 1 1 mask aligner The pattern was developed and dry etched in CHF$_3$/O$_2$ plasma through the poly-Si layer The pattern was thereby transferred to the wafer. The capillaries were then formed by wet chemical etching in concentrated HF solution, this etching is isotropic and the quartz is etched about 1 µm/min. The poly-Si layer was then totally oxidized and the holes were closed by LPCVD deposition of Tetra Ethyl Ortho Silicate, TEOS, oxide. Another photolithographic step was made for the contact holes, which were etched in the same manner as the poly-Si layer. The metal, gold with chromium as adhesion layer, was deposited through sputtering. The gold layer was also patterned photolithographically with photoresist and was then etched in iodine solution. To get larger metal areas electroplating was used. This also filled the contact holes.

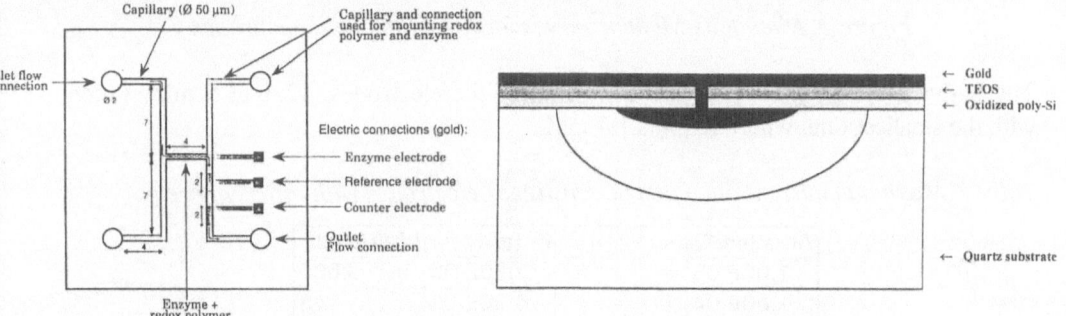

Figure 2. Capillary topview, a, and cross-section, b.

After dicing the wafer, each chip was mounted in a plastic measurement holder. The enzyme electrode was introduced in one of the channels through electroosmotic pumping. After completed crosslinking the glucose solution was introduced in the same way, but at the other end of the capillary system, thereby only passing the enzyme electrode in the middle section of the channel system, fig 2a. 80 V was applied over the channel, leading to a 1 cm/min flow rate for a 0.1 M KCl solution. With this kind of pumping there will be no pressure drop over the capillary.

3. Results and discussion

The results from the measurements with two of the gold electrodes are shown as Michaelis-Menten curves in figure 3.

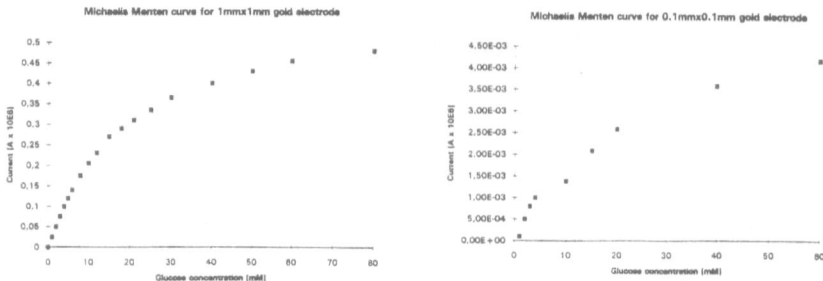

Figure 3. Michaelis-Menten curves for two of the gold electrodes.

The K_M values was almost the same with all of the electrodes, 12.5-13.5 mM, except with the smallest one, where K_M was 1.7 mM.

Table I. Maximum current and current density for electrodes with different areas.

Area [cm2]		i [uA]	j [uA/cm2]
1,00E-06		2,60E-03	2600
1,00E-04		4,30E-03	43
1,00E-02		0.48	48
0.12		9	75
0.35		13,5	39

From Table I, showing maximum current and current density for different electrode areas, it can be seen that the current density is almost the same for the lager electrode areas, but is much higher for the smallest electrode, which is in agreement with an earlier work [2]. This makes a higher degree of miniaturization possible, given a specific limit on smallest measurable currents. The current density for the glassy carbon electrode is slightly higher than the ones with gold.

The redox polymer has three functions, it communicates with the active site in the enzyme, transfers the electrons to the electrode surface and it forms a matrix in which the enzyme is immobilized [3]. This is the reasons which makes the osmium polymer such a successful mediator. A very sensitive and fast biosensor is achieved with the redox polymer bound enzyme. There are no membrane passages that will delay the reaction and the mediator is immobilized in the same matrix and will not diffuse away. An amperometric glucose sensor with the mediator electrostatically bound to a PVP polymer has been studied in a previous paper [4], where the mediator loss seemed to be a problem. The hydrophilicity of the osmium redox polymer also contributes to the rapid response because of the fast transport of water soluble substrates and products.

4. Conclusions

A novel μ–TAS for glucose measurements has been demonstrated. An enzyme electrode with the enzyme and redox centra immobilized in the same polymer matrix on an Au-electrode, has been evaluated for incorporation in a capillary system on quartz It has been shown from the dipstick measurements that the small responses from the miniaturized enzyme electrodes are detectable A fabrication method for the capillary system enabling the incorporation of enzyme electrodes has been developed It has also been shown that sample introduction can be done with electroosmotic pumping It should be possible to achieve a glucose sensor successfully incorporated in the μ–TAS we have developed in the near future

5. References

1   W Kaplan, H Elderstig and C Vieider, A Novel Fabrication Method of Capillary Tubes on Quartz for Chemical Analysis Applications, *IEEE, Micro Electro Mechanic Systems*, 1994, Japan
2   A Heller, Electrical Connection of Enzyme Redox Centers to Electrodes, *J Phys Chem* 1992, 96, 3579-3587
3   M Elmgren, Amperometric Biosensors Based on Enzymes Wired by Redox Polymers Acta Univ Ups , *Comprehensive Summaries of Uppsala Dissertions from the Faculty of Science* 471 55pp Uppsala ISBN 91-554-3175-5
4   S Haemmerli, A Schaeffler, A Manz and H M Widmer, An Improved Micro Enzyme Sensor for Bioprocess Monitoring by Flow Injection Analysis, *Sensors and Actuators B*. 7 (1992), 404-407

ELECTRIC FIELD MEDIATED CELL MANIPULATION, CHARACTERISATION AND CULTIVATION IN HIGHLY CONDUCTIVE MEDIA

G. Fuhr[1] and B. Wagner[2]
[1]Humboldt-University of Berlin, Department of Biology,
Invalidenstr 42, 10115 Berlin,
Germany
[2]Fraunhofer-Institute for Silicon Technology (ISiT),
Dillenburgerstr 53, 14199 Berlin,
Germany

0. Abstract

In this paper we show that, under appropriate conditions, ultra-micro-electrode systems fabricated on glass or silicon wafers are able to apply a c -electric fields to highly conductive media without electrolysis and strong heating A helpful result in relation to cell manipulation (trapping, positioning, cell sorting, cell cultivation) is the occurrence of exclusively repulsive forces independent of the applied frequency Surprisingly, under these conditions adherently growing animal cells, like mouse fibroblast (3T3, L929), can be cultivated in original culture media under permanent field application (frequency 10 MHz, field strength 10-100 kV/m)

1. Introduction

The manipulation of living cells or microparticles by applying high frequency electric fields requires field strengths of between 2 and several hundred kV/m [1-3] Although electrodes arrangements consisting of two or more electrodes were effectively used for cell separation since the pioneering work of POHL twenty years ago, up to now such intense fields could only be generated in purely non-conductive media (<0 05 S/m) Otherwise, strong heating and electrolysis accompanied by gas bubble formation prevented practical use However, most cell culture media, especially those for animal of human cells, exhibit conductivities between 1 and 5 S/m (e g DMEM, PPMI)

The conductivity of sea water ranges between 1 and 9 S/m Therefore, only a small number of medical, pharmacological, chemical and environmental applications could be developed (mostly brief field-pulse techniques, such as electrofusion or electropermeabilisation [4] Because, this situation was unchanged over more than twenty years, the questions arises, in what respect there is physical limitation?

A van den Berg and P Bergveld (eds), Micro Total Analysis Systems, 209–214

2. Physical description of the problem

Increasing the conductivity of the solution increases the current flow between the electrodes The current through a solution is also directly proportional to the field strength and the electrode area (Ohm`s Law) Therefore, raising the conductivity limit requires miniaturisation of the electrodes For cell manipulation and characterisation, it is sufficient to use electrodes and spaces of several microns up to hundreds of micrometers Compared with conventional electrode systems this is a increase by a factor between 100 and 10 000 Therefore, the conductivity can be increased by the same order of magnitude To reduce heating and electrolytic processes further, we developed the following design principles for ultramicroelectrode systems

a   Insulation (e g SiO_xN_y or even dielectrics) of all electrodes except those parts used for cell handling
b   Processing of multielectode arrays to distribute the power dissipation over a larger area
c   Processing of flat electrodes (micron- and submicron-range)
d   Application of high frequency electrical signals to decrease electrical loading of the cell membrane and to avoid low frequency electrode processes
e   Use of substrates with good thermal conductivity (e g silicon)

3. Microstructure fabrication

Microelectrodes with typical dimensions in the micron- and submicron-range were fabricated on quartz as well as on oxidised silicon wafers by electron-beam-lithography (see Fig 1A, B) The chips were mounted and bonded in standard ceramic carriers (LCC 90) The electrically connected chip could be inserted in a plastic ring system (Fig 1C) and driven from outside under sterile conditions (for more information see [5])

Fig   1A) Overview of a multi-electrode structure fabricated with e-beam lithography
        B) SEM of ultramicroelectrodes in the central part of A) (width 170 nm)
        C) Complete system for sterile cell cultivation

4. Results

4 1 HIGHLY CONDUCTIVE WATER SOLUTIONS

Flat electrodes (as shown in Fig 1) cover chip areas of several thousand of square micrometers and, allow the insertion of highly conductive media up to 7 S/m and the application of electrical signals of up to 5 V amplitude (frequency > 100 kHz)

In our smallest electrode system (width of electrodes 170 nm, spaced by the same distance) field strength's of more than 1 MV/m occurs Regardless of these high values cell culture media, especially that for animal and human cells, can be penetrated by such strong electric fields This is a result of the changed physical behaviour of extremely miniaturised electrodes [6] To get an impression about the chemical complexity of such a solution the constituents of DMEM are given in Table 1

4 2 CULTIVATION OF CELLS UNDER PROLONGED FIELD INFLUENCE

The best way to examine electric field effects on cell growth is to apply them directly to the culture medium Micro-structures allow this for prolonged periods (several days) We used adherently growing fibroblasts (3T3, L929) and exposed them for 3 days to field strengths of 50 kV/m (frequency 10 MHz) Typical growth characterising data in comparison to the controls without field are summarised in Table 2 Due to the high frequency no electrode reaction could be identified which influenced cell physiology and simple cell behaviour

4 3 NEGATIVE DIELECTROPHORESIS OF CELLS AND MICROPARTICLES

At conductivities greater than 1 S/m living cells show negative dielectrophoresis, independent of the frequency That means, they were repelled from the electrodes From a purely physical point of view this means that the real and imaginary part of the effective complex permittivity ($\varepsilon^* = \varepsilon + j\chi/\varepsilon_0 2\pi f$) of cells is smaller than that of the surrounding solution In the case of most plant and animal cells the conductivity of the cytoplasm ranges between 0 3 and 0 6 S/m and permittivity between 50 and 70 For frequencies up to 200 MHz the conductivity is the dominant quantity

One reason for the lower conductivity of the cytoplasm is the decreased mobility of ions compared to the outside medium Another reason is the pumping of ions across cell membranes We investigated algae yeast cells isolated plant protoplasts, animal cells like erythrocytes fibroblasts and hybridoma cells All cells exhibit exclusively negative dielectrophoresis This effect can be used to develop various cell manipulation principles

Table 1 composition of typical animal cell culture medium

Electrolytes mg/1000 ml	Sugars mg/100 ml	Amino acids and proteins mg/1000 ml	Complex substances % or mm/1000 ml
$CaCl_2\ 2H_2O$ 264 $Fe(NO_3)\cdot 2H_2O$ 0,1 KCl 400 $MgSO_4\cdot 7H_2O$ 200 NaCl 6400 $NaHCO_3$ 3700 $NaH_2PO_4\cdot 2H_2O$ 141	D-Gluc 4500	L-ARG 84 L-CYS 48 L-GLUT 584 Glycine 30 L-HIS 42 L-ILE 105 L-LEU 105 L-LYS 146 L-METH 30 L-PHA 66 L-SER 42 L-THR 16 L-TYR 72 L-VAL 94	foetal calf serum (10%) D-Phantothenic acid 4 Choline chloride 4 Folic acid 4 Myo-Inositol 7 2 Niacinamide 4 Pyridoxal HCl 4 Riboflavine 0 4 Thiamine 4

Table 2 Behaviour of fibroblast with and without field influence

Cell-type	anchorage time		lay phase		average time of cell cycles		speed of mobility	
	with	without	with	without	with	without	with	without
3T3	25 min ± 10	25 min ± 10	16h ± 2	17 h ± 2	11h ± 2,5	11h ± 2	0-40 µm/h	0-46 µm/h
L929	45 min ±15	60 min ±15	22h ±2	24h ±2	26h ±3	18h ±2	0-35 µm/h	0-25 µm/h

4 4 TRAPPING OF CELLS IN ELECTRIC FIELD CAGES

To measure characteristic parameters of single cells requires exact positioning (µm-range) However, most cells show physiological reactions after mechanical contact with artificial surfaces The occurrence of negative dielectrophoresis allows the development of field cages since cells are focused toward the central part of three-dimensional octupole-electrode configurations (for more details see [2, 7])

In such cages individual cells can be levitated in physiological solutions and held in a stable position or rotated by applying travelling electric fields [2] The field cage can be opened and closed by changing the amplitude, frequency of phase of the electrode signals [8] By this procedure cells can be released or trapped

4 5 ELECTRIC FIELD-MEDIATED CELL ADHESION CONTROL

In the case of stripwise arranged submicron electrodes as shown in Fig 1, the field strength near the substrate becomes very strong However, several micron away the influence of neighbouring electrodes is cancelled due to the superposition of field components of different phase Therefore, the smaller the electrodes and gaps, the stronger are the forces repelling cells near the surface On the other hand, the stronger the field, the smaller the particles that can be repelled There is a limit of view, particles smaller than 100 nm should be difficult to handle by polarisation forces

For cells, commonly several microns in size of more, dielectrophoretic forces can be used to prevent cell surface contact This is one of the key problems in medical material research, so called biocompatibilty If strong electric fields were applied, adherently growing cells could not occupy "field shielded" surfaces (Fig 2)

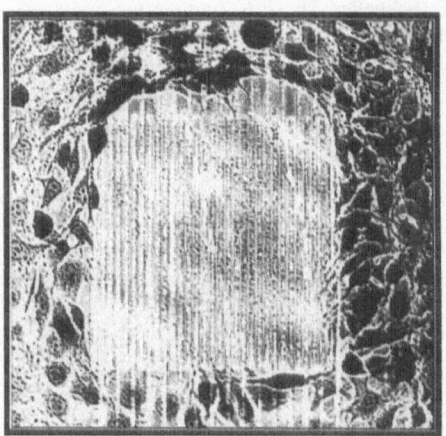

Figure 2. Ultramicroelectrode array after 28 h of cultivation of fibroblasts (dark areas) under prolonged field influence (1.5 V, 5 MHz). The electrode area is completely free of cells.

5. Perspectives

We have shown that planar and three-dimensional microelectrode systems, integrated in cultivation of flow through structures, are well suited to the manipulation of individual living cells and microparticles suspended in highly conductive solutions Overcoming the

214

conductivity limitation and introducing submicron electrodes opens up a large number of biotechnological, medical and also chemical applications

Additionally, the cultivation of cells under permanent radio-frequency field enabled us to investigate high-frequency electrical effects on living cells under defined condition, without change of solutions This may be also helpful in the discussion about the influence of "electrosmog" on human and animal physiology

Acknowledgement
We thank Mr Reimer from Fraunhofer-Institute for Silicon Technology (ISiT, Berlin) for fabrication of the ultramicroelectrode systems This work was supported by grants of BMFT (No 0310260A and No 13MV03032)

References
1 Pohl, H A Dielectrophoresis, Cambridge University Press, Cambridge, (1978)
2 Fuhr, G , Arnold, W -M , Hagedorn, R , Muller, T , Benecke, W , Wagner, B and Zimmermann, U , BBA 1108 (1992), 215-223
3 Pethig, R , Huang, Y , Wang, X -B and Burt, J P H , *J Phys P 24*, (1992)
4 Zimmerman, U , Electric-field mediated fusion and related phenomena,   *BBA 694* (1982), 227-277
5 Fuhr, G , Glasser, H , Muller, T , Schneller, Th , *BBA* (in press)
6 Aoki, K , *Electroanalysis 5* (1993), 627-639)
7 Schnelle, Th , Hagedorn, R , Fuhr, G , Fiedler, S and Muller, T , *BBA 1157* (1993), 127-140
8 Fuhr, G , Fiedler, S , Muller, T , Schnelle, Th , Glasser, H , Lisec, T , Wagner, B , *Sensors & Actuators A,* 41-42 (1994), 230-239

ELECTROCHEMICAL MICROANALYTICAL SYSTEM FOR IONOMETRIC MEASUREMENTS

W. Hoffmann, M. Bruns, B. Büstgens, E. Bychkov, H. Eggert, W. Keller,
D. Maas, R. Rapp, R. Ruprecht, W.Schomburg, W. Süß

Kernforschungszentrum Karlsruhe GmbH

0. Abstract

A better reliability of chemical analysis is expected by combining microsensors with microactuators and microelectronics representing complete microanalytical systems The ELMAS-concept (electrochemical microanalytical system) of the Nuclear Research Center Karlsruhe is presented in order to fabricate such a complete microsystem for the analysis of electrolytes It combines potentiometric microsensors with solid state ion sensitive membranes, active and passive microfluidic components fabricated by plastic molding and adhesive mounting, microelectronic components for signal processing and system management

1. Introduction

Beside the aspects of universal miniaturization, which is necessary especially for medical applications, and general cost reduction, the increase of reliability is one of the most striking credits given to the young emerging field of miniaturized Total Analysis Systems (μTAS) [1] General, chemical sensors and in particular, chemical microsensors suffer from drift phenomena or sudden device failure related mostly to contamination and corrosion phenomena This is also the reason for the limited acceptance of chemical microsensors at the market, despite promising advantages

Although a large number of chemical macrosensors for electrolyte measurements are commercially available, these sensors do not offer self-testing or self-calibration These functions are realized until now in large volume devices like flow-injection analyzers Just recently, several groups started to develop μTAS, using predominately silicon and glass structures [2-5]

2. The ELMAS Concept

Based on some specific microcomponent development at the Karlsruhe Nuclear Research Center, Institutes for Radiochemistry, Microstructure Technology and Applied Informatic

215

A van den Berg and P Bergveld (eds), Micro Total Analysis Systems, 215–218

216

a concept for a modular setup of electrochemical microanalytical system (ELMAS) has been created (fig.1).

Self-testing of the electronic and the chemical transfer functions are substantial features. Whereas electronic feedback tests can be realized by integrated electronics (like in „smart sensors") any check of the chemical transfer characteristics needs microactuators for controlled change of the chemical environment of the sensor. Micropumps are used for the alternating transport of measuring and calibrating solutions to the sensors. Then, the modulated sensor signal is a measure for the unknown species of the analyte. Using such procedures drift phenomena of the working point of the sensors can be compensated. Furthermore, in the same way, changes of the sensitivity may be indicated by a two-point calibration

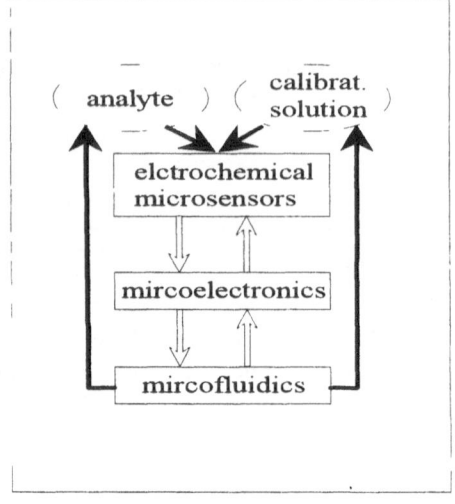

Fig. 1. ELMAS concept, schematically.

3. Chemical Microsensors

In a first µTAS approach electrochemical sensors are preferred because of their compatibility of their sensing and signal transfer principles and fabrication technology to microelectronics [6]. Ion sensitive field effect transistors (ISFET's) and ion selective electrodes (ISE's) will be used. Special attention is given to the fabrication of longlife and robust ion sensitive membranes (ISM's) as sensing interface in these potentiometric devices. Inorganic solid thin films were favored for this purpose. Examples for ISM materials, preparation methods and ions to be detected are given in Tab.1.

Tab. 1. solid state thin film ion sensitive membranes

Material	Thin Film Deposition	Detected Species	Reference
Silver iodide	HV evaporation	I^-	[7]
Sodium alumo silicate glass	RF cyl. magn. sputtering	Na^+	[8]
Cu doped arsenic selenide	RF magnetron sputtering	Cu^{2+}	[9]

The sensitivity and selectivity of these ISM's were after some technological process optimization comparable to those of the conventional macro ISE's.

The dynamic response of the sensors has to be considered with respect to the possible analysis frequency For the basic pH sensor a very fast response was found (fig 2), observed as a constant sensor signal within 1 second after fast exchange of the pH solution. In the case of Cu^{2+} ion sensitive thin film ISE a slower dynamic response has to be accepted.

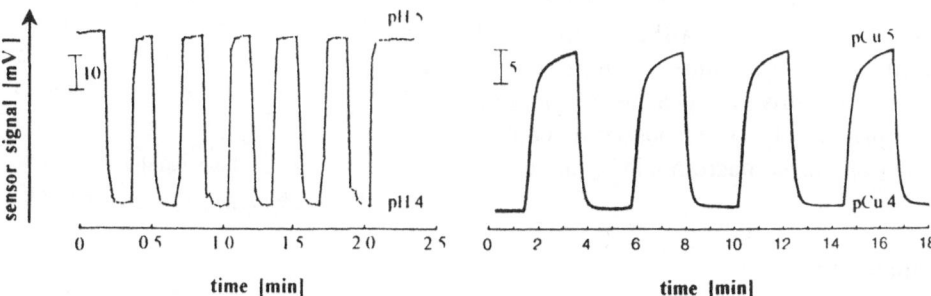

Fig. 2. dynamic response of a pH-ISFET (left) and a Cu-ISE (right) with thin film ISM.

4. Microfluidics

Plastic molded and adhesive mounted micropumps and microvalves will be used as active microfluidic components (see [10] this issue). The microchannel system for connecting all components (several sensors, the analyte to be measured, the calibrating solution reservoirs, the micropumps and valves) is under construction and will be fabricated with the standardized molding and mounting technology.

5. Electronic data handling and system management device

As a tool for the ELMAS development a computer-model was created which used specifications of the functional interactions within the system based on Statecharts ® Statecharts are extended finite automata The model was combined with a graphical user interface for investigating the system behavior during running simulations [11]

The simulated model components are stepwise substituted by real system modules. The model running on a workstation controls the sensor and actuator components of macro dimension. Going from macro to micro components the workstation will finally be replaced by a microcontroller

6. ELMAS functional model

A first functional model of ELMAS was realized by integrating two ISFET-chips, a micropump which transports the solutions indirectly by an air buffer and two miniaturized minivalves which control the flow direction The active elements are connected by microtubes (i d 0 5 mm) The principle ELMAS function of alternating measuring and calibrating cycles could be demonstrated, e g for the detection of sodium ions with a pNa-ISFET (fig 3) A reduced response time compared to macroactuator driven exchange of solutions (fig 2, pH-ISFET) was observed coming from not optimized microfluidic handling

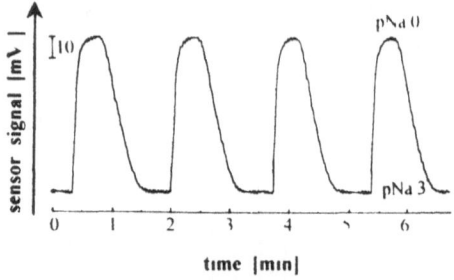

Fig. 3. changing pNa in ELMAS.

7. Conclusions

Directed at better reliability of chemical analysis by chemical microsensors, a concept of integrating ISFET's, micro ion selective electrodes, micropumps and microvalves as well as microchannels in a microsystem is followed Components have been prepared and combined to get a first functional model, which can be controlled by a computer model of the ELMAS

References

1 A Manz, N Graber H M Widmer, Miniaturized Total Chemical Analysis Systems A Novel Concept for Chemical Sensing, *Sensors and Actuators B,1* (1990) 244-248

2 D J Harrison, Z Fan, K Seiler, K Furri, Miniaturized chemical analysis System Based on ElectrophoreticSeparation and Electroosmotic Pumping, *Transducers '93, Digest of technical Papers* (1993) 403-406)

3 B H van der Schoot, S Jeanneret, A van den Berg, N F de Rooij, Microsystems for Flow Injection Analysis, *Analyt Meth and Instr 1* (1993) 38-42

4 S Howitz, M T Pham T Vogel A Steinbach, M Burger, H Fiehn, Mirkosysteme fur schwierige Messungen, *Chemische Industrie 6* (1994) 64-66

5 W Hoffmann H Eggert, W Schomburg, D Seidel, Elektrochemisches Mikroanalysensystem fur die Ionometrie von Flussigkeiten, *KfK-Report No 5238* (1993) 89-93

6 W Hoffmann, R Rapp H J Ache D stolze, D Neuhaus, D Hofmann, K H Freywald, Modular Potentiometric Measuring System for the Development and Comfortable Testing of Miniaturized Ion Sensors, *Proceed of the Mikro Total Analysis Systems Workshop.* (1994) this Issue

7 M J Schoning, M Bruns, W Hoffmann, B Hoffmann, H J Ache, Iodide ion-sensitive field effect structures, *Sensors and Actuators B 15-16* (1993) 192-194

8 R Becht, M Bruns W Hoffmann, H J Ache, Sodium Ion Sensitive Membranes for Microsensors Fabricated by R F sputtering technique *Abstract of 44th ISE Meeting,* Berlin ,(1993)

9 E Bychkov, M Bruns, H Klewe-Nebenius G Pfennig, K Raptis, W Hoffmann, H J Ache, Copper(II)ion response of Cu-As-Se thin film sensors in a flow-through microcell, *Abstract of 5th Int Meeting on Chem Sens* Rome, (1994)

10 W K Schomburg B Bustgens, J Fahrenberg, D Maas, Components for Microfluid Handling Modules, *Proceed of the Micro Total Analysis Systems Workshop,* Enschede, The Netherlands, (1994), xxxx

11 W Suss, H Eggert, M Gorges-Schleuter, W Jakob, W Hoffmann, Modelling of a microsystem for the detection of ions in fluids *2nd Annual Internat i-Logix Conf,* Burlington (1993)

MODULAR POTENTIOMETRIC MEASURING SYSTEM FOR THE DEVELOPMENT AND COMFORTABLE TESTING OF MINIATURIZED ION SENSORS

W. Hoffmann, R. Rapp, H.J. Ache
Kernforschungszentrum Karlsruhe GmbH, Institut für Radiochemie

D. Stolze, D. Neuhaus, D. Hofmann
Centrum für intelligente Sensorik Erfurt

K. H. Freywald
Sigma Electronic Erfurt

0. Abstract

An ionometric system for the analysis of electrolyte solutions has been developed Planar chip structures of ion sensitive field effect tansistors (ISFET's) and ion selective electrodes (ISE's) are used as sensors Their layout allows an easy preparation of the ion sensitive membrane and also a very simple electrical contacting The sensor chips can be clipped to small volume flow-through cells for dynamic measurements An advanced electronic device was developed for measuring both, the ISFET and the ISE signals This system is useful for basic investigations of ion sensitive materials and can be integrated comfortably into electroanalytical sensor/actuator microsystems

1. Introduction

Electrochemical sensors are of outstanding interest as signal converting components in miniaturized Total Analysis Systems (µTAS) [1] because they transform the chemical signal „species concentration" directly into an electronic signal which is used for further data handling , system control and management functions In this way the number of interfaces is minimized increasing the system reliability as one of the goals to meet with microsystems In the case of ion sensitive field effect transistors (ISFET's) [2] the basic element of integrated circuits - the MISFET - is modified to get a chemical sensor enabling an ideal adaptation of microelectronics and chemical microsensors

The sensor signal of ISFET's is modulated by an ion concentration dependent interface potential established at an ion selective membrane (ISM), similarly to conventional ion selective electrodes (ISE's) Material research for reaching high sensitivity, selectivity and stability for the ISM and technological investigations for miniaturizing the membrane dimension are currently conducted in many laboratories [3-8] for getting optimal sensor performance and for integrating these potentiometric sensors in microsystems

A van den Berg and P Bergveld (eds), Micro Total Analysis Systems, 219–222
© 1995 *Kluwer Academic Publishers*

220

A modular setup of a measuring system, where all the components are easily exchangeable, and where preparation and mounting of the components may be done not exclusively with the high sophisticated equipment of microelectronics fabrication techniques, can be helpful for basic investigations of membrane and sensor properties as well as for the construction of function models for μTAS. Therefore, a chip sensor, a flow-through cell and an electronic measuring device were developed, whose dimensions were not brought down to the lowest possible level; rather comfortable handling and possibilities for easy membrane preparation on cheap single chips were desired.

Sensor chip

As basic sensor a pH-ISFET chip was developed which can be used for different ISM depositions. It contains a dual ISFET structure (n-channel FET, LP-CVD Si_3N_4 gate membrane, gate size 16 x 400 μm) and a temperature sensitive diode. The chip is fabricated in a standard MOS-production line. For using the chip in a ISFET difference measuring mode without a conventional reference electrode an integrated pseudo reference electrode has been designed too.

The over all chip size is (5 x 8) mm which enables easy chip handling. ISE's prepared on chips of the same dimensions can be used alternatively. This size enables individual preparation of ion sensitive membranes on single chips without lithographic processes. For example, solid state membranes can be deposited on top of the basic pH gates to modify their sensitivity. This was demonstrated by evaporation [9] and sputtering [10], [11] methods. Furthermore, spin-on techniques and lithographic pattering of polymeric membrane materials are possible at this chip dimension.

The size of the contact pads on the chip is matched to the 0.5mm spacing of commercially available micro-connectors. The electric connections to the electronic device can be established simply by plugging the chip into the connector of the conductor cable. In this way a multichannel connection can be realized without any wire bonding processes (fig1)

Figure 1. ISFET with pads for direct electrical contacting (left), and flipped into an electric microconnector (right).

Flow-through cell

For testing the chemical function the chip can be clip-mounted in a flow-through cell (fig2) The cell body made of Plexiglas® is mechanically pressed against the chip surface sealed by an 0-ring The cell volume is about 1 µl Several cells can be stacked together to get a multisensor array

Again it is a rather simple mounting procedure No adhesive or welding processes have to be practiced and if necessary the sensors can be exchanged

Figure 2. ISFET chip mounted to a microflow-through cell

Electronic Measurement Device

For testing the sensor performance an ionometer device has been developed with the following features (fig3)

* measuring and calibration of 5 ISFET channels
* display of the difference signal of two channels
* two-point calibration
* temperature measurement and compensation
* automatic control of electronic sensor function (electronic transfer function, insulation)
* storage functions for operation conditions, calibrating factors and for measuring results (10 files with up to 12 000 data sets)
* RS 232 interface for PC combination, complete device control by PC
* analog output
* display mV (resolution 0 1 mV) or pX, °C
* measuring rate 1/sec
* external power or rechargeable battery
* handhold instrument, dimensions (195 x 100 x 40) mm

The basic ISFET-meter can be expanded by a module containing amplifiers enabling the direct potentiometric measurements of ISE's too

Figure 3. Electronic device. for measuring ISFET's and simultaneously ISE's.

Conclusions

A miniaturized modular potentiometric measuring system has been developed which can be used for basic investigations of ion sensitive membrane materials The advantages include simplicity of preparation of membrane materials on top of the sensors for tailoring the sensitivity and selectivity, easy mounting of the sensors and the fluidic components, comfortable signal handling for measuring both, ISFET's and conventional ISE's The measuring system can be used also for basic investigations of elctroanalytical microsystems with integrated actuators for realizing chemical control functions of the sensors performance

References

1 A Manz, N Graber, H M Widmer. Miniaturized Total Chemical Analysis Systems A Novel Concept for Chemical Sensing, *Sensors and Actuators B,1* (1990) 244-248

2 J Janata, R J Hubert, Ion Sensitive Field Effect Transistors, *Ion-Selective Rev , 1* (1979) 31-79

3 D N Reinhoudt, Application of supramolecular chemistry in the development of ion-selective ChemFET's . *Sensors and Actuators B, 6* (1992) 179-185

4 P D van der Wal. A van den Berg, N F de Rooij, Universal approach for the fabrication of Ca -, K -, and NO3-- sensitive membrane ISFET's, *Sensors and Actuators B, 18-19* (1994) 200-207

6 V V Cosofret. T M Nahir, E Lindner, R P Buck, New neutral carrier based H$^+$ selective membrane electrodes, *J Electroanal Chem , 327* (1992) 137-146

7 E Davini, G Mazzamurro. A P Piotto. Lead-selective FET complexation selectivity of ionophores embedded in the membrane, *Sensors and Actuators B ,7* (1992) 580-583

8 C Dumschat. R Fromer. H Rautschek. H Muller. H J Timpe, Photolithographically patternable nitrat-sensitive acrylate-based membran, *Anal Chcim Acta, 243* (1991) 179-182

9 M J Schoning. M Bruns, W Hoffmann. B Hoffmann, H J Ache, Iodide ion-sensitive field effect structures, *Sensors and Actuators B, 15-16* (1993) 192-194

10 R Becht, M Bruns. W Hoffmann. H J Ache. Sodium Ion Sensitive Membranes for Microsensors Fabricated by R F sputtering technique. *Abstract of 44th ISE Meeting,* Berlin ,(1993)

11 E Bychkov. M Bruns. H Klewe-Nebenius. G Pfennig, K Raptis, W Hoffmann, H J Ache, Copper(II)ion response of Cu-As-Se thin film sensors in a flow-through microcell" *Abstract of 5th Int Meeting on Chem Sens ,* Rome, (1994)

A NOVEL SAMPLING TECHNIQUE FOR TOTAL ANALYSIS SYSTEMS

Wolfgang Künnecke and Ursula Bilitewski
Biosensors Group, Gesellschaft für Biotechnologische Forschung (GBF) mbH,
Mascheroder Weg 1, 38124 Braunschweig, Germany

0. Abstract

A novel method of valveless sample introduction in flow analysis systems called flow-diffusion analysis (FDA) is described Due to the reduction of mechanical parts, i e an injection valve is dispensable, the system becomes very simple and its transformation into the micro-scale should require less effort than the respective transfer of conventional flow-injection systems Additionally, due to the time based sampling, the method is more flexible in terms of sensitivity and linear range compared to established volume based methods

1. Introduction

The development of Micro Total Analysis Systems (μTAS) will require several transformations of existing macroscopic devices, such as pumps, valves and detectors into the microscopic scale However, even in the macroscopic world with lots of reliable components readily available, total analysis systems are not yet state of the art This is mainly due to the lack of sufficient sampling and preconditioning techniques

One of the most promising candidates amongst the analytical methods to be transferred into the microscopic scale is flow-injection analysis (FIA) This is due to its ease of automation, simplicity, high flexibility and compatibility with several standard detectors FIA is suitable for process and environmental monitoring and especially its combination with biosensors seems to be a realistic solution for several analytical problems However, also in macro-FIA, the problem of sampling is still unsolved

It was the aim of our work to optimize the sampling and preconditioning step of a flow analysis system in order to obtain a long-term-stable device, high flexibility in terms of range of measurement as well as simple and cheap construction (i e minimum parts) This paper describes the integration of sampling and separation in one step by using membrane separation and time based sampling In analogy to flow-injection analysis we choose the term flow-diffusion analysis (FDA) for this novel technique Basic characteristics of the method will be described and a first combination with a microcomponent (i e the sampler) will be presented

A van den Berg and P Bergveld (eds), Micro Total Analysis Systems, 223–226
© 1995 *Kluwer Academic Publishers*

2. Materials and Methods

Flow rates were adjusted to 1 5 ml min⁻¹ except for microdialysis experiments In this case the flow rate was 67 μl min⁻¹ Potassium phosphate buffer (pH 7 5, 0 1 M) was used as acceptor stream All experiments were performed at room temperature

2 1 INSTRUMENTATION

The configuration of the flow-diffusion analyzer is shown in figure 1 It is composed of only three elements pump, sampler and detector Two types of diffusion units were investigated a) a standard sandwich-type unit (channel 0 5 mm depth x 1 4 mm width x 21 mm length) as macroscopic device and b) a microdialysis probe (CMA/12, CMA AB, Stockholm, Sweden) representing a microscale element Another pump (not shown) was required to transport continuously sample or standard solutions as donor through the diffusion cell For experiments with the microdialysis probe, the tip of the probe was immersed in the sample solution A thick film glucose electrode in wall-jet formation incorporated in a flow cell [1,2] and a potentiostat (Meredos, Bovenden, F R G) were used for electrochemical detection at 700 mV The signals were recorded on a chart recorder (Kipp & Zonen, Delft, The Netherlands) Timing and data acquisition were performed by a personal computer (Hewlett Packard, Waldbronn, F R G) and home made software (FiaFox, GBF)

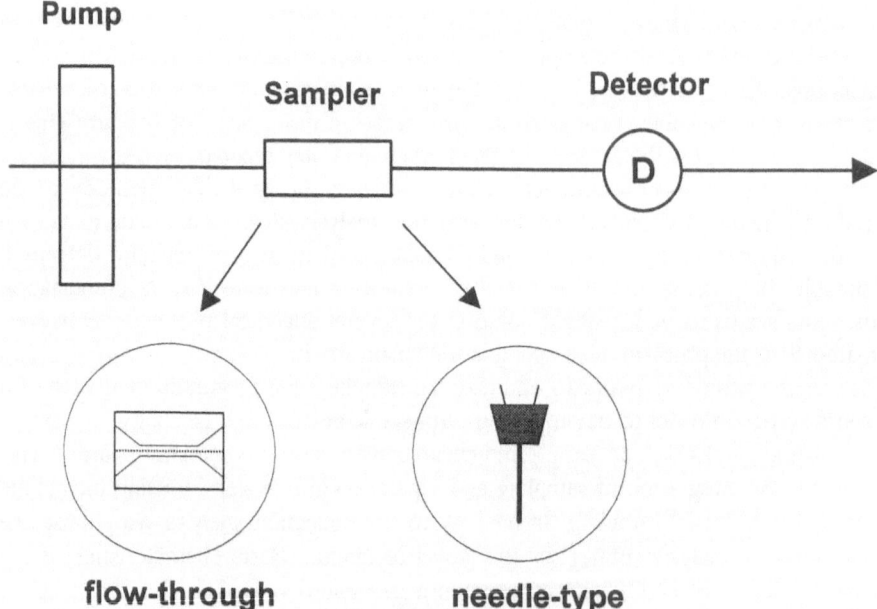

Figure 1. Flow-diffusion analysis (FDA) setup.

2 2 PRINCIPLE OF FDA

The pump propelled an acceptor stream through the diffusion unit located between the pump and the detector (see figure 1) By stopping the pump the acceptor stopped in the diffusion unit and the liquid element next to the membrane was accumulated with analyte The amount of analyte permeating the membrane into the acceptor was dependent on the stop time (accumulation time) Subsequently with the next switching an accumulated analyte plug was carried to the detector The peak shapes are similar to FIA-peaks, thus enabling evaluation by peak height measurement

3. Results and Discussion

By integration of separation and sampling into one device, three main improvements have been achieved first, the system became very simple, i e an injection valve was dispensable Second, almost no volume was extracted from the sample medium because the analyte was extracted by diffusion Third, time based sampling was more flexible compared to volumetric sampling using injection valves Different analyte concentrations could be accumulated by variation of sampling time (see figure 2), whereas an injection valve based method would require a manual exchange of the sample loop

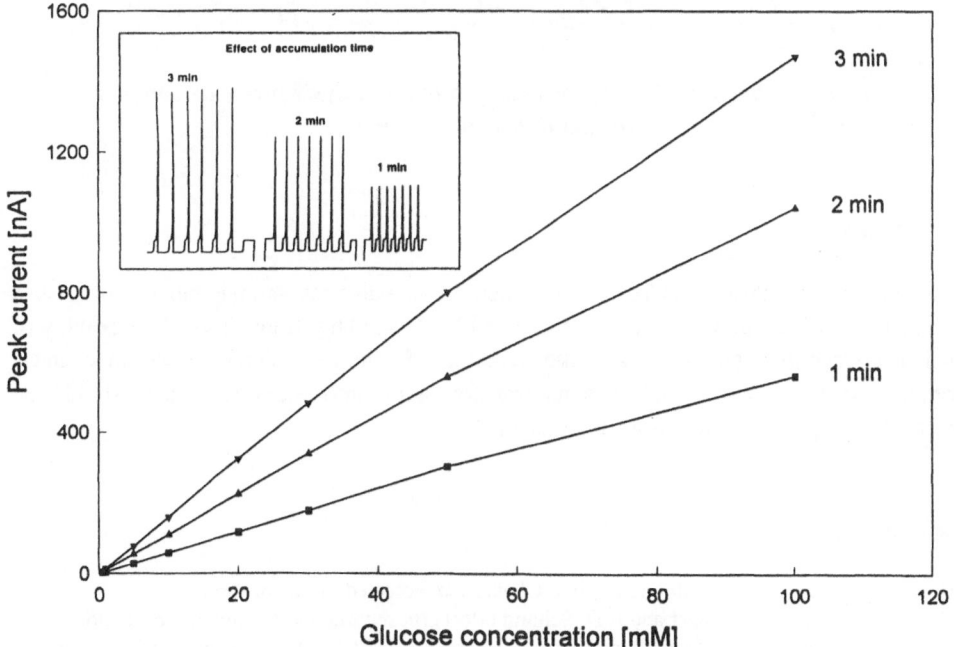

Figure 2. Effect of accumulation time. Glucose determination in mineral medium for E. coli cultivations.

226

For applications in fermentation monitoring glucose could be determined in the linear range of 0 5 mM to 100 mM, depending on the accumulation time. The screen printed glucose biosensor was stable for each day in continuous use without any decrease in activity

As a first step towards microsystems we used a microdialysis probe as the sampling device. Microdialysis is a method to sample the extracellular fluid of living tissue by means of an implanted dialysis tube[3] It was introduced a decade ago and became a standard technique in many neurochemical laboratories[4]. Therefore a range of different microdialysis probes is commercially available. Its performance in a flow-diffusion mode was investigated in order to prove the general applicability for microsystems. Pump and detector were the same as in all other experiments. As demonstrated in figure 3, reproducible signals could be obtained with this micro-sampling technique

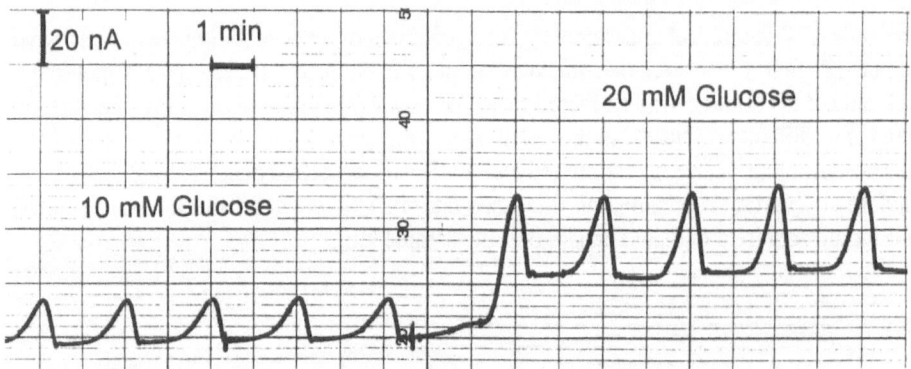

Figure 3. Original chart record using a microdialysis-probe as sampler (accumulation time: 1 min.)

4. Conclusions

Flow diffusion analysis (FDA), a novel method of valveless sample introduction, is a combination of membrane separation and time based sampling. It increases the flexibility of analytical readout and minimizes the amount of sample volume Less mechanical components are required and thus its transformation into the μTAS-scale should be favorable compared to other flow analysis techniques.

References

[1]   A Günther and U Bilitewski, *Anal. Chim. Acta*, accepted for publication
[2]   U. Bilitewski, P. Rüger and R.D. Schmid (1991) *Biosensors and Bioelectronics*, 6, 369
[3]   U. Ungerstedt, M. Herrera-Marchintz, U. Jungnelius, L. Stale, U. Tossmann and T. Zetterstrom (1982) In: M. Kotsiaka (Ed.) Advances in dopamine research. Pergamon Press, N.Y. , 219.
[4]   T.E. Robinson and J.B. Justice (1991) Microdialysis in the Neurosciences (Techniques in the Behavioral Neural Sciences, Vol. 7), Elsevier, Amsterdam.

A MICROMACHINED GLUCOSE OXIDASE ENZYME REACTOR

T. Laurell, L. Rosengren[1] and J. Drott
Department of Electrical Measurements, Lund Institute of Technology,
Lund University, P.O. Box 118, S-221 00 Lund, Sweden.
[1]Division of Electronics, Department of Technology, Uppsala University,
P.O. Box 534, S-751 21 Uppsala, Sweden.

0. Abstract

The design and fabrication of a flow-through cell in <110>-oriented silicon, which acts as a glucose oxidase enzyme reactor, is described. The reactor structure was made of several tall, parallel, standing walls, anisotropically etched in silicon, which occupies a wafer area of $3*15$ mm^2. Reactors of two geometries were fabricated on the designated area, holding either 30 lamellæ (165 µm tall, spaced 50 µm apart) or 75 lamellæ (235 µm tall, spaced ≈30 µm apart). Glucose oxidase was immobilised to the silicon surface available in the reactor. The lamella structure acted as an area enlarging geometry to achieve a high enzyme activity in the reactor. The two reactors exposed 172 mm^2 and 515 mm^2 of silicon surface respectively for the enzyme immobilisation. Enzyme activity determinations of the reactors showed that the reactor area increase yielded an increase in maximum glucose turnover rate from 10 nmol/min to ≈37 nmol/min. The 515 mm^2 reactor was also operated in a system for continuous glucose monitoring, based on continuous glucose sampling via microdialysis. At a perfusion rate of 25 µl/min, the system responded linearly to glucose levels between 0 and 5 mM after which the oxygen limitation affected the response.

1. Introduction

Components incorporated in chemical analysis systems are rapidly being developed in micro scale. Today microstructured pumps, valves and flow channels are already available, [1]. Micro scaled flow transducers and temperature sensors as well as chemical sensors i.e. pO_2, pCO_2, pH, [2], have also been developed. Flow injection analysis systems commonly use enzyme reactors for the analysis of more complex components, and until recently little work has been done on the wafer integration of enzyme reactors [3, 4].

This paper describes the development of a silicon wafer integrated enzyme reactor (ER). By using silicon as the substrate, methods for enzyme coupling commonly used for silica gels can be applied. In order to achieve a large amount of enzyme in the reactor volume, the design strategy focused on producing a structure which had a surface enlarging characteristic. Anisotropic etching of <110>-oriented silicon provided an array of deep narrow channels which comprised the reactor structure.

A. van den Berg and P. Bergveld (eds.), Micro Total Analysis Systems, 227–231.
© 1995 *Kluwer Academic Publishers.*

2. Materials and Methods

2.1 REACTOR FABRICATION

The reactor was designed to occupy a wafer area of 3*15 mm. By altering the lamella geometry two versions of the reactor were fabricated, offering either 172 mm^2 (lamella width: 50 μm, 165 μm tall, spaced 50 μm apart) or 515 mm^2 (lamella width: 20 μm, 235 μm tall, spaced 20 μm apart) of silicon surface for the enzyme coupling. The reactors were structured by anisotropic etching of <110>-oriented silicon [6]. As the lamellæ were etched from one side of the wafer, two holes were simultaneously etched from the rear side, adding flow connections to the structure. The lamella array was subsequently covered by a Pyrex glass lid, which was anodically bonded to the structure, to provide a sealed reactor structure with its only access via the rear side holes.

Since the silicon surface in the reactor was covered by a layer of native oxide, standard method for enzyme coupling to silica matrixes was used, [6].

2.2 ENZYME ACTIVITY DETERMINATION AND GLUCOSE MONITORING

The Trinder reagent [5] was used to determine the enzyme activity (EA) in the reactor. Trinder reagent solutions containing peroxidase and varying amounts of glucose (2, 5, 10, 50 mM) was sequentially pumped through the enzyme reactor. The glucose in the reagent solution was converted to gluconic acid and H_2O_2 according to the glucose turnover rate of the enzyme in the reactor. The produced H_2O_2 was in turn converted to a chinonimin colour compound by the Trinder reagent. The absorbance of the colour compound was monitored by a Waters 486 UV/VIS absorbance detector at 490 nm, yielding the glucose turnover rate of the reactor at each glucose concentration.

A 515 mm^2 reactor was connected to a system for continuous glucose monitoring, using microdialysis as the sampling technique, figure 1. The micro dialysis probe was immersed in glucose solutions ranging from 2 mM to 100 mM. As the dialysate entered the enzyme reacto:, a drop in dissolved oxygen was measured, caused by the enzymatic glucose breakdown.

3. Experimental and Results

The outcome of the anisotropic etching yielded a 172 mm^2 reactor with a lamella width of 50 μm and a channel width of 50 μm, figure 2. The 515 mm^2 reactor turned out to have a lamella width of 3-11 μm and a channel width of ≈33 μm. The non uniformities of the 515 mm^2 reactor was caused by minor problems in the lithographic processing.

The EA was estimated for three reactor structures, a 172 mm^2 reactor, a 515 mm^2 reactor and a 515 mm^2 reactor which was thermally oxidised (oxide thickness ≈ 280Å) before the bonding of the glass lid. The oxidation was done to investigate whether the native oxide was sufficient for the enzyme coupling or if a thick oxide layer would better serve as the base

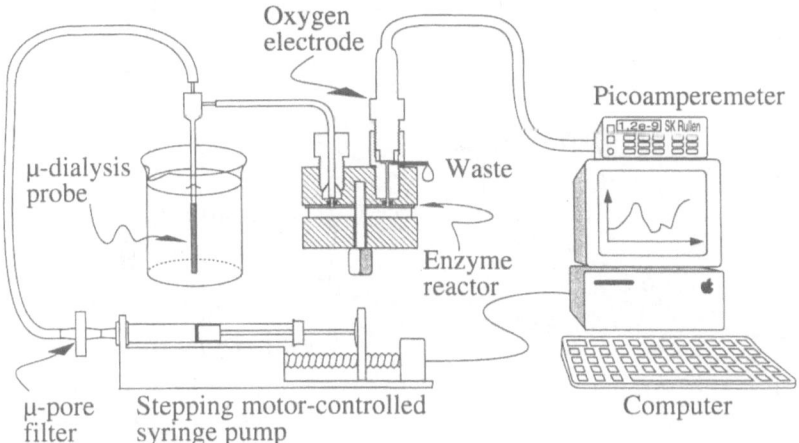

Figure 1. The microdialysis based glucose monitoring system with the silicon lamella structure operating as the enzyme reactor. The oxygen electrode monitored the changes in dissolved oxygen as the dialysis probe was immersed into samples of varying glucose concentrations.

Figure 2. A SEM-view of one end of the 50μm-reactor showing the flow entrance hole and the beginning of the lamella structure.

for the enzyme couplmg. A Lmeweaver/Burke plot of the glucose turnover rate at varying glucose concentrations gave the maximum turnover rate, V_{max}, as the x-axis intercept and the Michaelis constant, K_m, as the y-axis intercept, for each reactor, figure 3. Both the 515 mm^2 reactors gave a V_{max} of $\approx$37 nmol/min and a K_m of 20 mM, using the Lineweaver/ Burke approximation and consequently, an additional oxidation of the reactor did not affect the EA. The 172 mm^2 reactor displayed a V_{max} of 10 nmol/min and a K_m of 12 mM.

The 515 mm^2 reactor was also operated in a microdialysis based glucose monitoring system. The dialysis probe was immersed in glucose concentrations ranging from 2 mM to 100 mM. The glucose recovery across the dialysis membrane was 24% at the selected flow rate of 25 μl/mm and, thus, the glucose concentrations entering the enzyme reactor ranged

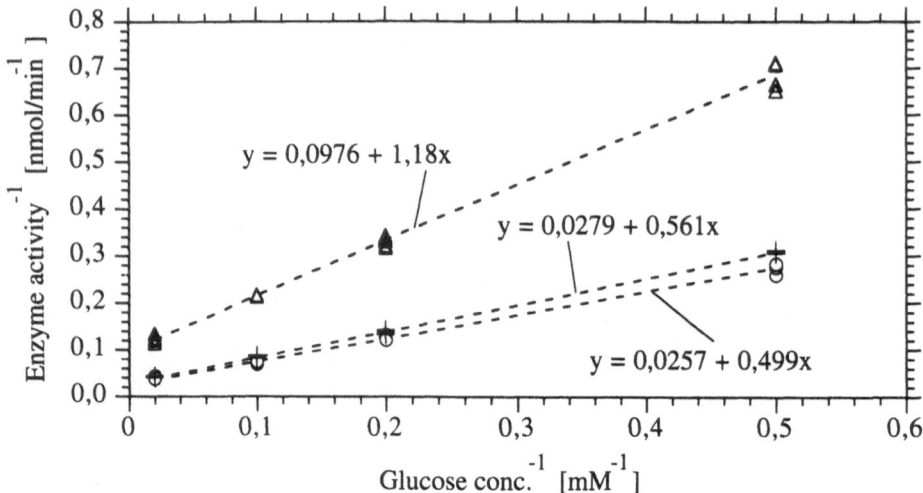

Figure 3. The enzyme activity plotted in a Lineweaver/Burke plot provided the V_{max} (x-axis intercept) and K_m (y-axis intercept) for the reactors in test. - △ - 172 mm² reactor, - +- 515 mm² reactor, - 0 - oxidised 515 mm² reactor.

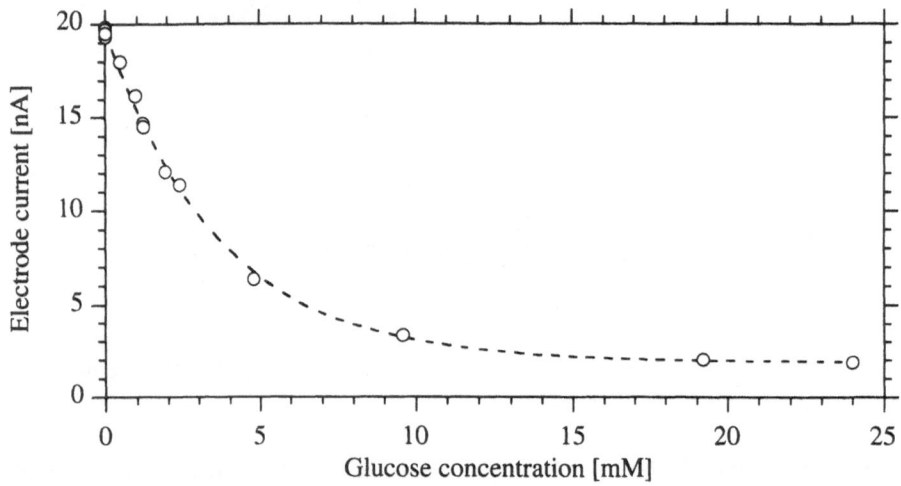

Figure 4. A calibration plot of the microdialysis system response to glucose, flowrate = 25 µl/min, shows a linear response to 5 mM before the oxygen limitation affects the response. The reactor in use was the 515 mm² reactor with native oxide.

from 0.5 to 24 mM, as seen in the calibration plot, figure 4. The system showed a linear response to glucose up to 5 mM after which the oxygen limitation affected the response characteristics.

4. Discussion and conclusions

The integration of ER s in silicon should follow simple process steps to allow for the cointegration with other micro structured chemical analysis components A major goal for the described ER design is to achieve a large silicon area and thus a high EA Tedious manual procedures such as packing enzyme carrying microbeads into a flow-through silicon cavity can be avoided if large areas can be achieved in the ER by other means Anisotropic etching of silicon offers a simple way with few process steps to achieve a structure with sufficient area Most previous reports on integrated ER s focus on V-grooved flow channels [3, 4], which do not offer any large area increase with respect to the wafer area used A vertically etched lamella array increases the reactor area to a much greater extent than the V-groove design does The employed strategy of increasing the EA in the ER by increasing the lamella frequency and the lamella height pays off approximately in proportion to the area increase

$$\frac{515 \text{ mm}^2}{172 \text{ mm}^2} = 3 \ 0 \quad , \quad \frac{V_{max} \ (515 \text{ mm}^2 \text{ reactor}) \ \ 37 \text{ nmol/min}}{V_{max} \ (172 \text{ mm}^2 \text{ reactor}) \ \ 10 \text{ nmol/min}} = 3 \ 7$$

The apparently higher gain in EA, 3 7, than in surface enlargement, 3 0, is partly an effect of the choice of approximation for V_{max} Another estimation yields an EA gain of 3 3 Also, the area at the entrance and exit holes are believed not to contribute as much to the EA as the lamella area due to the difference in flow distribution If only the lamella area gain is considered, the factor is 3 2 Furthermore, the yield in the enzyme immobilisation procedure may differ somewhat from one occasion to another These factors may explain the deviation

Although the achieved EA may seem low it should be noted that the sample volumes and the flow rates for micro structured systems are very low As an example, a flow rate of 2μl/min and a glucose concentration of 10 mM corresponds to a turnover rate of 20 nmol/min We therefore believe that a reactor with a lamella design will have the potential to fulfil the requirements for clinical applications

References

1 van der Schoot B H , Jeanneret S , van den Berg A , de Rooij N F , Microsystem for flow injection analysis , *Analytical Methods and Instruments*, 1 (1993), 38-42

2 Arquint Ph , van den Berg A , van der Schoot B H and de Rooij N F , Integrated blood-gas sensor for pO2, pCO2 and pH , Sensors and Actuators B, 13 14 (1993) 340-344

3 Xie B , Danielsson B , Norberg P Winquist F and Lundstrom I , Development of a thermal micro-biosensor fabricated on a silicon chip, *Sensors and Actuators B*, 6 (1992) 127 130

4 Murakami Y , Takeuchi T , Yokoyama K , Tamiya E , Karube I and Suda M , Integration of enzyme-immobilized column with electrochemical flow cell using micro machining techniques for a glucose detection system , *Analytical Chemistry*, 65 (1993) 2731 2735

5 von Gallati H Aktivitatsbestimmung von Peroxidase mit hilfe des Trinder Reagens, *J Clin Chem Clin Biochem* , 15 (1977), 699 703

6 Laurell T and Rosengren L , A micromachined enzyme reactor in <110>-oriented silicon, *Sensors and Actuators B*, vol 18 19 (1994), 614 617

FIRST STEPS OF µTAS IN LATVIA

A. Lúsis, J. Kleperis, V. Eglitis, A. Lloyd-Spetz[1], I. Lundström[1], H. Sundgren[1] and F. Winquist[1], G. Strautmanis[2], I.Slaidinš[2] P.Misâns[2], P. Rozukalns[3] and S. Sjulzics[3]

Institute of Solid State Physics, University of Latvia,
8 Æengaraga Street, LV-1063 Rìga, Latvia,
[1]Laboratory of Applied Physics, Linkoping University,
S-581 83 Linkoping, Sweden,
[2]Riga Technical University, Latvia,
[3]Microelectronic Comp A/S ALFA, Rìga, Latvia

0. Abstract

Owing to participation of Latvian and Swedish scientists in the field of chemically sensitive FET and MOS structures it was possible to utilize the technological facilities of the Microelectronics enterprise in Latvia in making an integrated smart sensor as a matrix of chemically sensitive FETs by using silicon based microtechnology Thereby the first steps of µTAS in Latvia are made in the direction of integration of solid state ionic materials with FET and MOS structures to develop new chemically sensitive structures and systems

1. Introduction

The electronic industry, which has been the dominating industry in Latvia over the last thirty years, is now under reconstruction One of the goals of that is the implementation of microsystems technology (MST) for production of new microdevices for different applications in an electronic instrumentation and equipment The base for restructuring is formed from a large group of researchers in solid state electronics and ionics, designers of integrated circuits and electronic equipment and Microelectronics Company A/S ALFA A number of different semiconductor devices and integrated circuits are produced by Microelectronics Company A/S ALFA, for which some foreign analogs are Integral Circuits (mA709 (A, HC, CN-14 etc), LM101H (CH, AM, A)), Voltage Comparators (mA710H (HC, NC-8) NC-14, N-8), mA711H etc), Voltage-Frequency-Voltage Converters (VFC32), Digital-Analog Converters (AD7520, AD7541, Hi562, AD558), Analog-Digital Converters (AD750, ICL7107IN, AD7574, AD7574kN, AD7581), Timers (NE555, NE555C), Semiconducting Devices (2N4260, BD136-6, PN2905A, BC556 etc)

The best tool for restructuring is international cooperation based on technology transfer through an innovation center and mobility of specialists

A van den Berg and P Bergveld (eds), Micro Total Analysis Systems, 233–236
© 1995 Kluwer Academic Publishers

We have initiated a cooperation program between Institute of Solid State physics of University of Latvia (ISSP of LU); Riga Technical University (RTC), A/S ALFA and Laboratory of Applied Physics of Linköping University (LAP of LU) for R&D regarding chemical sensors. The main themes of this cooperation are:

1. Gas and ion sensing elements based on physical, chemical and/or electrochemical phenomena.
2. Gas sensing gate materials for GASFETs.
3. Microminiaturization and integration of sensing elements.
4. Arrays of sensors for artificial olfactory systems.
5. Signal processing and interfacing methods and electronics (hardware - software) for arrays of sensors.

The first step of μTAS in Latvia is the research and development of chemical sensing elements. The next steps will be the miniaturization of them and integration with silicon technology to create arrays or matrices for an artificial olfactory system.

2. Gas sensing elements

As materials for sensing elements different catalytic metals, metal oxides and their composites are used in the form of thin and/or thick films obtained by different deposition, evaporation or sol-gel technologies. The electronic and ionic properties of the surfaces (interfaces) and bulk of such materials are sensitive to different gas molecules in the environment. The gas sensitivity is based on physical and chemical phenomena on the catalytic metals and solid state ionic materials.

On the first step catalytic metals and materials with high ionic (or mixed ionic-electronic) conductivity are studied, for example, thin film of Pd, ultra thin films of Pt, Ir; thick layers or bulk of hydrates of antimonic acid, Prussian Blue, beta-alumina, thin films of transition metal oxides etc.

2.1. ELECTROCHEMICAL ELEMENTS

Room temperature chemical sensors for alcohol, acetone, water vapor and ammonia detection were developed [1,2] by using beta alumina and xerogel of antimonic acid hydrate. The sensitivity and selectivity depend on ion-exchange and preparative methods. Ammonia-exchanged samples of beta alumina were sensitive to water and ammonia vapor. Potentiometric and amperometric sensors are made by using ion conducting material as the substrate, an Au electrode as the reference and Pd (or Ag, Ni, Pt) as the sensitive electrode.

2.2. RESISTIVE ELEMENTS

We prepared sensors based on resistivity changes from ion-exchanged beta alumina [2] and from Prussian Blue [3]. An ammonia-exchanged beta alumina with two Au electrodes (gap 0.1 mm) is sensitive to ammonia and humidity. The sol-gel produced polycrystalline

Prussian Blue thick film on ITO electrodes (the width of the gap is 0 1 mm, and the length is 15 5 cm) is sensitive to humidity only (and not to ammonia) at room temperature

2 3 GASFET ELEMENTS

In early 1975 an idea about new gas sensitive MOS device was presented [4] by I Lundstrom et al Since then in Sweden a great deal of work has been done on gas-sensitive field-effect devices (see, for example, [5-7]) Hydrogen sensors are prepared by using a Pd thin film electrode (about 200 nm) as the gate material By using ultra thin catalytic metal layers (about 10 nm thick), sensitivity to ammonia is created and the sensitivity to gases like alcohols and unsaturated hydrocarbons is increased

Gas sensors based on the MOS and FET structures is a new trend in ISSP of University of Latvia It is supported by cooperation with the Laboratory of Applied Physics of Linkoping University Our attention is given to the use of non-stoichiometric oxides of transition metals with mixed electronic-ionic conductivity (WO_x, MoO_x, IrO_x, NiO_x) instead of layers of metal and/or oxide in MOS structures

3. Sensor arrays

In a joint project between Linkoping University and A/S ALFA (Riga, Latvia) chips are created with CMOS arrays and diodes for temperature control Elements of different dimensions are collected on one chip to test the technical possibilities of A/S ALFA Changes in selectivity and sensitivity of different elements in the array to gases will be made by using different metals as gate materials and/or by using different temperatures of similar sensing elements

In Riga we are planning to use materials with mixed ionic-electronic conductivity as gate electrodes, thereby making the available region of selectivity and sensitivity of different elements in an array broader

In Linkoping alloys of different catalytic metals will be tested as gate materials to increase the selectivity and sensitivity to different gases

4. Signal processing methods and interfacing electronics

Researchers at Riga Technical University (groups of Prof G Strautmanis and Dr Doz I Slaidiðø) have more than 20 years experience of signal processing and design of LSIC for analog/digital signal processing That experience is now used for processing of signals from sensor arrays with elements of different sensitivities and selectivity patterns Several signal processing methods are now under development including statistical analysis and pattern recognition that have different properties regarding
1) precision of concentrations of gases,
2) possibility to distinguish between mixtures of gases,
3) complexity (linear processing or adaptive self-learning),

236

4) possibility of practical realization (hardware + software)
One of the main task in signal processing is to develop cost-effective optimization of the function of complex sensor arrays

5. Olfactory systems

The starting point for the development of an electronic nose in Latvia is the experience of the Linkoping's group more than 20 years with chemical sensors Several examples of the use of electronic noses have been used in evaluation of the sensor signals [8-10] In one case a sensor array consisting from six field-effect transistors, three with a thick Pd gate and three with a thin Pt gate, as well as its response to mixtures of hydrogen, ammonia, ethylene and ethanol was analyzed by the use of software based both on artificial neural nets and on the so-called abductory induction mechanism [10] Very simple networks can predict the hydrogen and ammonia concentrations in complicated mixtures, and the learning time was found to be very short In the case of recognition of different natural smells (from cheese, meat, grains etc) the learning time will be longer, but it is here not necessary to recognize individual components of the smells The electronic nose is planned to be used in the control of the quality of products in agriculture, food industry, pharmacy, for medical diagnosis and process control etc

References

1   J Kleperis, G Bajars, G Vaivars, A Kranevskis and A Lusis, gaseous sensors based on solid proton conductors, *Sensors and Actuators* **A 32** (1992) 476-479
2   J Kleperis, G Vaivars, G Bajârs, A Kranevskis, A Lúsis and G Vîtiðø, Solid proton conductors as room-temperature gas sensors, *Sensors and Actuators* **B 13-14** (1993) 269-271
3   G Vaivars, J Pitkev, iàs and A Lusis, Sol-gel produced humidity sensor, *Sensors and Actuators* **B 13-14** (1993) 111-113
4   I Lundstrom, M S Shivaraman, C M Svensson and I Lundkvist, Hydrogen sensitive MOS field effect transistor, *Appl Phys Lett* **26** (1975) 55-57
5   I Lundstrom, A Spetz, F Winquist U Ackelid and H Sundgren, Catalytic metals and field effect devices - a useful combination, *Sensors and Actuators* **B1** (1990) 15-20
6   I Lundstrom, C Svensson, A Spetz, H Sundgren and F Winquist, From hydrogen sensors to olfactory images - twenty years with catalytic field-effect devices, *Sensors and Actuators* **B 13-14** (1993) 16-23
7   A Spetz, F Winquist, H Sundgren and I Lundstrom, Field effect gas sensors In **Gas Sensors,** G Sberveglieri (ed), Kluwer Academic Publishers, Dordrecht, The Netherlands, 1992, chap 7, 219-279
8   H Sundgren, F Winquist, I Lukkari and I Lundstrom, Artifical neural networks and gas sensor arrays - quantification of individual components in a gas mixture, *Meas Sci Technol*, **2** (1991) 464-469
9   F Winquist, E G Hornsten, H Sundgren and I Lundstrom, Performance of an electronic nose for quality estimation of ground meat, *Meas Sci Technol* **4** (1993) 1493-1500
10 H Sundgren, I Lundstrom and H Vollmer, Chemical sensor arrays and abductive networks, *Sensors and Actuators* **B 9** (1992) 127-131

MICROREACTOR WITH INTEGRATED STATIC MIXER AND ANALYSIS SYSTEM

H. Mensinger, Th. Richter, V. Hessel, J. Döpper and W. Ehrfeld
IMM Institut fur Mikrotechnik GmbH, Carl-Zeiss-Str 18-20
55129 Mainz-Hechtsheim, Germany

0. Abstract

A modularly constructed, continuous flow microreactor has been developed containing a mixer unit, reaction unit and an analysis system Examples of these different units are demonstrated The mixer unit has been realized by micromachining with an excimer laser, demonstrating the possibility of shaping materials laterally as well as into the depth of the material The reaction unit demonstrates the ability to control the reaction parameters in microchannels e g by adjusting the flow velocity An optical detection system allows the use of several detection techniques such as UV-absorption and fluorescence measurements These units have been realized by combining deep X-ray lithography or excimer laser structuring with electroforming and plastic molding The embossing technique used allows the low cost replication of the structures in mass production

    The system presented demonstrates the possibility of low cost microfabrication for continuous flow microreactors with integrated mixing unit and analysis system This opens up completely new developments concerning chemical reactions in micro-scale devices

1. Introduction

New processes in microtechnology such as the LIGA technique [1], the structuring of photosensitive glass [2] or Laser LIGA, the combination of excimer laser structuring with electroforming and plastic molding [3], now allow the low cost mass fabrication of three dimensional microstructures from a wide variety of materials, such as metals, ceramics, polymers and glass

    These new methods open up completely new possibilities concerning chemical analysis and chemical reactions It is now possible to fabricate highly efficient analysis systems with short analysis times, high efficiency and low reagent consumption [4] Miniaturized analysis systems enable field analysis and additionally it is possible to develop disposable systems which are highly significant for medical applications

    Microreactors are characterized by small dimensions, an extremely large surface to volume ratio and, therefore, different reaction conditions compared to large scale reactors Using continuous flow microreactors, fast and highly selective reactions

A van den Berg and P Bergveld (eds), Micro Total Analysis Systems, 237–243

can now be improved and multistep synthesis is possible in one device The large surface to volume ratio allows a very fast interruption of the process, e g by rapid cooling after passing a short reaction channel, thus improving the purity of the reaction products Because of the small amount of chemicals and the high rate of heat transition the systems are inherently safe and chemical feedback systems in analogy to nature can be realized Microreactor arrays can be used for screening purposes and process optimization

2. Principle Process Steps of a Microreactor

Figure 1 shows the principle process steps of a continuous flow microreactor Two reactants are mixed in a static mixer unit followed by a reaction channel and an analysis system The modular construction of the system allows the exchange of individual units in order to optimize the system to a particular application

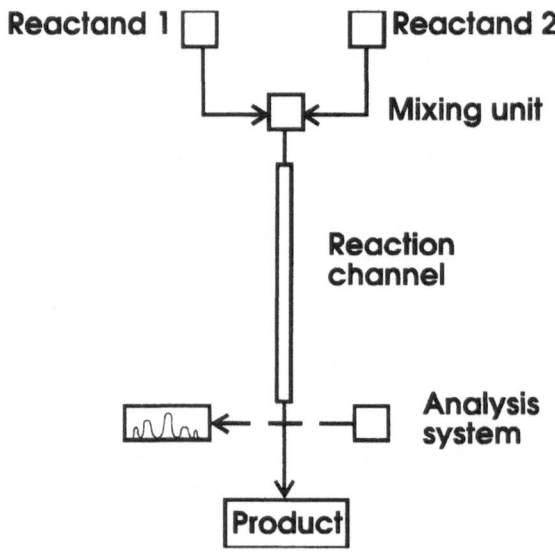

Figure 1. Continuous-flow reactions in microchannels.

Such a system can be realized by a sandwich system consisting of a stack of layers Figure 2 shows the arrangement of the modules of the stack The mixing unit is placed at the top, the detector unit at the bottom of the system

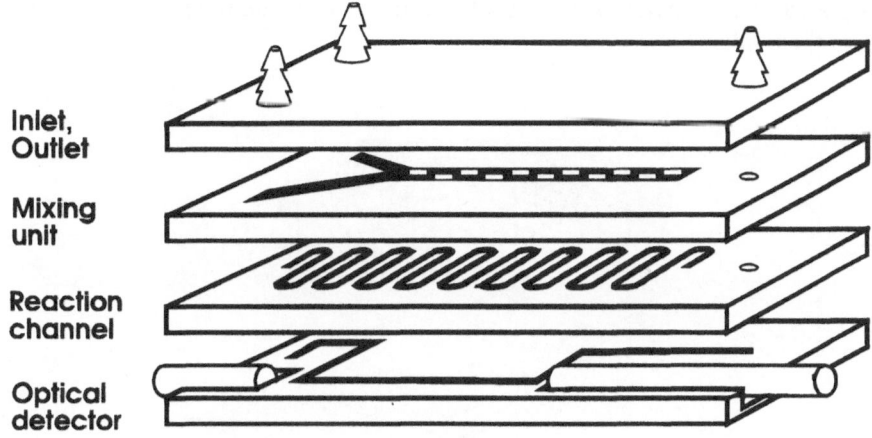

Inlet,
Outlet

Mixing
unit

Reaction
channel

Optical
detector

Figure 2. Layer-structure of the microreactor-system.

3. Experimental results and discussion

3 1 PACKAGING AND INTERCONNECTION TECHNIQUE

The systems have been realized by embossing using polymer materials such as PMMA and PC The sealing of different layers is achieved by a thermal diffusion process In this process the layers are oriented in respect to each other, heated up to a temperature above the glass transition temperature of the material employed and welded together under a slight positive pressure

Silicon tubes were connected to tubing adapters made of polypropylene These were glued to the top layer directly over a hole, 500 μm in diameter, by means of a two-component epoxy-glue The driving force for the solutions to be pumped through the system was achieved by applying a slight vacuum to the system outlet

3 2 STATIC MIXER

A static mixer has been developed for use either as a mixing unit e g for phase transfer catalysis or as a heat exchanger within the reaction channel This mixer improves the heat transfer perpendicular to the flow, thus achieving more uniform residence times and avoiding local overheating

A mixing unit consists of a twisted band, either left- or right-handed, which is similar to the so-called Mobius band, a ring-shaped structure with a one-sided surface Therefore this type of mixer is called Mobius mixer The principle is illustrated in figure 3 The layers of two non-miscible fluids are separated perpendicular to the boundary layers, subsequently twisted and reunited, thus doubling the exchange surface between the liquids. An array of several of these units leads to a complete mixing of the fluids.

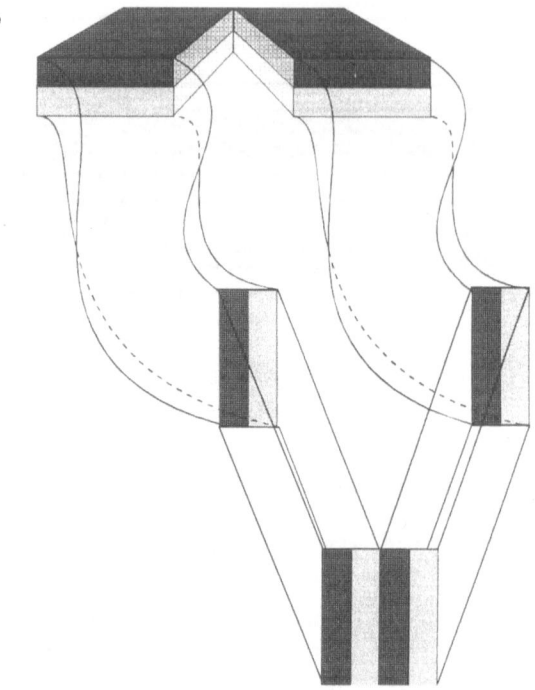

Figure 3. Principle of a static mixer (Möbius type).

3.2.1 *Experimental*

The static mixer device has been realized by excimer laser ablation, which can be used to shape materials laterally as well as into the depth of the material. There are several possibilities for varying the in-depth dimension across the lateral shape, e.g. by modulating light intensity across the sample to be produced. Another method is to step the beam across the surface within the frame of interest, thereby controlling the fluence (energy density per shot) and number of pulses applied to each spot. Figure 4 shows a static mixer containing three-dimensional surfaces fabricated as described above. This proves, that laser ablation is a very flexible method, allowing structuring of any geometry, unless undercutting is necessary.

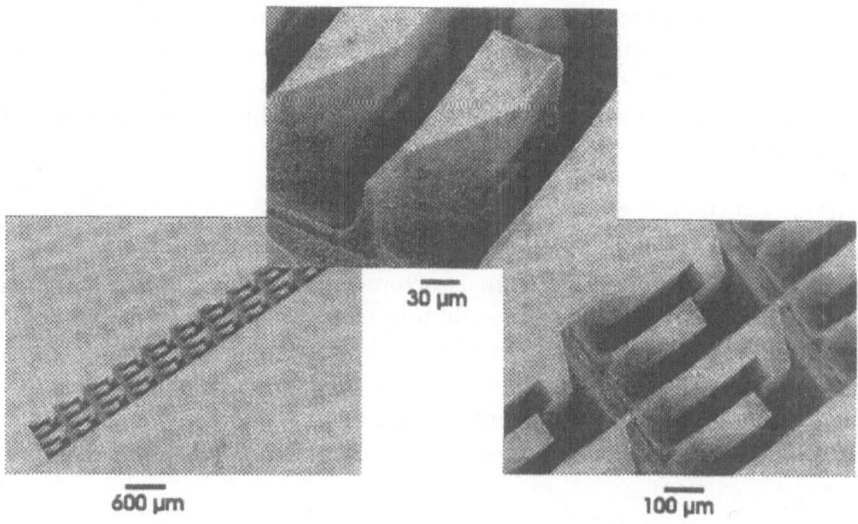

Figure 4. Static Mixer (Möbius-Type) pictured by scanning electron microscopy .

3.3 CONTINUOUS FLOW REACTION IN MICROCHANNELS

3.3.1 *Experimental*

Channels, 275 μm wide, in 1 mm thick PC-foil were sealed by applying a 250 μm sheet of PMMA by thermal diffusion and tubing adapters applied as described above.

The reduction of a solution of potassium permanganate with alkaline ethanol was used as a test reaction. The reaction process can be observed easily because the reactants as well as intermediates and products can be distinguished by their respective colors. The purple of the permanganate solution (MnO_4^-) is changed to a blue color due to a transition zone and mixing of the purple permanganate and the green color of manganate (MnO_4^{2-}). At slower flow velocities the green of manganate is visible. For very long reaction times this is followed by a yellow to brown color due to manganese dioxide (MnO_2). The brown precipitation can be removed completely by flushing the system with sulfuric acid immediately. The flow rate within the continuous flow system was in the range of a few ml/min.

3.3.2 *Results*

The continuous flow microreactor has been tested showing different phases of the reaction process between potassium permanganate and an alkaline solution of ethanol as a function of the flow rate. At fast flow rates a laminar flow pattern of the separated liquids, which do not react with each other during the reaction time defined by the length of the reaction channel, was observed. Reducing the flow velocity the reaction now takes place within the reaction channel forming manganate. At very low flow rates the reaction time in the system is sufficient for the formation of manganese dioxide. This example demonstrates that within a microchannel reaction conditions other than temperature can be defined in a very accurate way, thus enabling good control of the reaction process.

3.4 OPTICAL ANALYSIS SYSTEM

Fiber optical detection systems require structures with extremely high precision. Therefore the LIGA technique is a powerful tool to fulfill these requirements. LIGA structuring enables the design of a miniaturized "optical bench system". Precise positioning structures for ball lenses or fibers, for polarizers or other elements of various materials can be achieved. Figure 5 shows an optical detection system which can be used for measurements of UV-absorption or fluorescence.

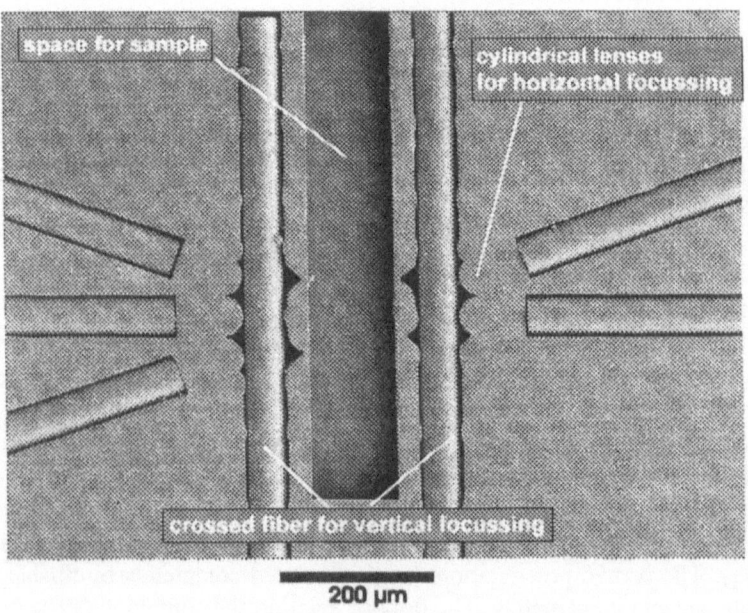

200 μm

Figure 5. Optical detection system.

4. Conclusion

The microreactor presented illustrates the possibility of using modularly constructed systems, which can be adjusted for specific applications. The combination of deep X-ray lithography together with excimer laser structuring provides greater advantages in microstructure fabrication, for instance, the high precision necessary for fiber optical units can be realized by means of the LIGA technique, whereas three dimensional structures necessary for the mixing unit can be obtained by micromachining with an excimer laser Packaging and interconnection techniques enable the combination of these units to a complete microsystem Using both X-ray lithography and excimer laser structuring only for a master the embossing technique allows low cost mass fabrication of such systems

A modularly constructed, continuous flow, microreactor has been developed containing a mixing unit, reaction unit and an analysis system A mixing unit has been presented which can be used either for mixing or for enhancing heat exchange within the system A chemical reaction has been performed in a continuous flow system whereby the flow rate allows control of the reaction process On-line analysis of the products can be achieved by integrating an optical detection system

Acknowledgments

We are grateful to Dr J Arnold for fabricating the excimer laser structures and H Leist for testing the systems

Literature

1a W Ehrfeld, H Lehr Deep X-ray lithography for the production of three-dimensional microstructures from metals, polymers and ceramics, Application Synchroton Radiation, Special edition of Radiation Physics and Chemistry, P Barnes, A Charlesby eds , in press

b E W Becker, W Ehrfeld, P Hagemann, A Maner, D Munchmeyer, *Microelectronic Engineering, 4* (1986) 35

2 T R Dietrich, M Abraham, J Diebel, M Lacher, A Ruf, *J Micromech Microeng , 3* (1993) 187-189

3 J Arnold, U Dasbach, W Ehrfeld, K Hesch, H Lowe, paper presented at E-MRS Spring Meeting, 24 - 27 May, 1994, Strasbourg, to be published in Applied Surface Science

4 A Manz, D J Harrison, E M J Verpoorte, J C Fettinger, H Ludi, H M Widmer, *Chimia, 45* (1991) 103 - 105

MICROELECTRODE ARRAYS AS TRANSDUCERS FOR MICROANALYSIS SYSTEMS

H. Meyer, B. Naendorf, M. Wittkampf, B. Gründig, and K. Cammann
Institut für Chemo- und Biosensorik e.V., Lehrstuhl für Analytische Chemie
Westfälische Wilh.-Univ. Münster, Wilh.-Klemm-Str. 8, D-48149 Münster, Germany

R. Kakerow, Y. Manoli, W. Mokwa, and M. Rospert
Fraunhofer-Institut für Mikroelektronische Schaltungen und Systeme
Finkenstr. 61, D-47057 Duisburg, Germany

0. Abstract

Two different microelectrode arrays are presented: the first array consists of 100 parallel connected ultramicroelectrodes. Counter as well as reference electrodes are integrated on the same chip. Although still exhibiting microelectrode features, this array yields analytical currents very similar to those of macroelectrodes. The second array consists of 400 individually addressable microelectrodes. It enables redundant as well as multi-analyte measurements. Here, we demonstrate the electrochemical imaging of two-dimensional distributions of ammonium chloride as well as urea.

1. Introduction

Microelectrodes are well-established instruments in medical science and diagnosis. Due to their small size, they are always used when bulky instruments fail. However, the current of a single microelectrode is very small. For that reason, we designed microelectrode arrays in silicon thin-film technology combining large overall currents with typical microelectrode features [1].

2. Chip-Layout and Results

2.1. ARRAY OF 100 PARALLEL CONNECTED ULTRAMICROELECTRODES

Ten arrays different in number, shape, size and arrangement of the electrodes have been designed and tested [2]. The optimized ultramicroelectrode array (UMA) consists of 10 x 10 platinum electrodes of a size of 2 μm in diameter and 20 μm interelectrode distance. In addition, a platinum counter electrode and two Ag/AgCl-reference electrodes were integrated on the chip. The chip with an overall size of 4.5 x 9 mm was fixed on an epoxy substrate, bonded and encapsulated with epoxy resin. AgCl was deposited with good adhesive strength on the sputtered silver electrodes by anodic treatment in 0.1 M KCl at a potential of +150 mV vs. Ag/AgCl/3M KCl for 60 s.

A. van den Berg and P. Bergveld (eds.), Micro Total Analysis Systems, 245–248.
© 1995 *Kluwer Academic Publishers.*

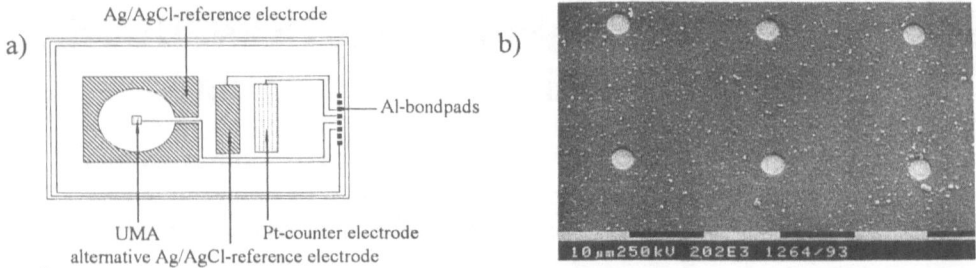

Figure 1. a) Schematic layout of the electrode arrangement.
b) SEM-picture of a part of the UMA.

Before starting the experiments the arrays were cleaned with dichromate-sulfuric acid followed by reduction of the formed platinum oxides at a potential of −160 mV vs. Ag/AgCl/3M KCl in 0.1 M KCl. The following figures demonstrate the different electrochemical behaviour of an UMA compared to a conventional macroelectrode.

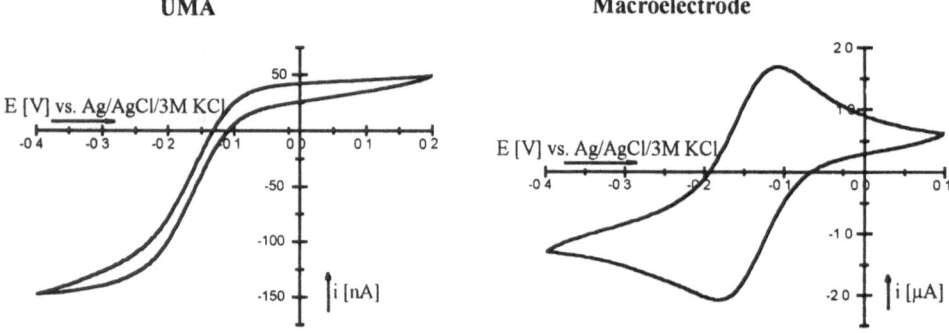

Figure 2. Cyclic voltammograms of an UMA and a 0.5 mm Pt-disk electrode, 5 mM $[Ru(NH_3)_6]^{3+}$ in nitrogen saturated 0.1 M KCl, 100 mV/s, T=298 K.

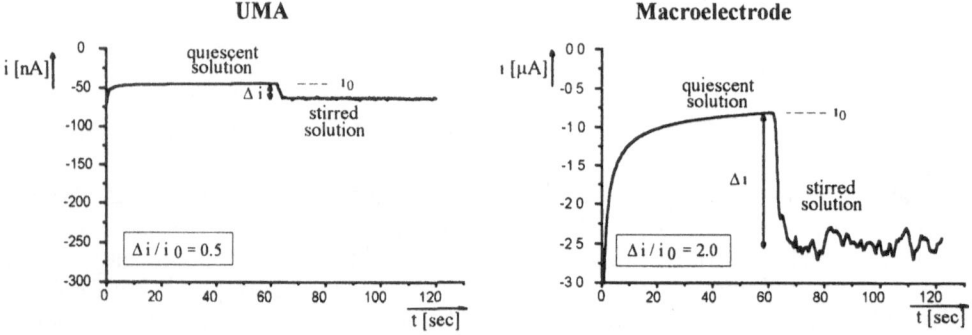

Figure 3. Stirring influence of an UMA and a 0.5 mm Pt-disk electrode on the reduction of dissolved oxygen in air saturated solution, T=298 K.

The experimental results confirm the characteristics of microelectrodes theoretically predicted In comparison with a conventional macroelectrode the array exhibits a decreased influence of stirring and an increased current density resulting in an excellent signal to noise ratio Furthermore, there is no peak-shaped cyclic voltammogram allowing measurements in quiescent as well as in stirred solutions

2 2 ARRAY OF 400 INDIVIDUALLY ADDRESSABLE MICROELECTRODES

Arrays of individually addressable microelectrodes offer some more applications Thus, redundant measurements of a single analyte will be possible as well as multi-analyte sensing Moreover, such an array can be used for electrochemical imaging of an analyte distribution over a two-dimensional area Therefore, a monolithic sensor array has been fabricated in a modified CMOS-process [3] It consists of 400 individually addressable sensor cells, arranged in a square matrix of 20 rows and 20 columns Individual addressing of each cell is possible by use of a horizontal and a vertical shift register and a control logic The size of a single sensor cell is 500 x 500 μm resulting in an array size of totally 1 x 1 cm Each cell contains a platinum electrode of 50 x 50 μm, a read-out amplifier and a sensor cell control unit The read-out amplifier converts the electrode potential to an analog output current The control unit switches the sensor cell into one of the three possible modes test mode, potentiometric mode or amperometric mode

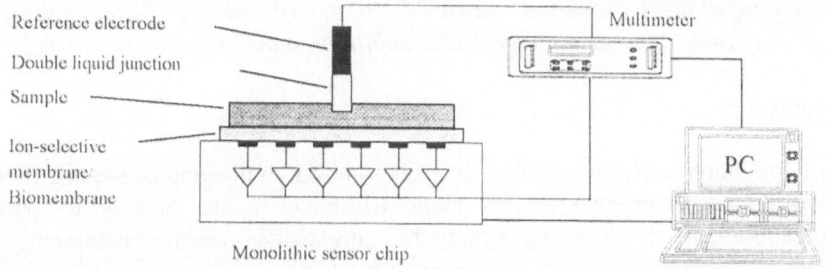

Figure 4. Schematic view of the experimental setup.

Chemical sensing and biosensing were performed using an ammonium-selective PVC-membrane and a urease-membrane, additionally [4] Both membranes were successively coated onto the sensor array The membranes were completely covered with pure buffer solution, whereas the double liquid junction was filled with 1 0 M ammonium chloride or urea in the same buffer solution Thus, after the contact of the double liquid junction and the buffer solution on the chip ammonium chloride or urea, respectively, diffused into the pure buffer solution on the chip, finally resulting in a homogeneous distribution all over the sensor array The distribution of ammonium chloride as well as urea was imaged by successive scans in potentiometric mode The scan time for a complete potentiometric scan was always 2 5 min

248

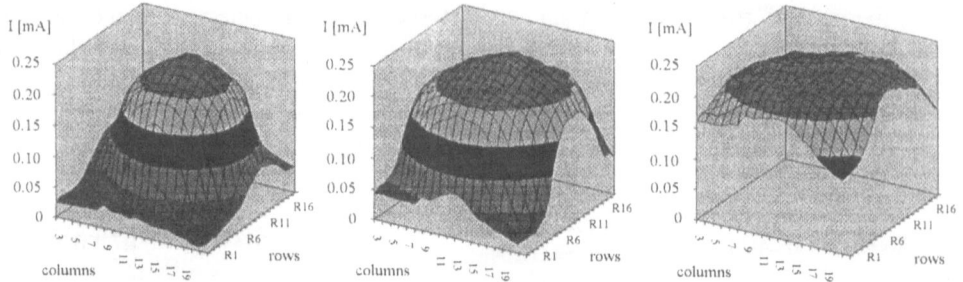

Figure 5   Imaging of ammonium chloride Scans were started 4 7 and 19 min after the contact of the double liquid junction and the buffer solution on the chip

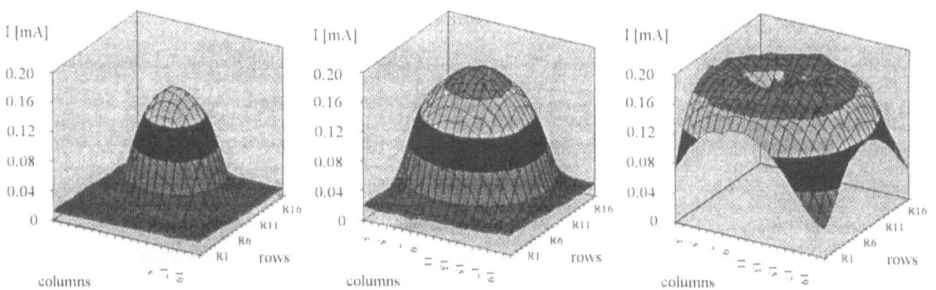

Figure 6   Imaging of urea Scans were started 7 10 and 19 min after the contact of the double liquid junction and the buffer solution on the chip

3. Conclusions

The array of 400 individually addressable electrodes allows an imaging of the distribution of some further analytes in amperometric mode Moreover it may be used for pattern recognition experiments Both arrays modified by any analyte-selective membrane can be used as chemical sensors or biosensors in micro total analysis systems

References

1   R M Wightman Voltammetry at ultramicroelectrodes
     in A J Bard (ed) Electroanalytical Chemistry Vol 15 Marcel Dekker New York Basel 1989
2   B Roß Ultramicroelectrode arrays as transducers for new amperometric oxygen sensors
     Sens Actuators B7 (1992) 758-762
3   R Kakerow A monolithic sensor array of individually addressable microelectrodes
     Sens Actuators 145 (1994) 296-301
4   H Meyer Chemical and biochemical sensor array for two-dimensional imaging of analyte
     distributions Sens Actuators B18 (1994) 229 234

Acknowledgement
This work was supported by the Bundesministerium fur Forschung und Technologie (BMFT Grant No 414-4013-13 MV 0357) and the Ministerium fur Wissenschaft und Forschung des Landes Nordrhein-Westfalen

A STACKED MULTICHANNEL AMPEROMETRIC DETECTION SYSTEM

M.Paeschke, U.Wollenberger[1] , A.Uhlig, U.Schnakenberg,
B.Wagner and R.Hintsche
Fraunhofer-Institute for Silicon Technology,
Dillenburger Str.53, D-14199 Berlin, Germany

0. Abstract

An universal stacked electrochemical detection system is described consisting of a noble metal electrode array, a flow chamber, and a miniaturized microprocessor controlled potentiostat. The electrode array and the flow chamber have been fabricated in silicon technology. The electrodes are arranged in four interdigitated electrode arrays (IDA). These IDA electrodes are controlled independently with a multichannel potentiostat (multipotentiostat). They are suited for the detection of redoxmediators. The combination of these components and biologic recognition elements in combination with various voltammetric procedures results in an electrochemical system for immunoanalysis.

1. Introduction

There is an increasing interest to integrate chemical and biological sensors in microsystems. The miniaturization of analytical systems allows rapid and efficient detection and separation. For the measurement samples are used with volumes down to the nano and femtoliter range.

The electrochemical detection with microelectrodes is introduced as a promising technique for microanalytical techniques. Microelectrodes have received a great interest for these systems because they exhibit a fast establishment of steady-state or quasi-steady currents, diffusion-controlled currents, low charging currents, and reduced solution resistance effects [1]. The microfabrication of electrodes in silicon technology opens new dimensions in structuring of transducer elements for microsize sensor probes. The arrangement of more than one microelectrode in an array of discs or bands can be easily designed and fabricated. Independent microelectrode bands arranged as an interdigitated array allow novel electrochemical measurements. One method is the amperometric amplification through redox recycling of mediators, which are e.g. analyte itself (dopamine) or reactions product of enzyme reactions (p-

---

[1] present adress University of Potsdam, c/o MDC 13122 Berlin

A. van den Berg and P. Bergveld (eds.), Micro Total Analysis Systems, 249–254.
© 1995 *Kluwer Academic Publishers.*

aminophenol) [2,3] The measurement is based on the redox cyclization between the adjacent microband electrodes of an interdigitated array electrode An other amperometric application is the detection of current of the individual electrodes to which different potentials were applied in order to obtain three dimensional results (time, current, potential) This is particular interesting for liquid chromatography [4]

Up to now miniaturized immunoanalytical systems based on electrochemical transducers did not succeed We combined electrochemical microelectrodes with micromachined flow components and multichannel detection in a novel complete system

2. Experimental

A multielectrode amperometric detection system has been developed as a part of a miniaturized chemical system for immunological sensing This system consists of a thin film electrode array, a flow chamber, and the structured inlet The microelectrode array is arranged as an assembly of independent interdigitated microband electrodes This array has been fabricated by photolithographic and electron-beam lithographic techniques on silicon substrates Thin-film Pt-(or Au-) electrodes have been evaporated on a Ti-adhesion layer The dimensions of the electrode width and space varied between 2 μm and 0 3 μm [5] The electrode chip has been pasted on a chip carrier which contained the connecting wires This arrangement forms an exchangeable unit

The flow cell on top of the electrodes was fabricated using double side anisotropic silicon etching in KOH The etched window of the measurement channel has the dimensions of the active area of the electrode chip (3 x 0 9 mm²) The volume of the channel can be adjusted between 0 027 and 1 8 mm^3 The bottom of the resulting cell was covered with a silicon rubber film that serves as packing to the electrode chip The inlet is structured as grating with holesizes between 20 x 100 μm² and 40 x 40 μm² The inlet tube works as a stainless steel auxiliary electrode, the outflow tube as a silver-/silverchloride reference electrode, respectively Both tubes were pasted in an acrylic holder, wich can be attached to a chip-carrier The channel of the flow cell serves as reaction chamber for biological reactions This chamber was filled with polymeric beads (0 02 - 0 1 mm diameter) on which enzymes or antibodies was immobilized (Fig 1) The beads are held back in the reaction room due to the gratings

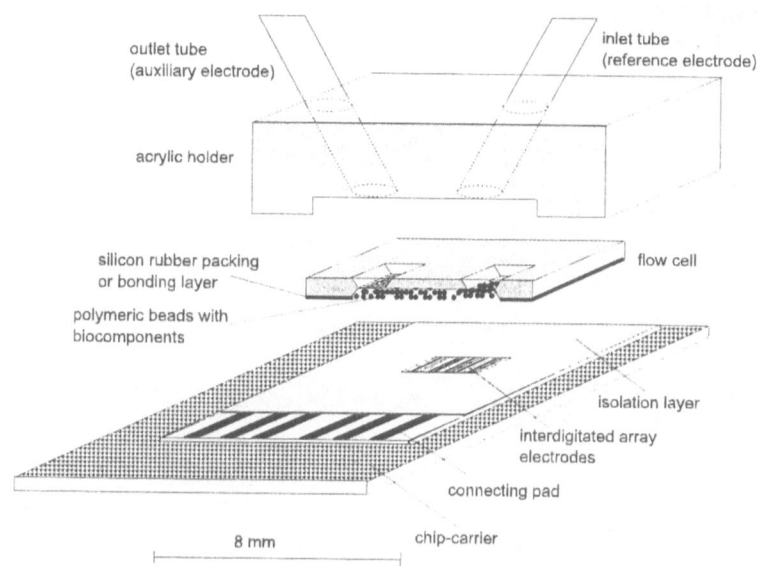

Figure 1 Schematic view of the amperometric detection system

The measurements were performed using a microcontroller equipped multipotentiostat The multipotentiostat has a capacity for up to 16 independent working electrodes providing the different potentials to the electrodes The microcontroller is able to perform all electrochemical measurement procedures, such as cyclic voltammetry, difference puls voltammetry, chronoamperometry and also control the analyte dosage and transport via pumps and valves

3. Results and discussion

The amplification of the amperometric response of redox species has been studied for p-aminophenol The sensitive detection of this compound is of particular interest since it is the basis for the measurement of enzymes such as alkaline phosphatase and β-galactosidase, which are frequently used in enzyme immunoassays The electrode active species (p-aminophenol) is liberated in the enzyme catalysed hydrolysis of its aminophenylated substrate which itself exhibit an irreversible electrochemical behavior The utility of p-aminophenyl phosphate in enzyme immunoassays in combination with IDA electrodes has been demonstrated in [6] The unfavorable pH-optimum of alkaline phosphatase is the drawback of this system Because of its neutral pH-optimum β-galactosidase appeared to be more advantageous An illustration of the proposed electrochemical enzyme immunoassay is given in Fig 2

252

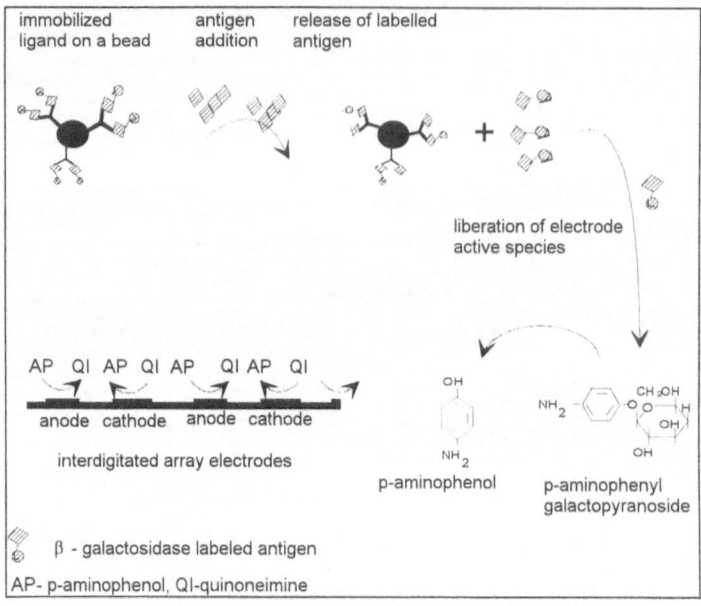

Figure 2. Principle of electrochemical immunoassay with redoxcycling amplification

p-Aminophenol enzymatically liberated from a p-aminophenyl-β-galactoside is oxidized to the respective quinoneimine at the anode, yielding an oxidation current. At the cathode the formed quinoneimine is reduced to p-aminophenol (cathodic current). In contrast, the aminophenylated substrate doesn´t give a response at the potential applied. In order to define the optimum potentials for oxidation and reduction of the mediator, cyclic voltammograms in single mode (Fig. 3b, without cathode) and in recycle mode (Fig. 3a) were recorded. Fig. 3c shows the dependence of the anodic and cathodic currents at one pair of IDA electrodes on p-aminophenol concentrations when the anode potential was fixed at +350 mV and the cathode potential was held at -150 mV. The p-aminophenol is recycled between the electrodes and an electrochemical amplification is observed The addition of 10 nmol/l p-aminophenol results in an anodic current of ≈0 4 nA within 1 s. A similar response is registrated at the cathode. These first measurement were performed with the flow cell under continuous flow realized with a peristalic pump. In batch the sensitivity for this mediator in recycle mode is about 16.4 nAl/μmol (to compare in single mode 0.2 μAl/μmol) [3]. Furthermore experiments are being carried out to investigate the redox recycling effects with respect to flow rate and flow cell volume.

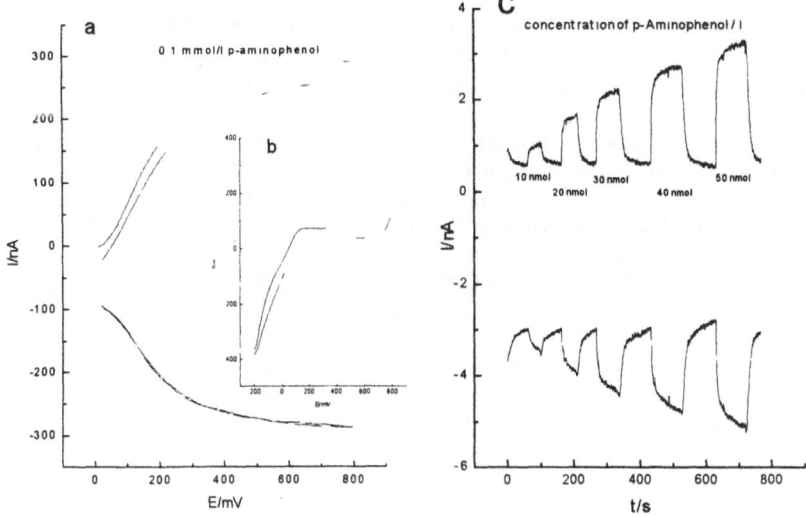

Fig 3.Cyclic voltammogram of p-aminophenol in 0.1 mol/l phosphate buffer in recycle mode (a) and single mode (without cathode) (b). The sweep rate was 20 mV/s versus Ag/AgCl. The used IDA electrode had a electrode width of 1 μm , a gap between two electrode fingers of 500 nm, and 70 electrode finger per band. The concentration dependent flow measurement is shown in Fig. 3c. Interdigitated electrodes were polarized at +350 and -150 mV versus Ag/AgCl.

4. Conclusions

The principle of the sensitive detection of mediators by redox recycling was demonstrated for measurements of p-aminophenol on interdigitated electrodes.

The microfabrication of electrodes is very promising for electrochemical probes. By means of silicon technology the integration of an interdigitated electrode array into a complete device is useful. The high integration of electrode structures is attractive for enhancing the sensitivity and selectivity of sensors. The combination of sensor chips and microfluidic elements using common production and mounting techniques promise to create novel modules for analytical microsystems

References

1 K Aoki, Theory of ultramicroelectrodes, *Electroanalysis*, 5 (1993), 627-639

2 O Niwa, M Morita, H Tabei, Highly sensitive and selective voltammetric detection of dopamine with vertically separated interdigitated array electrodes, *Electroanalysis*, 3 (1991), 163-168

3 U Wollenberger, M Paeschke, R Hintsche, Interdigitated array microelectrodes for the determination of enzyme activities, *Analyst*, 119 (1994), 1245-1249

4 J C Hoogvliet, J M Reijn, W P van Bennekom, Multichannel amperometric detection system for liquid chromatography and flow injection analysis, *Anal Chem*, 63 (1991), 2418-2423

5 M Paeschke, U Wollenberger C Kohler, T Lisec, U Schnakenberg, R Hintsche, Properties of interdigital electrode arrays with different geometries, accepted for *Analytica Chimica Acta*, (1994)

6 O Niwa, Y Xu, H B Halsall, W R Heinemann, Small-volume voltammetric detection of 4-aminophenol with interdigitated array electrodes and its application to electrochemical enzyme immunoassay *Anal Chem* 65 (1993), 1559-1563

COMPONENTS FOR MICROFLUIDIC HANDLING MODULES

W.K. Schomburg, B. Büstgens, J. Fahrenberg, D. Maas
Karlsruhe Nuclear Research Center
Institut für Mikrostrukturtechnik

0. Abstract

Microanalysis systems which are built up in a modular way can be adapted to various applications by exchanging individual modules. Thus, a bigger number of units can be sold and prices reduced. At the Karlsruhe Nuclear Research Center a microfluidic handling module is being developed which will be used in two different microanalysis systems joined with sensor modules. A batch fabrication process has been developed which was used to manufacture micropumps and microvalve systems by assembly of two molded thermoplastic parts and transfer of a polyimide diaphragm from a silicon wafer.

1. Introduction

Microsystems are expected to become a reliable, quick, portable and cost effective tool in the chemical analysis of gases and liquids [1 - 8 and this issue]. Such microanalysis systems will become an economic success, if sold in great numbers. Therefore, a modular setup of microanalytical systems is desirable which allows different sensor and actuator modules to be arranged so as to satisfy the needs in various applications. In addition, the output can be enhanced in this way, because modules tested separately can be combined to microsystems. For these reasons, several groups are working on modular concepts for microanalysis systems [6-9].

One of the most important advantages of the modular design is that a market for micro-analytical systems can develop. A single small or medium sized enterprise need neither to be capable of fabricating all components of a microanalysis system nor to have all the fabrication processes advantageous to microanalysis systems such as silicon etching, surface micromachining, electroplating in thick resist, LIGA etc. available in-house. It is possible to participate in the market by offering just a module for a special purpose fabricated with a special process. To achieve the development of such a market a standardisation is essential which enables the combination of modules fabricated with very different techniques and does not exclude a further development.

A. van den Berg and P. Bergveld (eds.), Micro Total Analysis Systems, 255–258.
© 1995 *Kluwer Academic Publishers.*

The diaphragm stress can be controlled by the temperature during bonding, because the difference in thermal expansions of the silicon wafer and the thermoplastic material has an influence on the stress. The first molded part and the polyimide are mechanically removed from the wafer (cf. Fig. 2b). Small diaphragms are removed from the substrate very easily, but even large membranes for X-ray masks are routinely removed with the help of separation layers [13]. Orifices are not damaged during diaphragm removal if their size is on the order of some hundred micrometers or less. Pins and corresponding orifices facilitate the alignment of the second molded part which is adhesively bonded similar to the first one (Fig. 2c). It is not necessary to pattern orifices in the diaphragm, for accommodation of alignment pins, because they are punched out very easily while the two microstructures are joined. All fabrication steps up to this point are carried out in parallel. Then samples are cut into pieces and the individual components are provided with electrical and fluidic contacts.

3. Applications

Figure 3 shows a micropump fabricated by thermoplastic molding and diaphragm transfer. The overall dimensions are $7 \cdot 10 \cdot 2 \, mm^3$. The 1 μm thin polyimide serves as diaphragms for pumping and for the check valves. A resistive copper heater is periodically powered to drive the pump. 1.7 ms short current pulses of 100 mA, frequency 30 Hz, were used to deliver unfiltered air at rates of up to 220 μl/min, and a maximum pressure of 130 hPa was generated. A micropump did not fail in a continuous run of 50 hours at 30 Hz.

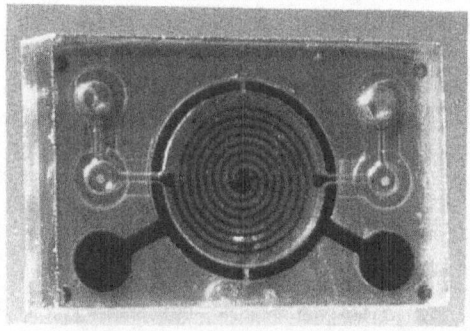

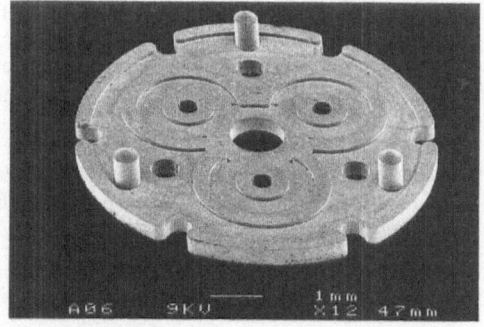

Figure 3. Micropump fabricated by molding and diaphragm transfer.

Figure 4. Lower part of a microvalve system consisting of three individual valves with a common inlet in the center. The design includes pins for alignment and fluidic ports for injection of adhesive.

Microvalve systems were fabricated from PMMA in a similar process. Figure 4 shows the complex structure of the upper and lower parts of the valve system. The diameter of

258

the system and of individual valves are 7 mm and 2 mm, respectively, while the depth of the flow channels and the thickness of the system are 70 μm and 1.9 mm, respectively. With an electrical power of 116 mW the microvalves closed a nitrogen flow supported at 130 hPa. At a pressure difference of 100 hPa a water flow of more than 50 μl/min was attained.

The micropump and microvalves are driven at low voltages (typically 15 V and 3 V, respectively) which are easily obtained from small batteries. This facilitates integration into microfluidic handling modules for microanalysis systems.

4. Conclusions

Micropumps and microvalves have been fabricated in a batch fabrication process by thermoplastic molding and transfer of a thin diaphragm. These examples show that microfluidic handling modules for microanalysis systems can be fabricated in a similar way. The success on the market of microanalytical systems is dependent on the development of a standard that allows modules fabricated with different techniques to be combined.

References
1 S. Shoji, M. Esashi, T. Matsuo, Prototype Miniature Blood Gas Analyser Fabricated on a Silicon Wafer, *Sensors and Actuators 14* (1988) 101 - 107.
2 A. Manz, N. Graber, H.M. Widmer, Miniaturized Total Chemical Analysis Systems: A Novel Concept for Chemical Sensing, *Sensors and Actuators B1* (1990) 244 - 248.
3 D.J. Harrison, Z. Fan, K. Seiler, K. Furri, Miniaturized Chemical Analysis Systems Based on Electrophoretic Separation and Electroosmotic Pumping, *Transducers '93, Digest of Technical Papers* (1993) 403 - 406
4 W. Hoffmann, H. Eggert, W. Schomburg, D. Seidel, K.-H. Freywald, Elektrochemisches Mikroanalysesystem (ELMAS) fur die Ionometrie von Flussigkeiten, *Abstract of papers presented at Symposium Mikrotechnik, Achema '94* (1994).
5 W.K. Schomburg, R. Rapp, B Bustgens, J. Reichert, O. Fromhein, Mikromembranpumpen als Elemente eines optochemischen Mikroanalysesystems, *KfK-Report No 5238* (1993) 78 - 84
6 B. H. van der Schoot, S. Jeanneret, A. van den Berg, N. F. de Rooij, Modular Setup for a Miniature Chemical Analysis System, *Sensors and Actuators B6* (1992) 57 - 60
7 A. van den Berg, P Bergveld, D.N Reinhoudt, J.H.J. Fluitman, From microsensors to micro total analysis systems (μTAS), *Abstract of papers at Symposium Mikrotechnik, Achema '94* (1994).
8 M.T. Pham, S. Howitz, T. Vopel, A. Steinbach, H. Hanke, Entwicklung eines Fluidik-ISFET-Mikrosystems zur chemischen Analyse, *Abstract at Symposium Mikrotechnik, Achema '94* (1994).
9 W.K. Schomburg, J Vollmer, B. Bustgens, J. Fahrenberg, H. Hein, W. Menz, Microfluidic Components in LIGA Technique, *J Micromech Microeng , vol 4/3* (1994).
10 B. Bustgens, W.Bacher, W. Bier, R Ehnes, L. Keydel, D. Maas, R. Ruprecht, W.K. Schomburg, Micromembrane Pump Manufactured by Molding, *Proc Actuator'94* (1994) 86 - 90.
11 J Fahrenberg, W Bier, D Maas, W Menz, R. Ruprecht, W.K. Schomburg, Microvalve System Fabricated by Thermoplastic Molding, *Proc Micro Mechanics Europe '94* (1994).
12 D. Maas, B. Bustgens, J. Fahrenberg, W. Keller, D. Seidel, Application of Adhesive Bonding for Integration of Microfluidic Components, *Proceedings Actuator'94* (1994) 75 - 78.
13 W.K. Schomburg, H.J. Baving, P Bley, Ti- and Be- X-Ray Masks with Alignment Windows for the LIGA Process, *Microelectronic Engineering 13* (1991) 323 - 326.

DEVELOPMENT OF A MICRO FLOW-SYSTEM WITH INTEGRATED BIOSENSOR ARRAY

G.Urban, G. Jobst, P.Svasek, M. Varahram, I. Moser, E.Aschauer
Institute für Allgemeine Elektrotechnik und Elektronik and Ludwig Boltzmann
Institut für Biomedizinische Mikrotechnik und Hirnkreislaufforschung,
Gußhausstr 27, A-1040 Vienna, Austria

Abstract

A micro flow-system was developed which consists of a chip with an integrated sensor array and a photo-patterned seal The flow channel was produced using hybrid technologies which allow the forming of flow chambers with less than 1µl internal volume The used integrated sensors for measuring glucose and lactate concentrations were produced using thin film technology and were evaluated in whole blood The underlying measuring principle is the electrochemical oxidation on a modified platinum working electrode of H_2O_2 which is produced by the enzymes glucose oxidase and lactate oxidase These enzymes were immobilised in a pHEMA membrane which can be photo patterned and additionally covered by a spacer and a catalase containing cover layer These biosensors were tested in undiluted blood with a linear range up to 40 mmol/l for glucose and 25 mmol/l lactate, a 95% rise time of 20 sec and an operational life time in undiluted serum of two weeks at ambient temperature The influence of interferents and flow conditions are negligible Such a micro flow-system was used as a laboratory on chip for decentralised acute analysis and as micro flow analysing system for ex-vivo monitoring during diabetological tests

1. Introduction

The demand for fast and reliable measurements in analytical chemistry, medicine, biotechnology and environmental sciences has evolved the need for small, easy to handle and inexpensive analytic devices For clinical purposes different parameters have to be measured at once [1,2] One possibility to develop such micro devices is the use of integrated chemo- and biosensors implemented in a micro flow system The urgency to use micro systems is caused by the fact that an analysing system has to handle analytes as serum or whole blood in a practical way for clinicians

259

A van den Berg and P Bergveld (eds), Micro Total Analysis Systems, 259–262
© 1995 *Kluwer Academic Publishers*

Different problems have to be overcome

1 To develop microsensors capable of measuring metabolic parameters in undiluted biological fluid which can be produced in large scale and
2 To implement microsensor arrays in a micro flow-system

Microelectronic sensors for pH determination and glucose measurements were produced utilizing ISFET-technology and thin film technology [3-7] Now none of the cited microelectronic sensors are able to work well in undiluted blood and even few conventional sensor systems are able to measure glucose and lactate concentration in undiluted media [9]

To overcome the described problems the solution is, beneath the dilution method, to utilize complex membrane systems [8,9] A big advantage is the UV initiated free radical crosslinking of the polymer directly on the substrate thereby designing the physical-chemical properties of the biosensor membrane

A thin film process was introduced which fulfills all technological requirements and overcome the mentioned drawbacks of blood measurements by using an electropolymerized semipermeable membrane on electrochemical Pt-working electrodes and different functionalized pHEMA-hydrogel layers that were structured by photolithography [8] The resulting glucose and lactate sensor device was tested in buffer solution and undiluted serum and whole blood To implement such integrated sensor arrays in a micro flow-system a hybrid solution was chosen and the whole system was tested during in vitro and ex vivo clinical trials

2. Experimental

The used integrated sensors for measuring glucose and lactate concentrations were produced using thin film technology The underlying measuring principle is the electrochemical oxidation on a modified platinum working electrode of H_2O_2 which is produced by the enzymes glucose oxidase and lactate oxidase These enzymes were immobilised in a photo patterned pHEMA membrane and additionally covered by a spacer and a catalase containing cover layer which was described in [8]

The micro flow-system consists of a chip with an integrated sensor array and a photo patterned seal The chip consists of five electrodes with a measuring area of 0 5 mm x 0 5 mm and interdistances of 800 μm Four Pt-electrodes act as working electrodes measuring concentrations of glucose, lactate, glutamate and glutamine [10] The fifth electrode is a Ag/AgCl-reference electrode [8]

The flow channel was produced using hybrid technologies which allow the forming of flow chambers with less than 1 μl internal volumes

On a base printed circuit board the sensorchip was mounted via a spacer layer This spacer was realised by a 100 μm thick photopatterned VACREL^R layer (Fig 1)

μ-FLOW-CELL

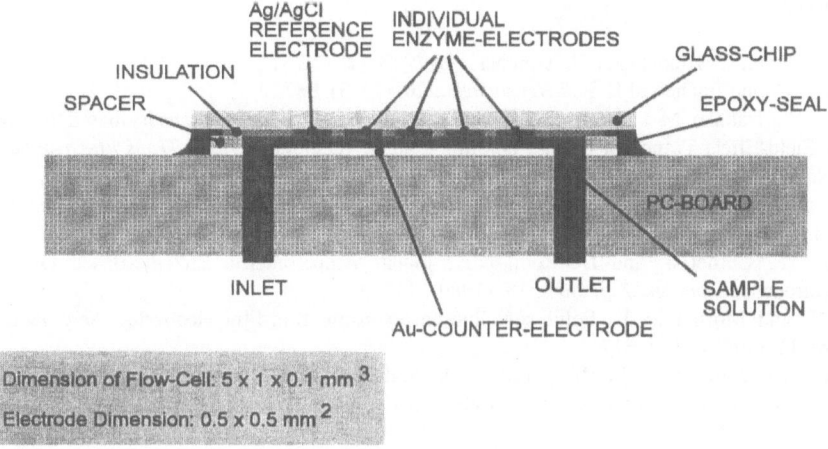

Figure 1. Cross section of the micro flow-cell showing the sensor array, the flow channel and the base plate.

3. Results and Discussions

The biosensors were tested in undiluted blood with a linear range up to 40 mmol/l for glucose and 25 mmol/l lactate, a 95% rise time of 20 sec and an operational life time in undiluted serum of two weeks at ambient temperature. The influence of interferents and changes of flow conditions are negligible [8]

The desired application is the laboratory on chip for decentralised acute analysis and the use as micro flow analysing system for ex-vivo monitoring during surgery or in intensive care units The integrated micro flow-chip was tested in fermentation broth of mammalian cell cultures and with undiluted heparinized whole blood during oral glucose tolerance tests continuously performed on volunteers for six hours The correlation between sensor reading and a clinical glucose analyser is $r^2 = 0.98$

This results highlights the possibility to handle biological fluids with microsystems and to measure undiluted biological fluid continuously with high precision Using hybrid technology an inexpensive and easy accessible tool for creating microsystems was successfully established

4. References

1 Astrup J , Symon L Branston N Lassen Na , *Stroke*, 8, (1977), 51

2 Strang R H C and Bachelard H S *J Neurochem* , 20, (1973), 987

3 Hanazato Y , Nakako M , Satorus S , Mitsuo M , Integrated Multi-Biosensors based on Ion-sensitive Field-Effect Transistor Using Photolithographic Techniques, *IEEE Trans Electron Dev* , Vol 36,No 7, (1989) 1303-1310

4 Shoji S and Esashi M , Micro flow cell for blood gas analysis realizing very small volume, *Sensors and Actuators B* 8, (1992) 205-208

5 Koudelka M , Gernet S and De Rooij N F , Planar Amperometric Enzyme-Based Glucose Microelectrode, *Sensors and Actuators*, 18 (1989), 157-165

6 Takatsu I and Moriizumi T Solid state biosensors using thin-film electrodes, *Sensors and Actuators* 11 (1987) 309-317

7 Mastrototaro J Johnson K Morff R Lipson D , Andrew C and Allen D , An electroenzymatic glucose sensor fabricated on a flexible substrate, *Sensors and Actuators B5*, Nr 1-4, (1991), 139-145

8 G Urban, G Jobst, E Aschauer, O Tilado, P Svasek and M Varahram, Performance of integrated glucose and lactate thin film microbiosensors for clinical analyzers, *Sensors and Actuators B*,Vol 19, No 1-3 (1994) 592-596

9 Pfeiffer D Scheller F W and Setz K Amperometric enzyme electrodes for lactate and glucose determinations in highly diluted and undiluted media, *Analytica Chimica Acta*, in press

10 I Moser, G Jobst E Aschauer P Svasek and G Urban, Miniaturized thin film glutamate and glutamine biosensors *Biosensors & Bioelectronics*, in press

TEMPERATURE CONTROLLER
FOR µTAS APPLICATIONS

P. van Gerwen, K. Baert, T. Slater[1], L. Hermans and R. Mertens
IMEC, Kapeldreef 75, 3001 Leuven, Belgium
[1]Wildcat Micromachining, 1032 Irving Box 227, San Fransisco, CA 94122, USA

0. Abstract

A temperature controller is a basic component of a µTAS. This controller regulates the temperature of a fluid, including cooling and heating. The advantages of a Peltier element, a thermal switch and a fluid cycle are compared taking into account parameters as volume that can be handled, temperature range that can be achieved and response time. Also the possibility for integration is studied.

1. Introduction

Miniaturization of systems for chemical analysis offers an improved efficiency with respect to sample size, response time and reagent consumption [1,2]. In microreactors, temperature control can be very important. Integrated cooling and heating has however hardly been studied. In this work we compare three possible techniques for integrated temperature control: a Peltier element, a thermal switch and a fluid cycle.

2. Peltier element

A Peltier element cools and heats, based on the Peltier effect. Whenever a circuit composed of two dissimilar conductors carries an electric current, heat is evolved at one junction and absorbed at the other (fig. 1). The rate at which heat is absorbed is proportional to the current and depends on the nature of the two materials (e.g. a p- and n- type semiconductor) comprising the junction. We can write $Q_{np} = S_{np}TI$, where S_{np} is the relative Seebeckcoefficient between the materials n and p [3]. Taking Joule heat production (RI^2) and conduction heat ($K\Delta T$) into account, heat absorbed at the cold junction and evolved at the hot junction is given by [3]

$$Q_c = ST_cI - K\Delta T - \tfrac{1}{2}RI^2,$$

$$Q_h = ST_hI - K\Delta T + \tfrac{1}{2}RI^2. \tag{1}$$

263

A. van den Berg and P. Bergveld (eds.), Micro Total Analysis Systems, 263–266.
© 1995 *Kluwer Academic Publishers.*

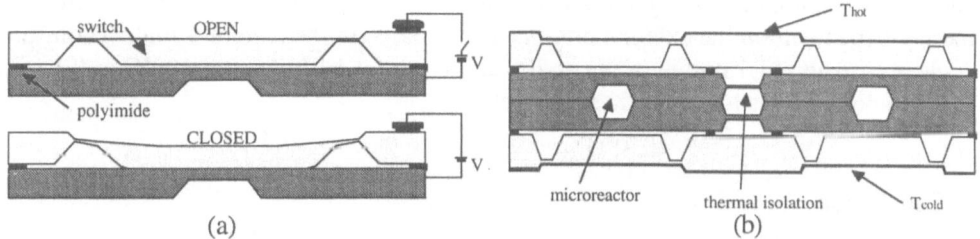

Figure 3 Thermal switch; (a) basic design; (b) integration.

through two thermal switches. When a thermal switch is closed there is almost no thermal resistance and when the switch is open there is a much larger thermal resistance. Heating is done by closing the switch to the hot reservoir and opening the switch to the cold reservoir, cooling is done by the inverse configuration. Defining $p = R_{open}/R_{close}$, the temperature in a reactor can be regulated between:

$$\frac{p \cdot T_{cold} + T_{hot}}{1+p} \leq T \leq \frac{T_{cold} + p \cdot T_{hot}}{1+p} \tag{2}$$

This device allows to achieve different temperatures in adjacent reactors (fig. 3b) and thus to control the temperature locally. The performance of this device can well be described by p and by $R_{tot} = R_{close} + R_{open}$. p should be high because than T_{max} and T_{min} are close to T_{hot} and T_{cold} (eq. 2); Rt_{ot} should be high to reduce the total amount of heat flowing from the hot to the cold reservoir: $Q_{tot} = R_{tot} \cdot (T_{hot} - T_{cold})$.

Thermal switches can be integrated (fig. 3). To surround a reactor of 1 μl, our square device has a surface of 7 mm^2. The switch is actuated electrostatically. It is not so difficult to make a good contact, the thermal contact resistance can be kept low [8]; R = 0.12 K/W. It is more difficult to get the resistance high in the open state because the difference between the switch and the base plate should be below 20 μm not to have to high actuation voltages. In the open state, the thermal resistance from the switch to the channel varies from 91 K/W (open) to 0.45 K/W (closed). Then p = 201 which allows a big temperature difference; for a cold reservoir at 0°C and a hot at 100°C, the temperature can vary between 0.5°C and 99.5°C. There is still a rather big conduction from the hot to the cold reservoir when one switch is in the open-state and one switch is in the close-state (R_{tot} = 91 K/W). For one reservoir at 0°C and one at 100°C there is a continuous heat flow of 1.1 W. This device can cool and heat 1 μl water with a time constant of 4 ms.

4. Fluid cycles

Cooling and heating can also be done by cycling a fluid. In a two-phase Rancine cycle, condensation heats and evaporation cools. In a Joule-Thomson cycle it is expansion of a gas which cools [9]. It is still impossible to completely integrate these devices because some parts (e.g. pumps, compressors) are hard to integrate. However condensors, evaporators or expansion valves can be integrated. This makes that the cooling or heating part can be integrated. The performance of these devices is however merely dependent on the external non-integrated parts. In the case of a Joule-Thomson cycle, expansion of 18l/min of N_2 at 200 atm can cool 1 μl of water 100°C in 0.1 sec [9]. Temperatures as low as 77 K can be achieved.

5. Conclusion

It is possible to completely integrate a Peltier element. When working with good elements ($z > 3 \ 10^{-3}$ K) and a clever design, good temperature ranges (10 - 80°C) and fast response times (3 ms) can be achieved for sample volumes of not more than 1 nl water. The disadvantage is that only small volumes can be handled with these smaller Peltier elements, only larger elements (> 1 cm long) can comprise volumes of some μl

In the design with two thermal switches only the control is integrated, two external heat reservoirs are needed. Fast response times (4 ms) can be achieved for as much as 1 μl of water A drawback is that there is a continuous heat flow of more than 1 Watt to achieve temperature ranges from 0 to 100°C

Fluid cycles allow only a very minimal integration. Temperature range and response time are completely dependent on external parameters.

Acknowledgement

The author would like to thank the IWONL (Institute for promotion of scientific research in industry and agriculture) for financially supporting this work

References

1   D J Harrison and P G Glavina, Towards Miniaturized Electrophoresis and Chemical Analysis Systems on Silicon an Alternative to Chemical Sensors, *Sensors and Actuators B*, 10 (1993) 107-116

2   B H van der Schoot, S Jeanneret, A van den Berg, N F de Rooij, A modular miniaturized chemical analysis system, *Sensors and Actuators B*, 13-14 (1993) 333-335

3   A F Ioffe, *Theory of Thermoelectric Cooling*, Infosearch, London, 1957

4   R R Heikes and R W Ure, *Thermoelectricity Science and engineering*, Interscience, New York, 1961, 458-517

5   G V Samsonov, N F Podgrushko, N F Selivanova, M I Lesnaya, L A Dvorina, *Test Method and Properties of Materials*, Plenum Publishing Corporation, New York, 1975

6   H R Meddins and J E Parrot, The Thermal and Thermoelectric Properties of Sintered Germanium-Silicon Alloys, *J Phys C Solid State Phys*, 9 (1976) 1263-1276

7   T Slater, R Prinz, P Van Gerwen, K Baert, E Masure and F Preud'homme, ART A Novel Thermal Valve for Temperature Control Applications, *MME '94 Workshop* (1994)

8   H Yuncu, S Kakaç, Thermal Contact Conductance - Theory and Applications, *Proceedings of the NATO Adv Study Institute on Cooling of Electronic Systems*, (1993) 29-54

9   W A Little, Microminiature Refrigeration, *Rev Sci Instrum*, 55 (1984) 661-680

REDOX-SENSITIVE FIELD-EFFECT TRANSISTORS AS TRANSDUCERS FOR MICRO-ANALYSIS SYSTEMS

T. Vering, D. Seiwald, W. Schuhmann, H.-L. Schmidt
Lehrstuhl fur Allgemeine Chemie und Biochemie
Technische Universitat Munchen, Vottinger Straße 40
D-85350 Freising-Weihenstephan, Germany

0. Abstract

A new type of (bio)chemical sensor, the redox-sensitive field-effect transistor is described It consists of a conventional ISFET with a noble metal added on top of the gate insulator The gate electrode is modified with a redox polymer containing osmium complexes The potentiostatic multi-puls method is introduced which allows the adjustment of the redox potential of the gate to a desired value in a stepwise way It is shown that the open circuit potential after switching off the potentiostat is a good measurement of the presence of the redox active species NADH

1. Introduction

The modification of the ISFET gate by immobilization of various enzymes results in a number of enzyme-FETs (EnFETs) [1] Generally hydrolytic enzymes e g urease or penicillinase were applied Principal problems of these sensors rise from the dependence on the buffer capacity and the pH-sensitivity of the enzyme´s activity

As many biologically relevant intermediates are converted by nicotinamide-(NAD$^+$-) dependent dehydrogenases, they are especially interesting for the application in biosensors The measurement of redox equivalents transferred during an dehydrogenase catalyzed reaction could thus largely extend the applicability of FETs In contrast to the pH-change measured with ISFETs, a shift of the local redox potential should be measured An additional layer of a noble metal on the gate insulator renders the FET sensitive to redox potentials Redox reactions catalyzed by immobilized enzymes could be potentiometrically monitored However, problems arise from the slow electron transfer kinetics of the products of enzymatic reactions The addition of mainly soluble so-called mediators is inevitable Recently, amperometric sensors were developed based on redox polymers to regenerate PQQ-dependent dehydrogenases [2] We could show that this polymer is also capable to catalyze the oxidation of the cosubstrate NADH Thus the redox polymer plays the role of an

A van den Berg and P Bergveld (eds), Micro Total Analysis Systems, 267–271

immobilized mediator. To measure the potential of the redox polymer as immobilized mediator, we developed a dynamic potentiostatic multi-puls method.

2. Results and discussion

2.1 LAYOUT OF THE FIELD-EFFECT TRANSISTOR

The field-effect transistors (schematically shown in fig.1) were produced in the Fraunhofer Institut, Arbeitskreis für Integrierte Schaltungen (FhG-AIS), Erlangen [3]. The FET was controlled by a custom designed feedback circuit and was driven in a constant charge mode [4]. The meander-shaped gate has a width of 5.000 µm and a length of 10 µm Onto the SiO_2 gate a 200 nm thick platinum layer was deposited where titanium was used as adhesion metal.

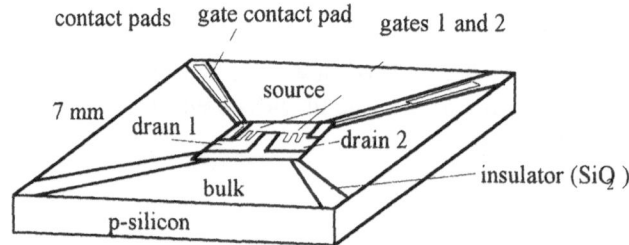

Figure 1. Design of the field-effect transistor. The contact pad for the gate electrode is shwon between those of the drain and source pads.

Each of the two gates can be contacted via an additional aluminum pad, which is positioned between the source and drain pads. This allows a current to be applied to the gate electrode by means of a potentiostat in order to control the redox state of a redox couple nearby the electrode surface.

2.2 MULTI-PULS METHOD FOR MODIFIED REDOX-FETS

A general problem of the potentiometric measurement is the low exchange current density of enzymatic substrate redox couples. Therefore additional mediators have to be applied with a high exchange current densitiy and fast electrode kinetics, which can react with the enzymatic products. Most often, these mediators are soluble low molecular weight substances, e.g. hexacyanoferrate. Until now, redox polymers based containing covalently bound osmium complexes [5] were only used in combination with oxidases for amperometric sensors. We could show that this redox polymer (E^0 =

290 mV vs SCE) catalyzes the electrochemical oxidation of the enzymatically produced coenzyme NADH [6]

By means of a potentiostat (fig 2) the redox state of the polymer-modified FET can be adjusted to any redox state (e g 400 mV vs SCE), which corresponds to a determined concentration ratio of oxidized and reduced osmium redox centres according to the NERNST equation The gate electrode is used as working electrode in a three electrode arrangement If we apply several pulses of a potentiostatic current to the polymer film, it will be charged stepwise to reach the potentiostatically given potential (e g 400 mV, fig 3a)

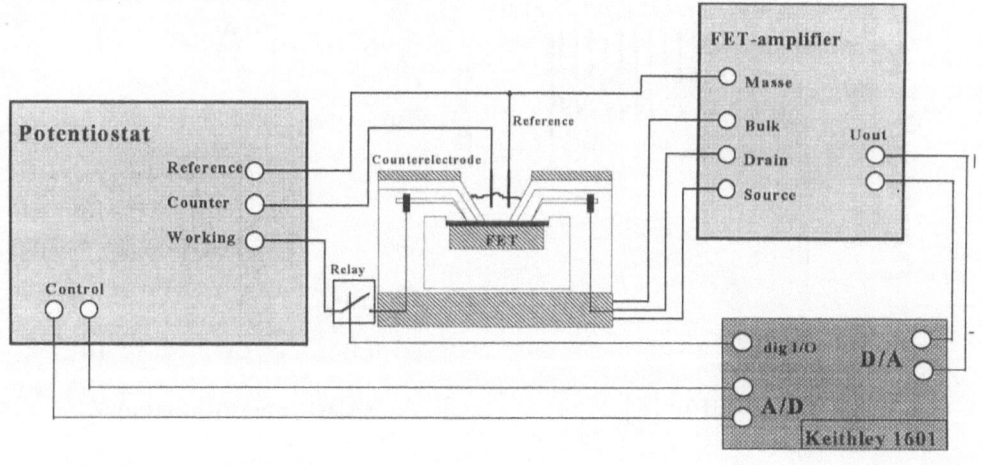

Figure 2. Setup for multi-puls measurement with the modified redox-FET. The Keithley-card allows to control the experiment with the software of a personal computer.

Between the pulses the gate is disconnected from the potentiostat, (open-circuit phase) and the potential of the platinum gate is dictated by the redox-state of the polymer The potential is approaching its equilibrium value, while the electrons spread from the electrode to equilibrate throughout the polymer film The redox polymer attached to the platinum surface functions as a condenser

If there is another redoxactive substrate in solution a homogeneous redox reaction is additionally coupled to this charging process The redox polymer works like an electron reservoir at a determined potential which provides electrons to or accepts electrons from a soluted redox species, e g enzymatically produced NADH In the case of NADH added to the buffer, the redox polymer is reduced, accepting electrons from NADH The effect resulting from this catalytic reaction could be observed during the open circuit phases (fig 3a)

The decrease of the equilibrium potential due to the chemical reaction was potentiometrically monitored by means of the redox-FET. As can be seen in fig 3, the effect depends on the concentration of NADH. After several pulses an equilibrium between the charge delivered by the redox reaction and the charge accepted from the potentiostat is established. Thus several pulses show the same value of the final potential at the end of the open-circuit phase. This value is plotted against the concentration in fig. 3 (b).

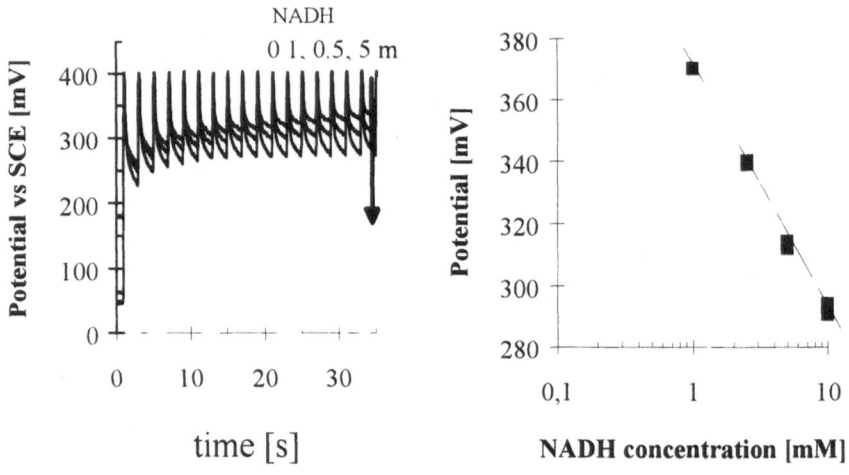

Figure 3. (a) Potentiostatic multi-pulse potentiometry for the determination of 0.1, 0.5 and 5 mM NADH in buffer solution with a redox-polymer modified FET, (b) calibration curve for a measurement of NADH in 0.1 M phosphate buffer, pH 7.5.

3. Conclusion

The potentiostatic multi-pulse potentiometry described here allows the dynamic measurement of potentials. The advantages of this method are the short time required for the analysis and the low noise of the signal. The "ancestor" of this technique, enzyme chronopotentiometry [7], posed problems of reproducibility when it was applied to the immobilized redox polymer. The excellent reproducibility of our method is clearly shown in fig. 3b. These techniques were the fundamental developments to conceive redox-FETs for the first time. After immobilization of NAD^+-dependent dehydrogenases covalently on the surface of the transducer the enzymatically produced NADH would be catalytically oxidized in situ by the polymeric mediator. To this very compact combination the substrate and NAD^+ as cosubstrate have to be applied externally. The coimmobilization of the coenzyme NAD^+ would lead to reagentless sensors. This is a subject of forthcoming investigations

4. References

1   S Caras, J Janata, *Anal Chem* **52** (1980) 1935-1937
2   L Ye, M Hammerle, A J Oolsthoorn, W Schuhmann, H -L Schmidt, J A Duine, A Heller, *Anal Chem* **65** (1993) 238-241
3   T Falter, T Mikolajick, H Ryssel in F Scheller, R D Schmid, GBF Monograpies, Vol 17, VCH Weinheim, p 345-347
4   J Janata, R H Huber, Solid State Chemical Sensors, Academic Press Inc , Orlando 1985
5   B A Gregg, A Heller, *J Phys Chem* **95** (1991) 5970-5975
6   T Vering, W Schuhmann, D Seiwald, H -L Schmidt, *J Electroanal Chem* **364** (1994) 277-279
7   B Uhe, G Janker, W Schuhmann, H -L Schmidt, J Janata, *Sens Act* **B 7** (1992) 389-392

PERFORMANCE OF THE COULOMETRIC SENSOR-ACTUATOR DEVICE IMPROVED BY μTAS

Wouter Olthuis and Piet Bergveld
MESA Research Institute, University of Twente
P.O. Box 217, 7500 AE Enschede, The Netherlands

0. Abstract

The ISFET-based coulometric sensor-actuator device turns out to be well suited for μTAS applications, because there is no need for a reference electrode. It is shown that μTAS essentially can improve the performance of the sensor-actuator device, with respect to its ease of operation (no interfering mass transport, an excess concentration of supporting electrolyte is available and signal processing is simpler) and its measuring range (controlled dilution).

1. Introduction

Acid or base concentrations can be determined by performing an acid-base titration with coulometrically generated OH^- or H^+ ions at an actuator in close proximity to the pH-sensitive gate of an ISFET. In figure 1 the basic elements of such a device are shown. The titrant can be generated either by the electrolysis of water at a noble metal actuator electrode [1]:

$$2H_2O \rightarrow 4H^+ + 4e + O_2 \quad \text{at the anode, and}$$
$$2H_2O + 2e \rightarrow 2OH^- + H_2 \quad \text{at the cathode}$$

or by the reversible redox reaction at an iridium oxide actuator electrode, which is accompanied with the release or uptake of protons [2]:

$$IrOOH \Leftrightarrow IrO_2 + H^+ + e$$

The ISFET is used as indicator to detect the end-point in the titration curve. The time needed to reach this end-point, t_{end}, depends on the acid or base concentration of the sample [3]. In figure 2 a typical measuring result is shown.

A van den Berg and P Bergveld (eds), Micro Total Analysis Systems, 273–277
© 1995 *Kluwer Academic Publishers*

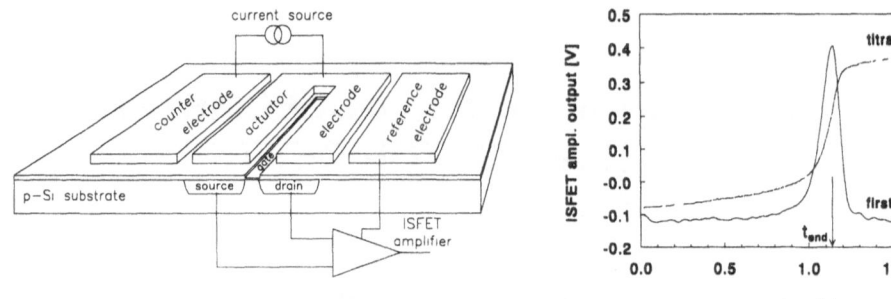

Fig.1 Basic elements of the coulometric device.   Fig. 2 Result of a typical titration

1.1. ADVANTAGES

When the coulometric sensor-actuator chip is glued on a small printed-circuit board carrier, a dipstick-like titrator is obtained, capable of performing fast titrations (seconds) in a small sample volume (microliters). Operation in a larger volume has the advantage that the bulk of the solution is not affected by the titration. The titration is fully computer controllable and does not require extra chemicals.

An additional important advantage arises from the fact that a separate reference electrode is not necessary, which can be explained by referring to the measurement set-up, shown in figure 3.

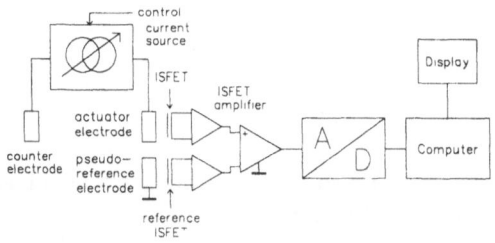

Fig. 3 Measurement set-up

The second ISFET of figure 3 serves as a reference, which is possible because the bulk pH does not change during the experiment. The actuator current is applied to only one of the sensor-actuator chips and the electrode shaped around the reference ISFET can now be used as a pseudo-reference electrode. A differential measurement of the ISFET output signals results in the desired signal as shown in figure 2. The counter electrode, shown in figure 3 is just a thin gold film applied on the reverse side of the dipstick.

1.2. DISADVANTAGES

Unfortunately, the coulometric sensor-actuator device shows some disadvantages, which can be made clear by presenting the result of a series of measurements in figure 4.

In figure 4, the square root of t_{end} is shown as a function of a series of different acid concentrations both of HNO_3 and of acetic acid (HAc). It is obvious that the curves for HNO_3 and HAc do not coincide. This can be understood from the operational principle of free diffusion; the slope of the curve is described by [3]:

$$\frac{\partial \sqrt{t_{end}}}{\partial C_{acid}} = \frac{F\sqrt{\pi D_{acid}}}{2 j_c} \qquad\qquad \text{eqn. 1}$$

where j_c represents the cathodic current density through the actuator, and C_{acid} and D_{acid} are the concentration and the diffusion coefficient of the acid, respectively. F is Faraday's constant. Eqn. 1 shows that the measurand of the titration, t_{end}, is not only a function of the analyte C_{acid}, but also of its diffusion coefficient, D_{acid}. In some practical applications, when the type of acid is not known, or when a mixture of several acids is present, this can be a disadvantage.

It is also clear from figure 4 and the corresponding theoretical eqn. 1 that a linear curve is only obtained when the *square root* of t_{end} is presented as a function of the acid concentration. For some applications, especially with respect to the subsequent signal processing, this can be disadvantageous.

The upper limit of the acid concentration, as shown in figure 4, is of the order of 10 mM. Although this limit, and with that the total measuring range, can somewhat be increased by changing the current density j_c, this upper limit turns out to be problematic with respect to the much higher total acid concentrations in several food products, such as fruit juice or wine (50 to 150 mM).

The operational principle of the device fully relies on only one type of mass transport: diffusion. The presence of either one of the two remaining mass transport mechanisms, convection or migration, will cause erroneous measuring results. Under laboratory conditions, the avoidance of convection can be obtained in a small-volume beaker by a 30-seconds pause after stirring, but it is clear that for larger volumes or in-line measuring, the necessary absence of convection forms a practical problem. The effect of migration can only be avoided, when an excess concentration of neutral supporting electrolyte is present in the analyte. Obviously, this also might be a disadvantage in some practical applications.

It will be shown in the following section that all the disadvantages, as mentioned in this section, disappear, when a suitable µTAS approach is chosen. The advantages of the device as mentioned in the previous section will be kept, of which the differential mode of operation, discarding the need for a reference electrode is the most important one, making this device exceptionally well suited for µTAS applications.

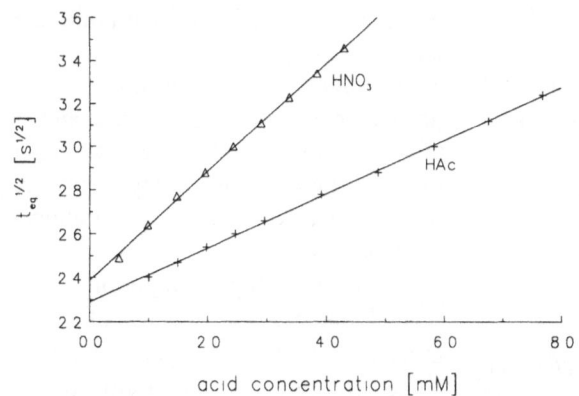

Fig. 4 Square root of t_{end} as a function of both HAc and HNO₃ concentrations, j_c=20 µA/mm².

2. Improvements of the performance by μTAS

In this section, the basic building blocks of μTAS are considered to be available or at least, the fabrication of these blocks is considered to be feasible. With basic building blocks, subsystems of μTAS are meant that are capable of performing a well-defined function, like inlet, selection valve, micro-pump, injector, mixing chamber, reaction chamber, outlet, etc. It will be shown that with the availability of these subsystems, the performance of the coulometric sensor-actuator device can be essentially improved.

As a start, let us consider the sensor-actuator device in differential mode to be placed in a small-volume reaction chamber, as schematically shown in figure 5. The height h of the reaction chamber is chosen small (typically 25 μm) with respect to the mean diffusion layer thickness after a few seconds of titrant generation (typically 100 to 500 μm). In addition, the actuator area A is chosen large (e.g. 1 to 5 mm^2) with respect to the mean diffusion layer thickness. Consequently, a homogeneous coulometric titration can be considered to take place in the volume formed by A·h, as indicated in figure 5 [4]. The slope of the curve, resulting from a series of measurements in different acid concentrations, C_{acid}, can now be expressed as

$$\frac{\partial t_{end}}{\partial C_{acid}} = \frac{FAh}{I_c}$$

eqn. 2

with I_c is the current through the actuator.

When comparing eqn. 1 with eqn. 2, it is clear that in the latter equation the diffusion coefficient is absent: regardless of any difference in the diffusion coefficient, the time to reach the end point t_{end} in the titration curve will, for a constant current I_c, depend only on the acid concentration (expressed in its corresponding normality). This is an important advantage over the former operation principle: that of free diffusion.

It is also clear from eqn. 2 that the relation between the acid concentration and t_{end} is linear, which may simplify subsequent signal processing.

Because lateral mass transport still would influence the measuring result, convection must be avoided, which is very easy in the small-volume reaction chamber after the valve controlling the inlet flow is closed: the sample in the reaction chamber will be at rest almost immediately.

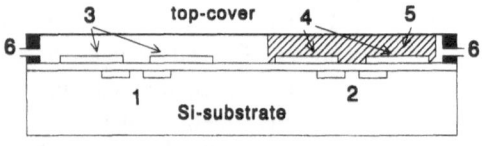

Fig. 5 Reaction chamber with (1) reference ISFET, (2) indicator ISFET, (3) pseudo-ref. electrode, (4) actuator, (5) titration volume and (6) spacer with inlet and outlet.

Considering migration, the second undesirable means of mass transport, another advantage of μTAS comes in view: the possibility to add supporting electrolyte. Both the proper control of the titrant-generating actuator current and the avoidance of migration requires a lower limit for the ion strength of the sample. With µTAS, a very small volume of a high concentration of neutral supporting electrolyte can be added and mixed with the analyte in a mixing chamber, thereby hardly changing the concentration of the acid.

Finally, the injection and mixing of moderate volumes of supporting electrolyte can be used to deliberately dilute the sample in order to be able to measure high acid (or base) concentrations, e.g. in fruit juices.

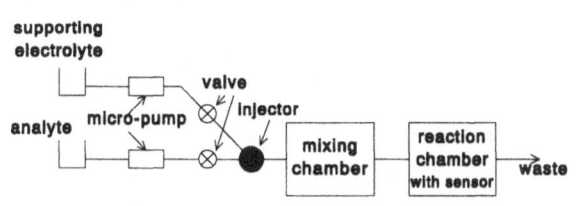

The µTAS thus proposed is schematically shown in figure 6. The subsystems of the µTAS are on purpose shown separately in figure 6; in reality the mixing and reaction chamber might be one chamber only. Also, the micro-pump itself might function as a valve.

Fig. 6 The proposed µTAS for improved coulometric sensor-actuator performance.

3. Concluding remarks

It has been the purpose of this paper to show that the performance of an already existing sensor can essentially be improved by the subsystems that are present in a µTAS. Concerning the coulometric sensor-actuator device, which is very well suited for integration in a µTAS by its lack of need for a reference electrode, the following is gained in combination with a µTAS:
- the measuring result no longer depends on the diffusion coefficient of the analyte,
- the measuring result depends linearly on the concentration of the acid (or base),
- both undesirable means of mass transport, convection and migration, can be cancelled,
- the measuring range can be expanded, especially to higher concentrations of the analyte.

Currently, the integration of the sensor-actuator device with a suitable reaction chamber is investigated.

References
1. W. Olthuis, B.H. van der Schoot, F. Chavez and P. Bergveld, A dipstick sensor for coulometric acid/base titrations, Sensors and Actuators, 17 (1989) 279-283.
2. W. Olthuis. J.G. Bomer, P. Bergveld, M. Bos and W.E. van der Linden, Iridium oxide as actuator material for the ISFET-based sensor-actuator system, Sensors and Actuators B, 5 (1991) 47-52.
3. W. Olthuis, J. Luo, B.H. van der Schoot, P. Bergveld, M. Bos and W.E. van der Linden, Modelling of non-steady-state concentration profiles at an ISFET-based coulometric sensor-actuator system, Anal. Chim. Acta, 237 (1990) 71-81.
4. B.H. van der Schoot and P. Bergveld, An ISFET-based microlitre titrator: integration of a chemical sensor-actuator system, Sensors and Actuators, 17 (1989) 279-283.

CONCEPT OF A MINIATURISED SYSTEM FOR MULTICOMPONENT GAS ANALYSIS BASED ON NON-DISPERSIVE INFRARED TECHNIQUES

M. Gebhard, W. Benecke

Institute for Microsensors, -actuators and -systems (IMSAS), Dept. of Physics and El. Eng., University of Bremen, 28334 Bremen, Germany

0. Abstract

The set up of a miniaturised multicompound gas analysis system based on the infrared absorption photometry is proposed. It seems to be a promising concept to apply the principles of the non dispersive infrared (NDIR) technique in a miniaturised analysis system using silicon based technologies. The fabrication of the optical components by silicon micromachinig technologies is discussed as well as different technological alternatives of the NDIR techniques.

1. Introduction

Gas sensors based on gas absorption of radiation in the IR spectral region offer advantages, especially concerning selectivity and long term stability, compared to gas sensors originating from other principles. Instruments based on dispersive techniques, using spectral decomposition of the IR radiation, are applied in the traditional laboratory analysis. They are large sized, complex, highly sensitive and expensive. Contrary, instruments working on NDIR absorption are proven to be robust, more compact and because of portability suitable for widespread use. Today they are utilised mainly in the environmental technique, especially in emission technique and process control for the detection of CO, CO_2, NO_x, SO_2, HCl and hydrocarbons. Within the last 15 years several technological alternatives of the NDIR technique have been developed in order to improve performance and universal employment. Following this development, it is a promising concept to apply the principles of NDIR techniques in a miniaturised analysis system for gases by using silicon based microsystem technologies. The content of this paper is to discuss the different technological alternatives of the NDIR technique for multicompound gas analysis with regard to a conversion in silicon based technologies.

2. Concepts for the construction of miniaturised systems for multicompound gas analysis

The fundamental absorption bands of the gases of interest, CO, CO_2, NO_x, SO_2, HCl and hydrocarbons are located in the wavelength region of 3 μm up to 7 μm. Fig. 1 shows

A. van den Berg and P. Bergveld (eds.), Micro Total Analysis Systems, 279–282.

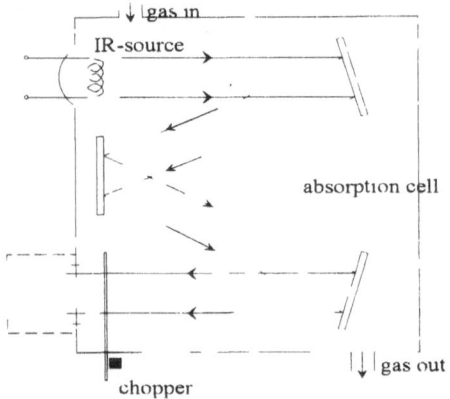

gas in

IR-source

absorption cell

chopper

gas out

Figure 1: Scheme of the optical set up of the NDIR gas analyser. The box contains filters and detectors.

the scheme of the optical set up of IR source, long path absorption cell and chopper. The box on the left of the exit window of the absorption cell includes the different filters for selective detection as well as the detectors. The three concepts to be discussed here are different with respect to this filter-detector part of the set up.

2 1 IR SOURCE, ABSORPTION CELL AND CHOPPER BASED ON SILICON TECHNOLOGIES

The fabrication of a poly-silicon IR source using IC technologies already has been demonstrated [1] The source consists of a poly-Si filament encapsulated in a vacuum sealed cavity for low electrical power consumption (5 mW). The source emitts broadband radiation with an emitted optical power for the total spectral region of 170 μ W The lateral size of the poly-Si filament is 510 x 5 μm^2 The filament emitter area of commercial available sources are larger by a factor of about 100. As the signal-to-noise ratio in background or intrinsic noise limited detection is proportional to the square of the optical power, which is proportional to the filament area, the emitting area of a Si microlamp has to be increased for use in the gas analyser. For example a meander filament structure with a total length of 5 mm and width of 50 μm should be suitable.

As there are deviations from the Lambert's law $I = Io \exp(-c \cdot l \cdot \varepsilon)$ for higher concentrations c, with ε the gas- and system specific extinction coefficient and l the length of the absorption cell, variation of the optical path length of the absorption cell is recommended. This is also inevitable, because the strengths of gas absorption are very different. For example CO_2 has line strength of about $S = 3 \cdot 10^{-18}$ cm²/mol cm and CH_4 of $S = 1 \cdot 10^{-19}$ cm²/mol cm [2] Total path lengths of 1 cm up to 30 cm are needed for gas detection in the 10`ths ppm range An absorption cell fabricated on or in Si brings up some problems, as the lateral dimensions for a cell on a typical Si chip (4x4 mm²) are too small. Even if a multifold channel cell is designed, the absorption losses of the high reflection material coated Si walls are improper A commercial available, variable, long path cell should be used in the gas analysers. Chopper and detector can be integrated, as this has been demonstrated for a chopper on the basis of piezoelectric bimorph vibration, to form one device [3]

2.2 MULTICOMPOUND GAS SENSOR WITH MULTICHANNEL DETECTOR AND FILTER ARRAY

Figure 2 shows the set up of a multichannel gas sensor. The IR multichannel detector consists of for example a pyroelectric detector array with n detectors. The interference filter array is made up of an arrangement of filters for each of the detectors n-1 channels are used for the detection of different gas components. One channel serves as the optical reference, where no gas absorption takes place.

The integration of filters on a silicon substrate has been demonstrated [4] Interference multilayer filters are deposited by vacuum thermal evaporation The lateral structuring was done by photolithography The filters are composed of a sequence of evaporated optical materials with high (Ge) and low (SiO) refractive index with an optical thickness of $\lambda/4$ or $\lambda/2$ respectively Technological problems are affected with the adherence of the multilayers, comprising up to 30 layers, during the subsequent processes A pyroelectric four channel detector with interference filter array packaged in a TO8 housing is commercial available [5]

2 3 MULTICOMPOUND WAVELENGTH MODULATED GAS SENSOR

Wavelength modulation and selection can be achieved by either a tuneable Fabry Perot interferometer, as shown in figure 3 or by a vibrating blazed diffraction grating Both devices have been fabricated on silicon wafers with IC processing technologies [6], [7] The diffraction grating is mounted on a rotation stage for wavelength scanning as well as for selecting different centre wavelengths for multicompound analysis Wavelength modulation and selection with the interferometer is achieved by electrostatic activation and capacitive control of the small gap between the highly reflective dielectric mirrors deposited on the interior surfaces of two bonded Si wafers Signal detection is done by phase sensitive amplification technique Signal analysis is performed by evaluation of the ratio of the first harmonic and the d c component The ratio is linearly related to the gas concentration and is independent from unspecified changes of the incoming radiation and detector sensitivity This technique shows high sensitivity Problems of the diffraction grating prototype are dealing with the low selectivity With a bandwidth of the prototype of 300 nm there is a partially overlap of the absorption band of CO and CO_2 and therefore bad resolution An increase in sensitivity can be established by the use of the interferometer For a finesse of 50 and a gap spacing of 20 µm the FWHM bandwidth at 4,26 µm (CO_2) is about 3 nm, which provides full resolution of CO to CO_2 .

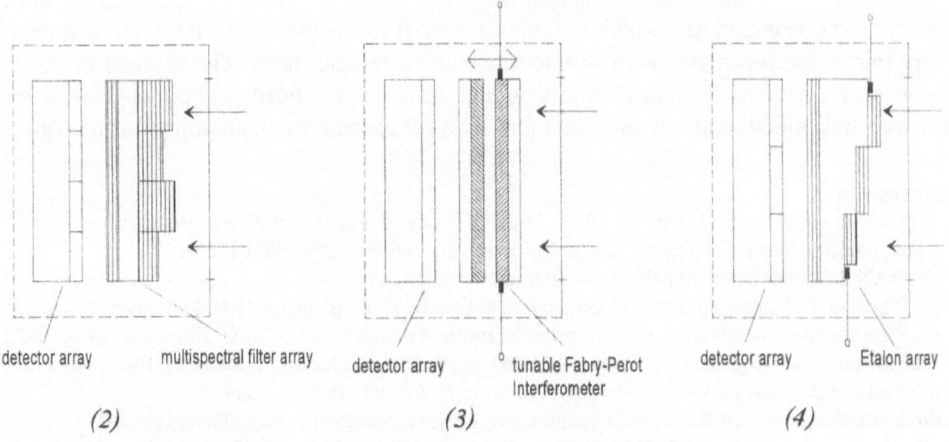

    detector array   multispectral filter array   detector array   tunable Fabry-Perot   detector array   Etalon array
                                                         Interferometer

          (2)   *(3)*   *(4)*

Figure 2: Set up of the multichannel detector
Figure 3: Set up of the wavelength modulated sensor
Figure 4: Set up of the etalon array with multispectral filters

2 4 MULTICOMPOUND GAS SENSOR BASED ON CORRELATION PHOTOMETRY

Figure 4 shows an etalon array with multispectral filters combined with a detector array The etalon array is an arrangement of etalons with different thicknesses and filters The etalon cavities for interference of the radiation are made of Si and the thickness d is larger by about one order of magnitude compared to the small gap of the Fabry-Perot interferometer (in 2 3) This means a very small free spectral range δv (distance of interference maxima) according to $\delta v = 1/ 2 \cdot n \cdot d$, with n the refractive index of Si The thicknesses of the etalons are matched in order to align the distance of interference maxima to the distance of the rotation lines (feinstructure) of the gases of interest This corresponds to about 400 μm for CO and to about 160 μm for CH_4 Coincidence of the interference lines with the rotation lines means a high degree of correlation and therefore sensitive detection The multispectral filters deposited on the etalons determine the transmitted spectral region for each gas to be detected Optical reference is given by time multiplexed detection of a reference phase In this phase the interference lines are phase shifted to the rotation lines by about a half of the distance of the lines This results in minimised coincidence Si etalons enable shifting by temperature dependent tuning of the optical thickness The difference in temperature for the shift is $\Delta T = 17°$ for CO and $\Delta T = 31°$ for CH_4 Selectivity of measurement is enhanced by adequate combination of measuring and reference temperature The performance of a correlation photometry gas sensor with one Si etalon, fabricated by thin film technology has been demonstrate [8] An etalon array with appropriate Si cavities and filters permits a multicompound analysis

3. Conclusion

Up to now micromachined IR sources with appropriate emitted optical power for use in the NDIR gas analysers have not been fabricated In the gas analyser a commercial available, variable, long path absorption cell with low cost optical components should be used For the wavelength modulated gas sensors it is of interest to fabricate a tuneable Fabry Perot interferometer with Si micromachining technologies The fabrication of a Si etalon array with multispectral filters would provide the head component for a very sensitive and selective multicompound gas analyser based on correlation photometry

References

1 C H Mastrangelo, J H J Yeh, and R S Muller, Electrical and optical characteristics of vacuum-sealed polysilicon micro lamps, *IEEE Trans Electron Devices, 39* (1992) 1363-1374

2 HITRAN database, molecular absorption parameters database

3 K Takeuchi,T Tanaka,M Ikeda,K Shibata,Y Sakauchi,Y Yamada,and S Nakano, Highly accurate gas sensor using a modulation-type pyroelectric infrared detector,*Jpn J Appl Phys ,32(1993)221-227*

4 J de Frutos, J M Rodriguez, F Lopez, A J de Castro, J Melendez and J Meneses, Electrooptical infrared compact gas sensor, *Sensors and Actuators B, 18-19* (1994) 682-686

5 Product information, IRIS infrared and inteligent sensors, pyroelectric four channel detector

6 J H Jerman, D J Clift and S R Mallinson, A miniature Fabry-Perot interferometer with a corrugated silicon diaphragm support, *Sensors and Actuators A, 29* (1991) 151-158

7 T Chen, Wavelength-modulated optical gas sensor, *Sensors and Actuators B, 13-14* (1993) 284-287

8 M Zochbauer, Fabry-Perot-Korrelationsphotometer fur die Gasanalyse, *Technische Messen, 61-5* (1994) 195-203

A DOUBLE CHEMFET FLOW CELL SYSTEM FOR DETECTION OF HEAVY METAL IONS AND INTEGRATION IN µTAS

J.F.J. Engbersen, P.L.H.M. Cobben, R.J.W. Lugtenberg and D.N. Reinhoudt

Faculty of Chemical Technology and MESA Research Institute, University of Twente, P O Box 217, 7500 AE Enschede, The Netherlands

0. Abstract

A new type of flow cell has been developed for detection of heavy metal ions in aqueous solutions, based on the flow injection method The cell contains two chemically modified field effect transistors (CHEMFETs) of which one is selective for the supporting electrolyte ion (in this case a 0 1 M potassium ion solution) and the other is selective for the heavy metal ion to be detected (Cd^{2+}, Pb^{2+}) The differential signal of the reference electrode CHEMFET and the heavy metal ion CHEMFET in the flow cell system which has been presently developed will be evaluated for miniaturization and further integration in a micro total analysis system

1. Introduction

A chemical sensor connects the chemical domain and the physical domain Our work is aimed at the design and synthesis of molecular receptors that can selectively recognize a guest species, and the development of sensor systems that incorporate these receptors When the guest species complexed by the receptor are charged, like cations, the chemical recognition process converts a neutral receptor into a charged species This allows the potentiometric detection by use of chemically modified field effect transistors for the transduction of the (chemical) complexation reaction into an electronic signal These sensors can provide direct information for localisation of environmental pollution and can be applied in water quality monitoring, biomedical analysis and biomonitoring The innovative technology from the semiconductor materials, micro-engineering, sensor and molecular engineering world shows great promise for the development of microsensors for the fabrication of multi-ion sensors by IC technology Miniature CHEMFETs compatible with IC techniques have been developed in our group recently and applied in a flow cell [1]

283

A van den Berg and P Bergveld (eds), Micro Total Analysis Systems, 283–288
© 1995 *Kluwer Academic Publishers*

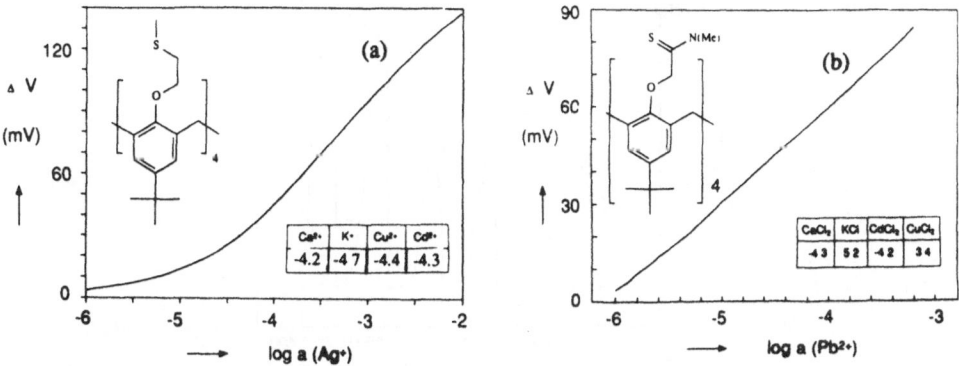

Figure 2. Response of Ag^+-selective CHEMFET (a) and Pb^{2+}-selective CHEMFET (b) in the presence of 0.1 M $Ca(NO_3)_2$ and 1 M KCl, respectively. Inset: Selectivity coefficients in the presence of 0.1 M interfering ion.

3. Development of a double CHEMFET flow cell system

Due to their small dimensions, chemically modified field effect transistors are particularly useful for application in flow sensor systems. A double sensor flow-through cell has been developed for detection of heavy metal ions in aqueous solutions (Figure 3). The cell contains two CHEMFETs of which one is selective for the supporting electrolyte ion (in this case a 0.1 M potassium ion solution) and the other is selective for the heavy metal ion to be detected (Cd^{2+}, Pb^{2+}).

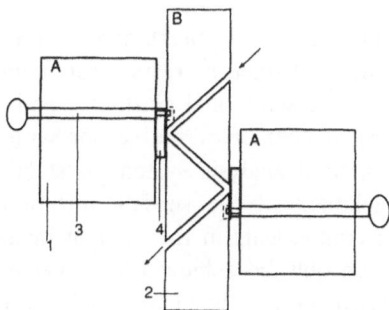

Figure 3. Side view of the detailed structure of the double sensor flow-through cell: perspex (1,2), contact wire (hook) (3), CHEMFET (4).

The differential signal of the reference CHEMFET and the heavy metal ion CHEMFET in the flow cell yields a stable signal, free of drift and noise effects (Figure 4).

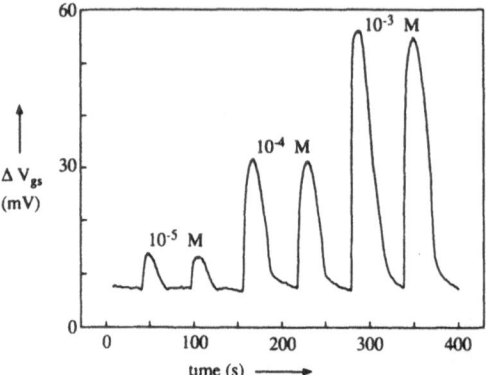

Figure 4. Response of Pb^{2+} CHEMFET measured differentially with respect to K$^+$ CHEMFET in the double sensor flow-through cell. (Q = 1.1 mL.min^{-1}; carrier: 10^{-6} M PbCl$_2$ + 100 mM KCl; injection: PbCl$_2$ + 100 mM KCl; V$_i$ = 220 µL.

4. Implementation of durable sensors in total analysis systems

A further step in miniaturization results in a micro total analysis system [7]. When incorporated in such a system, the requirements for the properties of the CHEMFETs change. First of all, the design of the CHEMFET must be in accordance with the micro fluid system as presented in [8]. Furthermore, drift is less critical, because automatic calibration can easily be incorporated in the micro system. On the contrary, the durability becomes more important if a continuous stand-alone operation of the µTAS is required.

Although plasticized PVC membranes with ionophores as sensing membranes give already quite satisfactory results for a limited period of time, the adhesion of the membrane is not stable and plasticizer and ionophore leach out from the membrane upon prolonged contact of the sensor with the aqueous solution. A sensor of practical use in a total analysis system must have a long lifetime, which requires the covalent attachment of the sensing membrane to the semiconductor surface. Also electroactive components in the sensing membrane must be prevented from leaching out, which can only be achieved by covalent attachment. Therefore a new membrane material has been developed to which the electroactive components, *viz.* the selective receptor and the anionic site, can be covalently bound. An example of such a modified CHEMFET architecture for detection of K$^+$ is depicted in Figure 5. Due to the covalent linkage of the ionophore via the methacrylate group leaching out from the membrane matrix is prevented and a stability of several months for the sensor in flow cell is obtained. In contrast, membranes with the same ionophore but without covalent attachment lose their activity quickly (Figure 6). In principle the above described architecture is generally applicable for the detection of charged species which can build up a membrane potential.

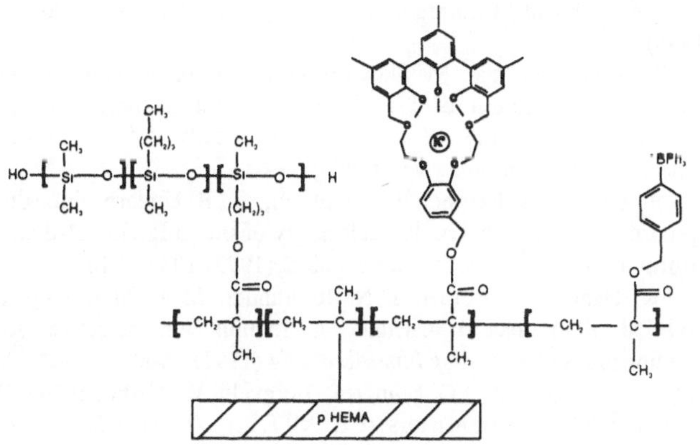

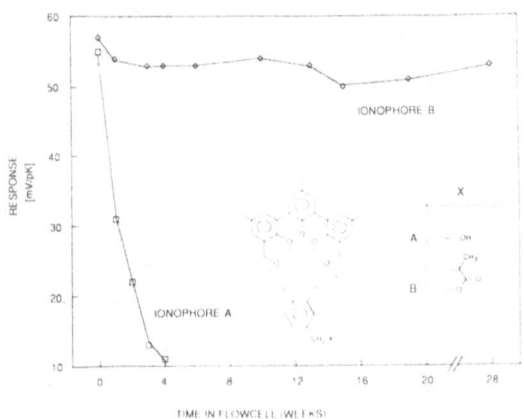

Figure 5. Schematic representation of the sensing membrane after photocrosslinking.

Currently we are also investigating the effect of covalent immobilisation of different ion receptors for heavy metal ions, which can be used in a micro flow system and can be implemented in a total analysis system circuit.

Figure 6. Sensor response characteristic in time. A: ionophore without covalent attachment. B: ionophore with covalent attachment.

References

1. P.L.H.M. Cobben, R.J.M. Egberink, J.G. Bomer, E.J.R. Sudholter, P. Bergveld, and D.N. Reinhoudt, Chemically modified ion-sensitive field-effect transistors: application in flow injection analysis cells without polymeric encapsulation and wire bonding, *Anal. Chim. Acta,* 248 (1991) 307-313.

2   P Bergveld, and A Sibbald, Comprehensive analytical chemistry, Vol XXIII, Elsevier, Amsterdam, (1988)

3   E J R Sudholter, P D van der Wal, M Skowronska-Ptasinska, A van den Berg, P Bergveld, and D N Reinhoudt, Modification of ISFETs by covalent anchoring of poly(hydroxyethyl methacrylate) hydrogel Introduction of a thermodynamically defined semiconductor-sensing membrane interface, *Anal Chim Acta, 230* (1990) 59-65

4   Z Brzozka, B Lammerink, D N Reinhoudt, E Ghidini, and R Ungaro, Transduction of selective recognition by preorganized ionophores, K+ selectivity of the different 1,3-diethoxycalix[4]arene crown ether conformers, *J Chem Soc Perkin Trans 2* (1993) 1037-1040

5   J A J Brunink, J R Haak, J G Bomer, D N Reinhoudt, M A McKervey, and S J Harris, Chemically modified field-effect transistors, a sodium ion selective sensor based on calix[4]arene receptor molecules, *Anal Chim Acta, 254* (1991) 75-80

6   P L H M Cobben, R J M Egberink, J G Bomer, P Bergveld, W Verboom, and D N Reinhoudt, chemically modified field effect transistors (CHEMFETs), *J Am Chem Soc* (1992) 10573-10582

7   J H Fluitman, A van den Berg, and T S Lammerink, Micromechanical components for μTAS, this volume, 73

8   P R Brown, and E Gruska, Advances in chromatography, Volume 33, Marcel Dekker, New York, (1993), 1-66

COMPONENTS AND TECHNOLOGY
FOR A FLUIDIC-ISFET-MICROSYSTEM

H. Fiehn, S. Howitz, M.T. Pham, T. Vopel, M. Bürger, T. Wegener
Research Center Rossendorf e V., Institute of Ion Beam Physics and
Materials Research Department for Microsystems/Sensors
P O Box 510119, 01314 Dresden, Germany

0. Abstract

We present a fluidic ISFET-microsystem (FIM) developed at the research center
Rossendorf during the last two years FIM represents a first step toward micro
instrumentations for chemical analysis FIM is based on a modular and planar concept
and composed of four main components, microfluidics, microsensors, micro actuators
and system electronics The feasibility of FIM is demonstrated by a complete
technology line for packaging and interconnection using Si-Si fusion, Si-glass bonding,
reflow soldering and adhesive techniques (developed by micro- and hybrid electronics)
FIM is featured by a) flexible exchange of the system modules due to a reversible
coupling, b) system compatibility for a variety of components by using a spacer chip
technology, c) an absolutely leakless fluid handling based on new microactuator
concept and d) dynamic differential measurement A FIM prototype is demonstrated
with ISFETs for H^+, Na^+, K^+, NH_4^+ and NO_3^-

1. Introduction

The μTAS idea was born in 1990 [1] The realization requires among other things
highly qualified microfluidic components and a technological concept for
micropackaging and interconnection suitable for processing on a wafer board [2] With
FIM an attempt was made to develop such components and microintegration
technologies The objective is a micro instrumentation for chemical analysis This paper
presents a FIM prototype with ISFETs as microsensors

2. Concept and realization of FIM

The FIM analyzer is fully planar organized by packaging the four main modules on a
Si-glass wafer structure with a network of microfluidic channels Fig 1 shows this
configuration for a FIM prototype realized on a 4 inch Si wafer with 8 x ISFETs, 2 x 2
fluidic injectors and 2 x fluidic diodes The number of modules and their configuration
are flexibly defined to meet different analysis tasks The modules microactuators,

289

A van den Berg and P Bergveld (eds), Micro Total Analysis Systems, 289–293
© 1995 *Kluwer Academic Publishers*

microsensors and microelectronics are micromachined by separate wafer processes and reversibly coupled to the fluidic wafer using both thermo-adhesive tape technique and thick film technique (metal mask) The adhesives have proven their suitability to assembly the sensor face down on the fluidic wafer

Electrical interconnections are made both by conductive adhesives and wire bonding The fluid handling is accomplished by means of piezoelectric micro fluidic injectors and micro fluidic diodes The function principle is explained in section 3 The main electronic functions are implemented on a programmable ASIC

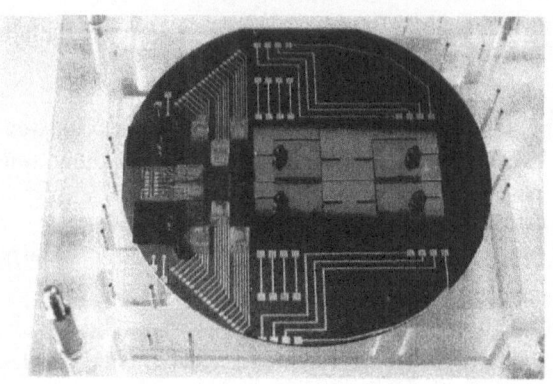

Fig. 1: Parts of the ISFET-microsystem: Fluidic wafer and containment with calibration fluids.

3. Integration of ISFETs in the fluidic concept

Measurements with ISFET require the sensors to be calibrated periodically This is done by switching the fluid flow between the sample and calibration fluid We have

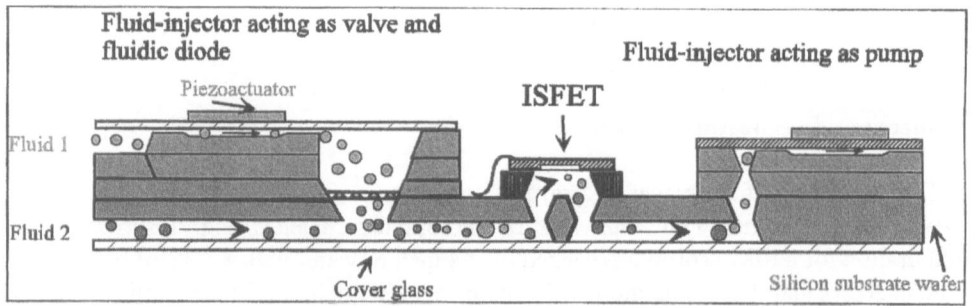

Fig. 2: Cross section through a microchannel with ISFET and actuators.

developed a microfluidic device called fluid injector (based on the working principle of ink jet printer heads [3]) to ensure proper fluid switching

The fluid injector is made from <100>-silicon and consists of a pump chamber with several in- and outlets A micro sieve with defined mesh size is used as a fluidic diode The outlet of the injector is separated from the micro sieve by an air space (see Fig 3 - left) so that in the off-state of the actuator the fluid 1 is kept totally decoupled from the carrier fluid 2 As to be seen in Fig 3 by computer-animation liquid-drops are injected to the carrier fluid without the carrier passes the sieve upward If the flow of the carrier is suppressed at the same time the injected drops will form a liquid volume below the sieve In this way, the fluid dosing can be performed without any leakage, even in the nl range The volume of the drop shown in Fig 3 (left) is about 1 5 nl The piezo-actuator (PZT-ceramic, 100 µm thick) is supplied with 30   100 V rectangular pulses The upper limit of the frequency is approximately 2 kHz

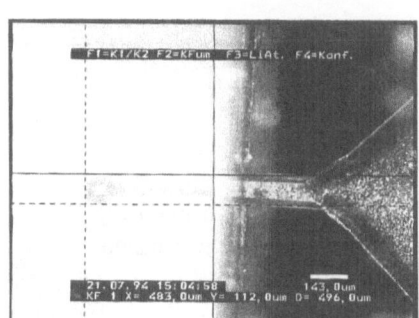 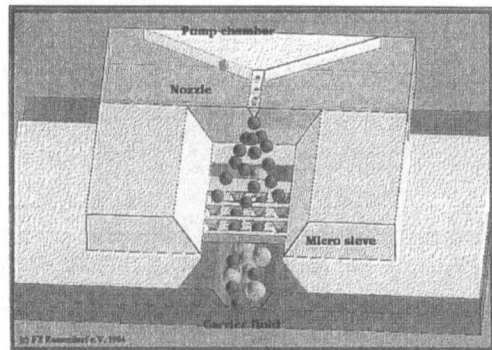

Fig. 3: Principle of fluid injector - the edge length of the outlet ranges from 50-100 µm.

One of the problems concerning microfluidic devices is to ensure a whirl free flow profile Otherwise disruptions may occur due to accumulated contaminants and air bubbles We have performed numerical calculations of the flow behavior of aqueous solutions using the finite element software FLOTRAN® The injector chamber we designed is based on the results obtained from this numerical simulation

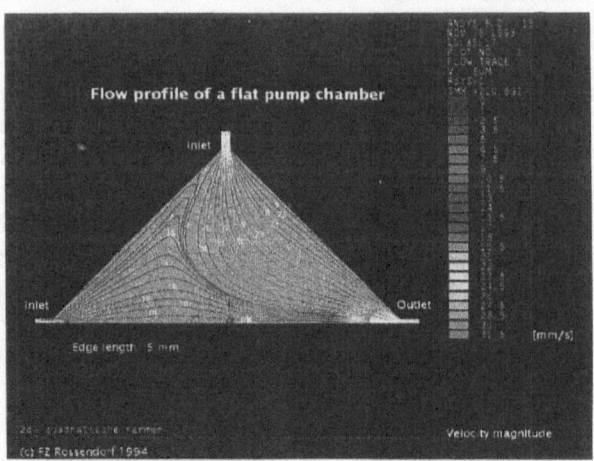

Fig. 4: Finite element analysis of the fluid flow in a flat silicon chamber.

4. Measurement results

Preliminary characterization of the FIM prototype was carried out using ISFETs as microsensors for pH, nitrate, ammonium, sodium and potassium. The sensibilization was made by either ion implantation or casting of a polymer solution.

Parameter	Sensibilization	Range
pH	Si_3N_4 membrane	pH = 1 ... 10
NO_3^-	pH-ISFET with polymer membrane: - 2-nitrophenyl octyl ether - tetradodecylammonium nitrate - PVC	$pNO_3 = 1 ... 5$
NH_4^+	pH-ISFET with polymer membrane: - bis(2-ethylhexyl) sebacate - nonactin - PVC	$pNH_4 = 1 ... 5$
Na^+/K^+	multiple ion implantation of Ca^+, Al^+, Na^+, and K^+ into SiO_2 top layer	pNa = 0 ...4 pK = 0 ... 4

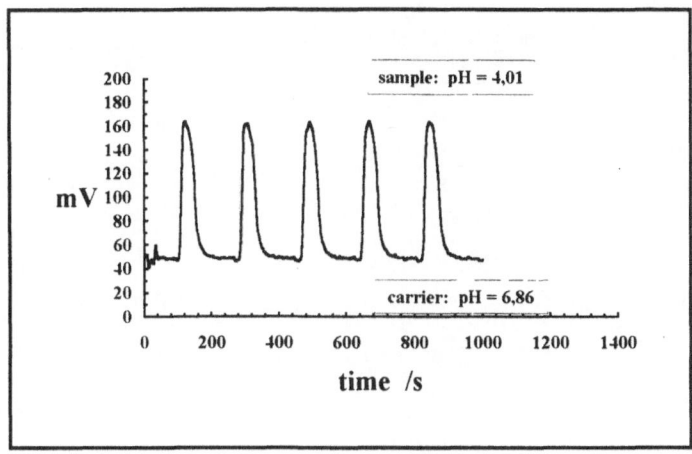

Fig. 5: Example of first measurements of pH with the FIM prototype.

5. Conclusions

The preliminary results obtained have demonstrated the feasibility of the FIM concept presented. Micro fluidic injectors and diodes represent an alternative approach to the microfluid handling and may be very useful with respect to dosing fluid in nanoliter range.

References

1   A. Manz, N. Graber and H.M. Widmer, Miniaturized total chemical analysis system: a novel concept for chemical sensing, Sensors and Actuators B, 1 (1990) 244-248
2   P. Gravesen, J. Branebjerg, O.S. Jensen, Microfluidics-a review, J. Micromech. Microeng. 3(1993) 168-182
3   N. Schwesinger, W. Bohmann, The behavior of liqiud jets escaping microchannels, , J. Micromech. Microeng. 3(1993) 210-213

AN ON-CHIP MINIATURE LIQUID CHROMATOGRAPHY SYSTEM: DESIGN, CONSTRUCTION AND CHARACTERIZATION

S. Cowen and D.H. Craston
Lab. of the Government Chemist, Queens Road,
Teddington, Middlesex TW11 0LY, United Kingdom

Abstract

Capillary liquid chromatography columns suitable for microchemical analysis have been fabricated on planar glass and silicon wafers, using standard microengineering techniques, including photolithography, wet chemical etching and anodic bonding. The columns are designed to allow the injection of sample solutions and elution solvents, and derivatisation of the inner walls. Columns modified with various stationary phases have been used to separate analytes, the detection of which is achieved by amperometric measurements.

The chromatography column forms part of an integrated system which, when completed, will comprise a miniature chemical measurement instrument, containing extraction and separation facilities. Fully portable, it will provide a cost-effective method of sample screening.

1. Introduction

At-site chemical measurement is an area of increasing importance in analytical chemistry, as it provides a basis for process control, continuous monitoring, and for cost-effective rapid screening of samples, allowing the elimination of 'negative' samples prior to laboratory analysis. Generally, instruments used for this purpose must satisfy several requirements if they are to be successful. They must provide a rapid response, have a high reliability factor, be simple to operate, and be small in size. Rapid screening devices are traditionally of chemical sensor type, but they are deficient in several areas: sensors are frequently extremely selective in their detection capabilities and have a lifetime limited by fouling of the sensor/environment interface. A more generic solution to the problem of effective rapid sample screening might be to develop a way of bringing the very non-selective methods of laboratory analysis

A. van den Berg and P. Bergveld (eds.), Micro Total Analysis Systems, 295–298.
© 1995 *Kluwer Academic Publishers.*

directly to the environment of interest, using miniaturised instrumentation. The additional problem of sensor lifetime is also solved. Such instruments would also be valuable for continuous monitoring and in process control, where the reduced requirement for analytical solvents provides additional benefits.

Some work in developing miniature analytical devices has already been published. The Stanford gas chromatograph [1] was fabricated almost entirely on a silicon wafer, and was the first attempt to produce a working hand-held device. More recently, work in microsystems has expanded, with capillary electrophoresis [2] and liquid chromatography [3] systems being produced.

In this paper we describe the first stages in the development of a complete, portable liquid chromatography instrument, which will be suitable for the rapid screening of samples in a wide variety of applications.

2. Experimental

Miniature chromatography columns were fabricated on polished, annealed Pyrex wafers using a standard photolithography process: gold and chromium masking layers were deposited onto the glass, followed by spin-coating of a $1\mu m$ layer of photoresist. The photoresist was then patterned with the column outline by selective UV exposure and developing, and the exposed metal layers etched with aqua regia and ammonium cerium nitrate solutions. A 20% HF, 14% HNO_3 glass etchant was used at temperatures of 50-70°C to etch the column pattern into the glass, to depths of 1-10μm and channel widths of 50-80μm. Following this, discharge drilling [4] was used to produce inlet and outlet holes in the column. Figure 1 shows a typical column layout. After the lithography stage, columns were bonded to <100>-oriented silicon wafers using the anodic bonding method [5].

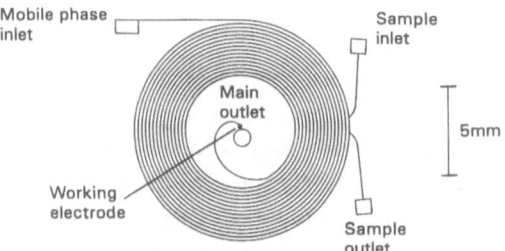

Figure 1. Miniature LC column design

Amperometric detection was incorporated into each miniature column by embedding a 50μm diameter platinum wire in a hole which had been drilled in the silicon prior to the anodic bonding stage. The wire was held in position with a low-temperature solder glass (Schott UK Ltd). This working electrode was situated at the mobile phase outlet, close to the drilled hole.

The open-tubular design of the columns allows a variety of stationary phases to be added, including octadecyltrichlorosilane and silicone polymers. The columns were modified by depositing the chosen stationary phase using the static coating method [6]. Injection of small sample volumes into the column was achieved by using a stainless steel mount containing three mechanically actuated valves at the inlets and outlets. This was attached to the column with a polyamide-based adhesive. Figure 2 shows a cross-section of the experimental setup, and also the position of the Ag/AgCl reference electrode.

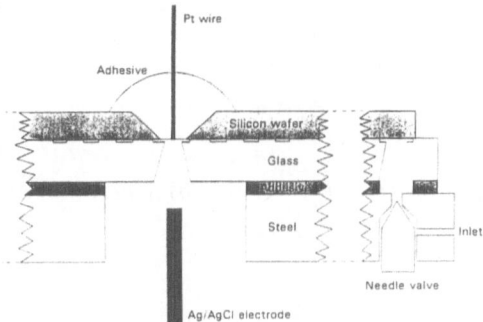

Figure 2. Cross-section of column and mount

Sample and mobile phase solutions were introduced into the column under pressure supplied by pressurised gas cylinders, and the valve system allowed the injection of samples.

3. Results and discussion

Initial tests of the system were performed with a 5mM solution of $K_4Fe(CN)_6$, chosen because of its reversible electrochemical behaviour. Injection of a small sample plug produced the peak shown in Figure 3(a).

Comparison of the retention time and peak width allows the sample volume to be estimated at 10nl, as the total column volume is known. The dead volume in the valve system places an upper limit on sample size, and can be reduced in order to reduce the volume of sample entering the column.

When a mixture of 5mM potassium ferrocyanide and four ferrocenecarboxylic acid (FAC)-derivatised amino acids was used as a sample, the chromatogram of Figure 3(b) was obtained. Clearly, there is a separation of FAC and potassium ferrocyanide, but the individual amino acids cannot be discriminated. To achieve acceptable separation efficiency, the sample/column volume ratio has to be reduced from 0.1 to around 0.01, which is at the focus of the present research work.

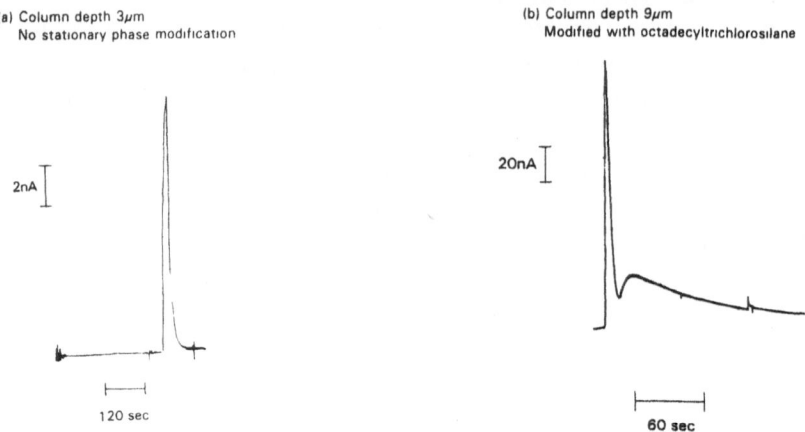

Figure 3. (a) Detection of 5mM K₄Fe(CN)₆. (b) Separation of K₄Fe(CN)₆ and FAC

4. Conclusions

A miniature chromatography system, suitable for rapid screening applications, has been constructed and initial tests performed. Results show that it is possible to inject a 10nl sample and separate two analytes contained with it. Modification of the column walls can alter the separation efficiency, as can the column dimensions.

In order to produce a complete chemical analysis system, miniaturisation of both extraction and clean-up stages is required, and work in this area is in progress.

References

1 S.C.Terry, J.H.Jerman and J.B.Angell, A gas chromatographic air analyzer fabricated on a silicon wafer, *IEEE Trans. Electron. Devices, ED-26* (1979) 1880-1886.
2 A.Manz, D.J.Harrison, E.M.J.Verpoorte, J.C.Fettinger, A.Paulus, H.Lüdi and H.M.Widmer, Planar chips technology for miniaturisation and integration of separation techniques into monitoring systems, *J. Chromatogr., 593* (1992) 253-258.
3 A.Manz, Y. Miyahara, J.Miura, Y.Watanabe, H.Miyagi and K.Sato, Design of an open-tubular column liquid chromatograph using silicon chip technology, *Sensors and Actuators, B1* (1990) 249-255.
4 M.Esashi, Y.Matsumoto and S.Shoji, Absolute pressure sensors by airtight electrical feedthrough structure, *Sensors and Actuators, A21-A23* (1990) 1048-1052.
5 G.Wallis and D.I.Pomerantz, Field assisted glass-metal sealing, *J. Appl. Phys., 40* (1969) 3946-3949.
6 J.Bouche and M.Verzele, A static coating procedure for glass capillary columns, *J. Gas Chromatogr., 6* (1968) 501-505.

A MICROSYSTEM MASS SPECTROMETER

A. Feustel, J. Müller, V.Relling
Department of Semiconductor Technology
Technical University of Hamburg-Harburg

0. Abstract

This paper presents the concept and a first realization of a miniaturized mass spectrometer fabricated as a microsystem based on anisotropic etching, thin film technology and anodic bonding The mass spectrometer includes a plasma chamber for electron generation, an ionization chamber and a new type of mass separator within a cube of 3x3x3 mm³

1. Introduction

Mass spectrometers are widely used in gas- and microanalysis Usually they are very bulky and expensive instruments which need high supply voltages and ultrahigh vacuum, which limits their application excessively If, however, a low cost, small size instrument would be available, a number of new applications , e g in environmental surveillance and process control could be in reach

To obtain such a device not only the way of its fabrication must be that of a mass product, but also the necessary periphery , especially with respect to supply voltages and the pump line have to be reduced All of these obstacles can be solved, if the appliance is realized as a microsystem This can be accomplished by using a combination of microsystem technology like anisotropic silicon etching, silicon to glass sealing by anodic bonding and thin film technology In addition new concepts for the key components of a spectrometer, the ionizer and the mass separator have to be applied

In order to reduce the size of the spectrometer, the reduction of the mean free path (i e an increase of the allowed pressure) and the supply voltages for the ionization and mass separating microstructures are necessary The mass separator is based on three chambers separated by small orifices from each other which are simultaneously used for charged particle extraction and as pressure stages

A van den Berg and P Bergveld (eds), Micro Total Analysis Systems, 299–304
© 1995 *Kluwer Academic Publishers*

2. The ionizer

Fig. 1 shows the cross-section of this component. The ionization of the measurand is accomplished by electrons extracted from an inert gas plasma (e. g. Ar) and accelerated in direction perpendicular to the flow of the measurand. The two cabinets are connected via a small gap of 300μm x 100μm. The electrode right above of the gap (area of bare Si) accelerates electrons. There is another gap between the upper chamber and the mass separator (not shown in fig. 1). Behind the second gap the accelleration electrode for the mass separator is positioned.

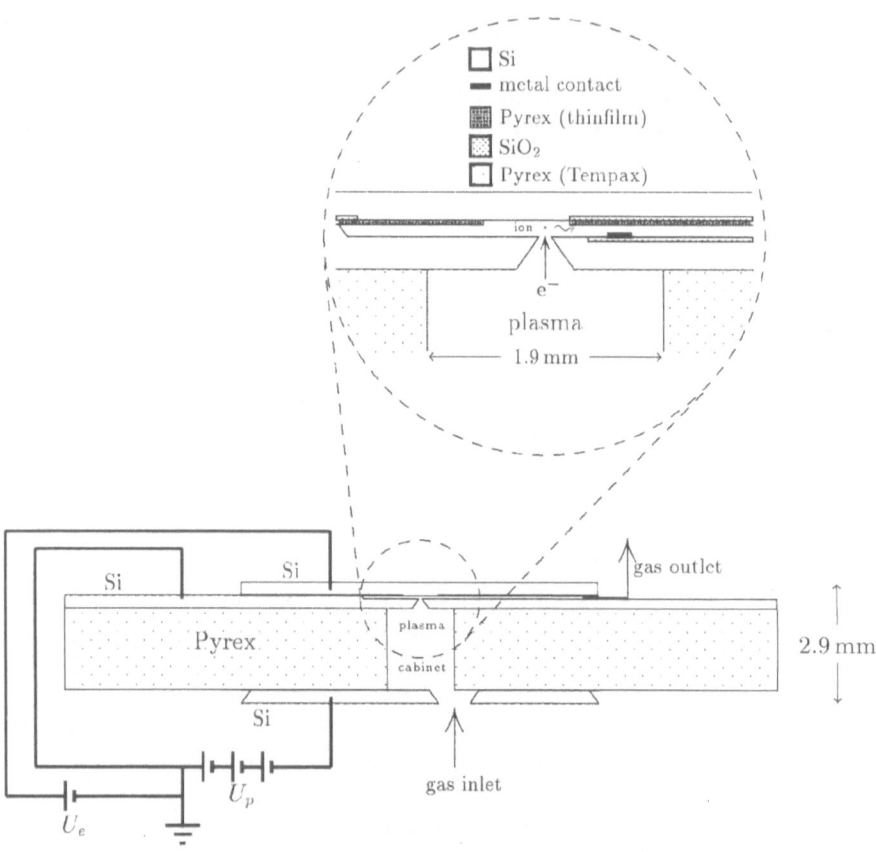

Fig. 1:Cross-sectional view of the ion source. The gap between the plasma cabinet and the upper chamber is 300μm to 100μm.

A rare gas plasma is ignited in the first cabinet (plasma cabinet). The electrons are accelerated through the first gap to the upward electrode. The design is similar to a Langmuir-probe, so a corresponding characteristics in respect to the electron current

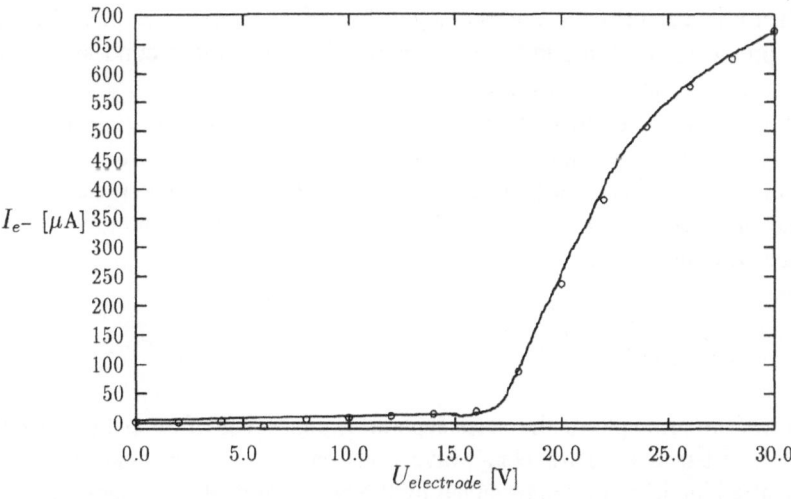

Fig. 2: Langmuir-probe like electron current in dependence on the electrode potential.

results. From an Ar-plasma in the first cabinet at a pressure of 400 Pa and a dc-potential of 220V electron currents up to 600μA can be extracted (fig. 2). This high electron density in the upper chamber allows for a high ionization rate even though the distance between the extracting gap and upward electrode is only 100μm. Ion currents up to 170μA were measured at a negatively biased accellerating electrode (fig. 3).

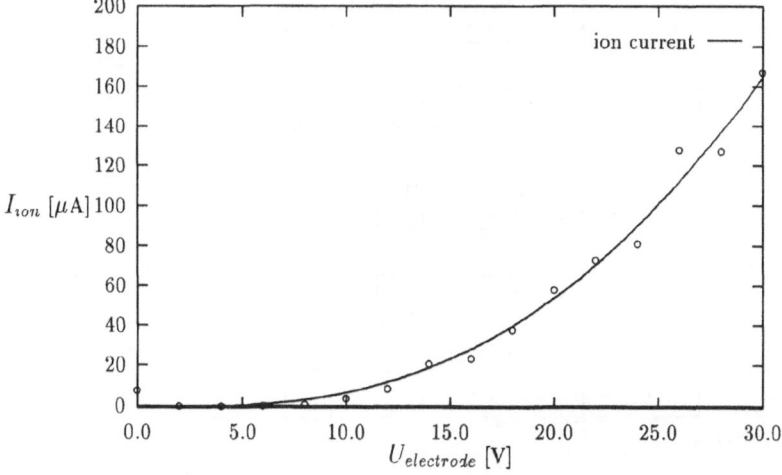

Fig. 3: Ion current in dependence on the potential of the ion-trap electrode.

Due to their effect as pressure stages at a gas flow of 0.1sccm at the inlet of the plasma chamber, pressures of 400Pa in the first cabinet, 200Pa in the second and below 2Pa behind the second gap are achieved.

With the remote location of the plasma cathode from the ionization chamber impurities caused by sputtering will have small influence on the measured spectra. The pressure in the plasma cabinet is determined by Paschen's law. The pressure of 400Pa marks the optimum for the geometries of the chamber for minimum voltage i. e. minimum sputtering effects.

3. The mass separator

Fig. 4 shows the principle of the mass separator, consisting of an array of electrodes. Its principle is based on a traveling wave. Ions enter the mass separator on the left hand side at a kinetic energy determined by the potential of the acceleration electrode, which results in a velocity of the ions which is inversely proportional to the root of their mass. Ions at a specific mass and an acceleration voltage will travel synchronously to the traveling alternating dipole field, as shown in fig. 4. In principle both the bold

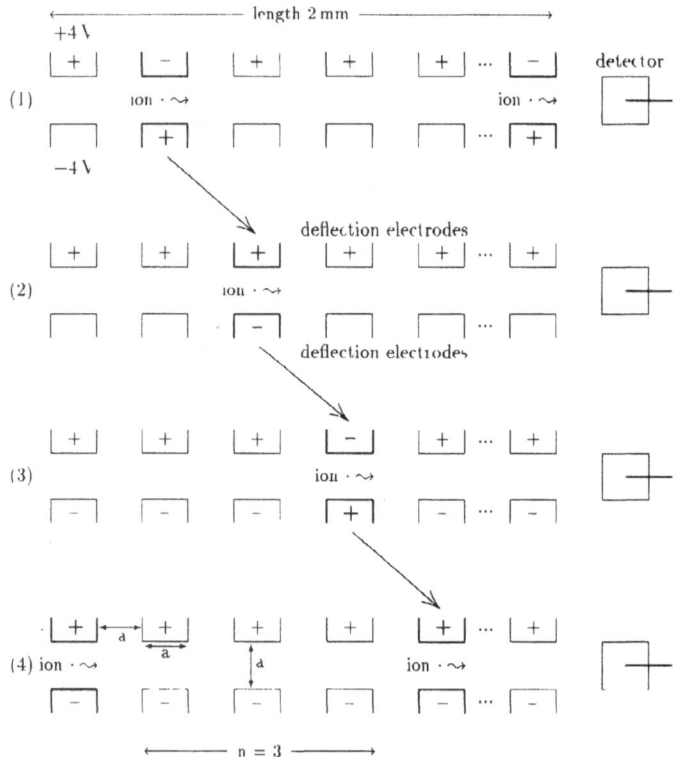

Fig. 4: Principle of the mass-separator. Distances of the electrodes are 100µm.

electrodes and these ions will travel through the separator. The ions reach the detector on a sinusoidal trajectory. Ions of lower and heavier mass will travel faster or more slowly, and will be subject to the constant electric field of the other electrodes and thus will be deviated to the wall.

Fig. 5 shows the simulated ion trajectories of two different ions, e. g. K^+ and Ar^+ at an acceleration potential of 25V. With a distance of 100μm between the electrodes even at a low potential of ±4V a high electric field of 80000 V/m can be achieved. To extract an ion of 40u (Ar^+) with a kinetic energy of 83eV an ac-field of 50MHz is necessary for the electrodes, which can be generated by common computer components.

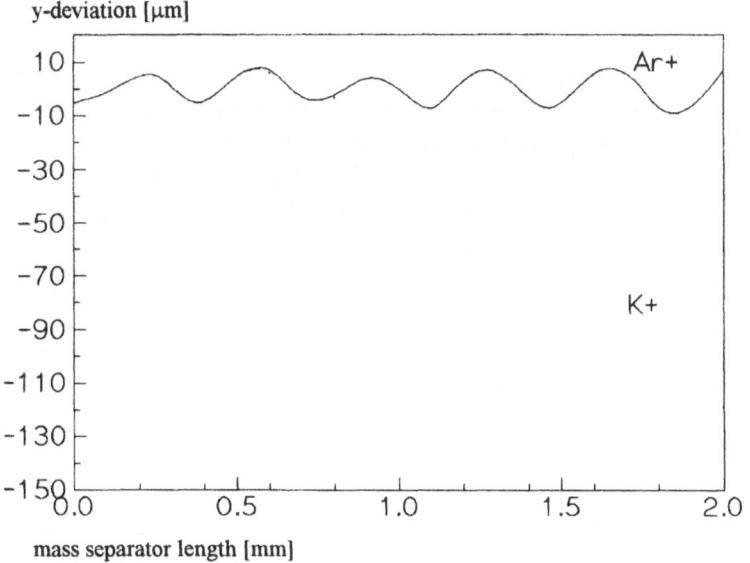

Fig. 5: Simulated trajectories of 25eV ions.

If one alternating is followed by three static dipoles the detector signal is an alternating ion current of 25MHz. A Faraday-cup coupled to a high sensitive, high frequency receiver, e. g. a short wave receiver, will be sufficient to detect the signal.

Changing the acceleration potential or the electrode frequency allows to vary the mass to be detected. From computer simulations ion currents of several 100nA up to 1μA are expected for pure gases and a resolution of m/Δm=18 for a separator of 2mm in length and the electrode dimensions as mentioned. Furthermore calculations show that the resolution is limited rather by geometry and available electrode frequency than by thermal motion of the ions. With respect to the mean free path a pressure in the separator below 4Pa will enable a collision free trajectory.

4. Fabrication technology

The mass spectrometer is assembled from four wafers, three silicon and one glass wafer, hermetically sealed by anodic bonding The geometries of the orifice and the plasma electrodes are fabricated via anisotropic etching of Si by KOH The plasma chamber for this first approach was drilled into the Pyrex wafer For the electrodes of the mass separator a deep anisotropic plasma etching process is used by which perpendicular trenches more than 200 μm deep with micrometer tolerances can be generated The electrodes are metallized for a defined potential within the separator

5. Outlook

Our first results of the device promise a very efficient and simple mass spectrometer device Further improvements are expected if an RF or microwave stimulated plasma (ECR condition) is used It would allow to reduce the pressure in the plasma chamber and hence an increase of the mean free path in the separator The more extended separator length, then available, will positively influence the resolution of the mass spectrometer

Authors index

Subject index